Stochastic Resonance

A Mathematical Approach in the Small Noise Limit

Mathematical
Surveys
and
Monographs

Volume 194

Stochastic Resonance

A Mathematical Approach in the Small Noise Limit

Samuel Herrmann
Peter Imkeller
Ilya Pavlyukevich
Dierk Peithmann

American Mathematical Society
Providence, Rhode Island

2010 *Mathematics Subject Classification.* Primary 60H10, 60J60;
Secondary 34D45, 37H10, 60F10, 60J70, 60K35, 86A10.

For additional information and updates on this book, visit
www.ams.org/bookpages/surv-194

Library of Congress Cataloging-in-Publication Data

Herrmann, Samuel, author.
Stochastic resonance : a mathematical approach in the small noise limit / Samuel Herrmann, Peter Imkeller, Ilya Pavlyukevich, Dierk Peithmann.
pages cm. — (Mathematical surveys and monographs ; volume 194)
Includes bibliographical references and index.
ISBN 978-1-4704-1049-0 (alk. paper)
1. Stochastic partial differential equations. 2. Diffusion processes. 3. Stability. I. Imkeller, Peter, 1951– author. II. Pavlyukevich, Ilya, 1974– author. III. Peithmann, Dierk, 1972– author. IV. Title.

QA274.25.H47 2014
519.2′3—dc23

2013034700

10 9 8 7 6 5 4 3 2 1   19 18 17 16 15 14

Contents

Preface

Stochastic resonance is a phenomenon arising in many systems in the sciences in a wide spectrum extending from physics through neuroscience to chemistry and biology. It has attracted an overwhelming attendance in the science literature for the last two decades, more recently also in the mathematics literature. It is generally understood as the optimal amplification of a weak periodic signal in a dynamical system by random noise.

This book presents a mathematical approach of stochastic resonance in a well defined framework. We consider weakly periodic systems in arbitrary finite dimension with additive noise of small amplitude ε. They possess two domains of attraction of stable equilibria separated by a manifold marking a barrier. Both the geometry of the attraction domains as well as the barrier height are not scaled with the amplitude parameter ε. Therefore, in contrast to other approaches, noise induced random transitions in our model happen on time scales given by the exponential of the quotient of barrier height and noise amplitude (Kramers' times), and are due to *large deviations*. Our analysis is therefore based on a new space-time large deviations principle for the system's exit and transition dynamics between different domains of attraction in the limit of small ε. It aims at the description of an optimal interplay between large period length T of the weak periodic motion and noise amplitude ε. Optimization is done with respect to appropriate measures of quality of tuning of the stochastic system to the periodic input.

The two principal messages of the book are these. First we show that—already in space dimension one—the classical physical measures of quality of periodic tuning such as the *spectral power amplification* or *signal-to-noise ratio*, due to the impact of small random oscillations near the equilibria, are not robust with respect to dimension reduction. Comparing optimal tuning rates for the unreduced (diffusion) model and the associated reduced (finite state Markov chain) model one gets essentially different tuning scenarios. We therefore propose—in arbitrary finite space dimension—measures of quality of periodic tuning based uniquely on the transition dynamics and show that these measures are robust. Via our central space-time large deviations result they are able to explain stochastic resonance as optimal tuning.

Concentrating on these more theoretical themes, the book sheds some light on the mathematical shortcomings and strengths of different concepts used in theory and application of stochastic resonance. It does not aim at a comprehensive presentation of the many facets of stochastic resonance in various areas of sciences. In particular it does not touch computational aspects relevant in particular in high dimensions where analytical methods alone are too complex to be of practical use any more.

With this scope the book addresses researchers and graduate students in mathematics and the sciences interested in stochastic dynamics, in a quite broad sense, and wishing to understand the conceptual background of stochastic resonance, on the basis of large deviations theory for weakly periodic dynamical systems with small noise. Chapter 1 explains our approach on a heuristic basis on the background of paradigmatic examples from climate dynamics. It is accessible to a readership without a particular mathematical training. Chapter 2 provides a self-contained treatment of the classical Freidlin–Wentzell theory of diffusion exit from domains of attraction of dynamical systems in the simpler additive noise setting starting from a wavelet expansion of Brownian motion. It should be accessible to readers with basic knowledge of stochastic processes. In Chapter 3 based on an approach from the perspective of semi-classical analysis, i.e. spectral theory of infinitesimal generators of diffusion processes, the conceptual shortcomings of the classical physical concepts of stochastic resonance are presented. In Chapter 4 the Freidlin–Wentzell theory is extended to the non-trivial setting of weakly time-periodic dynamical systems with noise, and concepts of optimal tuning discussed which avoid the defects of the classical notions. Both Chapters are accessible on the basis of the background knowledge presented in Chapter 2.

Introduction

Speaking about *noise* we usually mean something that deteriorates the operation of a system. It is understood as a disturbance, a random and persistent one, that obscures or reduces the clarity of a signal.

In nonlinear dynamical systems, however, noise may play a very constructive role. It may enhance a system's sensitivity to a small periodic deterministic signal by amplifying it. The optimal amplification of small periodic signals by noise gives rise to the ubiquitous phenomenon of *stochastic resonance* (SR) well studied in a plethora of papers in particular in the physical and biological sciences. This book presents a mathematical approach to stochastic resonance in a well defined particular mathematical framework. We consider weakly periodic systems with additive noise of small amplitude ε. The systems possess two domains of attraction of stable equilibria separated by a manifold marking a barrier. Both the geometry of the attraction domains as well as the barrier height are not subject to scalings with the amplitude parameter ε. Therefore, as opposed to other approaches, noise induced random transitions in our model happen on time scales of Kramers' law, i.e. they are exponential in the quotient of barrier height and noise amplitude, and are due to large deviations. Our analysis is therefore based on a new large deviations principle of the systems' exit and transition dynamics between different domains of attraction in the limit of small ε. It aims at the description of an optimal interplay between large period length T of the weak periodic motion and noise amplitude ε, where optimization is done with respect to appropriate measures of quality of response of the stochastic system to the periodic input. We will be uniquely concerned with the well founded and self contained presentation of this mathematical approach mainly based on a space-time extension of Freidlin–Wentzell's theory of large deviations of noisy dynamical systems, first on a heuristic and then on a mathematically rigorous level. The two principal messages of the book are these. First we show that — already in space dimension one — the classical physical measures of quality of periodic tuning such as the spectral power amplification, due to the phenomenon of the *small oscillations catastrophe*, are not robust with respect to dimension reduction. Comparing optimal tuning rates for the diffusion processes and the finite state Markov chains retaining the models' essentials one gets essentially different results (Chapter 3, Theorems 3.50, 3.53). We therefore propose — in arbitrary finite space dimension — measures of quality of periodic tuning based uniquely on the transition dynamics and show that these measures are robust and, via a crucial large deviations result, are able to explain stochastic resonance as optimal tuning (Chapter 4, Theorems 4.19, 4.29, 4.31). Concentrating on these more theoretical themes, the book sheds some light on the mathematical shortcomings and strengths of different concepts used in theory and application of stochastic resonance, in a well defined framework. It does not aim at a comprehensive presentation of the many facets of stochastic resonance in various areas of sciences (a sample will be briefly discussed in Chapter 1, Section 1.5). In particular it does not touch computational aspects relevant in particular in high dimensions where analytical methods alone are too complex to be of practical use any more (for an incomplete overview of stochastic resonance from a computational dynamics perspective see also Chapter 1, Section 1.5).

We now explain briefly our motivation and approach. The most prominent and one of the first examples in which phenomena related to stochastic resonance were observed is given by energy balance models of low dimensional conceptual climate dynamics. It was employed for a qualitative explanation of *glacial cycles* in earth's history, i.e. the succession of ice and warm ages observed in paleoclimatic data, by means of stochastic transitions between cold and warm meta-stable climates in a dynamical model. It will be discussed in more detail in Chapter 1. The model proposed by Nicolis [**83**] and Benzi et al. [**6**] is based on the balance between averaged absorbed and emitted radiative energy and leads to a deterministic differential equation for averaged global temperature T of the form

$$\dot{T}(t) = b(t, T(t)).$$

The explicit time dependence of b captures the influence of the solar constant that undergoes periodic fluctuations of a very small amplitude at a very low frequency. The fluctuations are due to periodic changes of the earth's orbital parameters (Milankovich cycles), for instance a small variation of the axial tilt that arises at a frequency of roughly 4×10^{-4} times per year, and coincide roughly with the observed frequencies of cold and warm periods. For frozen t the nonlinear function $b(t, T)$ describes the difference between absorbed radiative energy as a piecewise linear function of the temperature dependent albedo function $a(T)$ and emitted radiative energy proportional to T^4 due to the Stefan–Boltzmann law of black body radiators. In the balance for relevant values of T it can be considered as negative gradient (force) of a double well potential, for which the two well bottoms correspond to stable temperature states of glacial and warm periods. The evolution of temperature in the resulting deterministic dynamical system is analogous to the motion of an overdamped physical particle subject to the weakly periodic force field of the potential. Trajectories of the deterministic system relax to the stable states of the domain of attraction in which they are started. Only the addition of a *stochastic forcing* to the system allows for spontaneous transitions between the different stable states which thus become *meta-stable*.

In a more general setting, we study a dynamical system in d-dimensional Euclidean space perturbed by a d-dimensional Brownian motion W, i.e. we consider the solution of the stochastic differential equation

$$dX_t^\varepsilon = b\Big(\frac{t}{T}, X_t^\varepsilon\Big)\, dt + \sqrt{\varepsilon}\, dW_t, \quad t \geq 0. \tag{0.1}$$

One of the system's important features is that its time inhomogeneity is weak in the sense that the drift depends on time only through a re-scaling by the time parameter $T = T(\varepsilon)$ which will be assumed to be exponentially large in ε. This corresponds to the situation in Herrmann and Imkeller [**50**] and is motivated by the well known Kramers–Eyring law which was mathematically underpinned by the Freidlin–Wentzell theory of large deviations [**40**]. The law roughly states that the expected time it takes for a homogeneous diffusion to leave a local attractor e.g.

across a potential wall of height $\frac{v}{2}$ is given to exponential order by $T(\varepsilon) = \exp(\frac{v}{\varepsilon})$. Hence, only in exponentially large scales of the form $T(\varepsilon) = \exp(\frac{\mu}{\varepsilon})$ parametrized by an energy parameter μ we can expect to see effects of transitions between different domains of attraction. We remark at this place that our approach essentially differs from the one by Berglund and Gentz [**13**]. If b represents a negative potential gradient for instance, their approach would typically not only scale time by T, but also the depths of the potential wells by a function of ε. As a consequence, transitions even for the deterministic dynamical system become possible, and their noise induced transitions happen on time scales of intermediate length. In contrast, in our setting transitions between the domains of attraction of the deterministic system are impossible, and noise induced ones are observed on very large time scales of the order of Kramers' time, typically as consequences of large deviations. The function b is assumed to be one-periodic w.r.t. time, and so the system described by (0.1) attains period T by re-scaling time in fractions of T. The deterministic system $\dot{\xi}_t = b(s, \xi_t)$ with *frozen* time parameter s is supposed to have two domains of attraction that do not depend on $s \geq 0$. In the "classical" case of a drift derived from a potential, $b(t, x) = -\nabla_x U(t, x)$ for some potential function U, equation (0.1) is analogous to the overdamped motion of a Brownian particle in a d-dimensional time inhomogeneous double-well potential. In general, trajectories of the solutions of differential equations of this type will exhibit randomly periodic behavior, reacting to the periodic input forcing and eventually amplifying it. The problem of *optimal tuning* at large periods T consists in finding a noise amplitude $\varepsilon(T)$ (the *resonance point*) which supports this amplification effect in a *best possible way.* During the last 20 years, various concepts of measuring the quality of periodic tuning to provide a criterion for optimality have been discussed and proposed in many applications from a variety of branches of natural sciences (see Gammaitoni et al. [**43**] for an overview). Its rigorous mathematical treatment was initiated only relatively late.

The first approach towards a mathematically precise understanding of stochastic resonance was initiated by Freidlin [**39**]. To explain stochastic resonance in the case of diffusions in potential landscapes with finitely many minima (in the more general setting of (0.1), the potential is replaced by a quasi-potential related to the action functional of the system), he goes as far as basic large deviations' theory can take. If noise intensity is ε, in the absence of periodic exterior forcing, the exponential order of times at which successive transitions between meta-stable states occur corresponds to the work to be done against the potential gradient to leave a well (*Kramers' time*). In the presence of periodic forcing with period time scale $e^{\frac{\mu}{\varepsilon}}$, in the limit $\varepsilon \to 0$ transitions between the stable states with critical transition energy close to μ will be periodically observed. Transitions with smaller critical energy may happen, but are negligible in the limit. Those with larger critical energy are forbidden. In case the two local minima of the potential have depths $\frac{V}{2}$ and $\frac{v}{2}$, $v < V$, that switch periodically at time $\frac{1}{2}$ (in scale T accordingly at time $\frac{T}{2}$), for T larger than $e^{\frac{v}{\varepsilon}}$ the diffusion will be close to the deterministic periodic function jumping between the locations of the deepest wells. As T exceeds this exponential order, many short excursions to the wrong well during one period may occur. They will not count on the exponential scale, but trajectories will look less and less periodic. It therefore becomes plausible that physicists' *quality measures for periodic tuning* which always feature some *maximal* tuning quality of the random trajectories to the periodic input signal cannot be captured by this phenomenon of quasi-deterministic periodicity at very large time scales.

These quality measures, studied in Pavlyukevich [**86**] and Imkeller and Pavlyukevich [**59**] assess quality of tuning of the stochastic *output* to the periodic deterministic *input*. The concepts are mostly based on comparisons of trajectories of the noisy system and the deterministic periodic curve describing the location of the relevant meta-stable states, averaged with respect to the equilibrium measure (of the diffusion as a space-time process with time component given by uniform motion in the period interval). Again in the simple one-dimensional situation considered above the system switches between a double well potential state U with two wells of depths $\frac{V}{2}$ and $\frac{v}{2}$, $v < V$, during the first half period, and the spatially opposite one $U(\cdot)$ for the second half period. If as always time is re-scaled by T, the total period length is T, and stochastic perturbation comes from the coupling to a white noise of intensity ε. The most important measures of quality studied are the *spectral power amplification* and the related *signal-to-noise ratio*, both playing an eminent role in the physical literature (see Gammaitoni et al. [**43**], Freund et al. [**41**]). They mainly contain the mean square average in equilibrium of the Fourier component of the solution trajectories corresponding to the input period T, normalized in different ways. These measures of quality are functions of ε and T, and the problem of finding the resonance point consists in optimizing them in ε for fixed (large) T.

Due to the high complexity of original systems, when calculating the resonance point at optimal noise intensity, physicists usually pass to an effective dynamics description. It is given by a simple caricature of the system reducing the diffusion dynamics to the pure *inter well motion* (see e.g. McNamara and Wiesenfeld [**74**]). The reduced dynamics is represented by a continuous time two state Markov chain with transition probabilities corresponding to the inverses of the diffusions' Kramers' times. One then determines the optimal tuning parameters $\varepsilon(T)$ for large T for the approximating Markov chains in equilibrium, a rather simple task. To see that the Markov chain's behavior approaches the diffusion's in the small noise limit, spectral theory for the infinitesimal generator is used. The latter is seen to possess a spectral gap between the second and third eigenvalues, and therefore the closeness of equilibrium measures can be well controlled. Surprisingly, due to the importance of small *intra well fluctuations*, the tuning and resonance pattern of the Markov chain model may differ dramatically from the resonance picture of the diffusion. Subtle dependencies on the geometrical fine structure of the potential function U in the minima beyond the expected curvature properties lead to quite unexpected counterintuitive effects. For example, a subtle drag away from the other well caused by the sign of the third derivative of U in the deep well suffices to make the spectral power amplification curve strictly increasing in the parameter range where the approximating Markov chain has its resonance point.

It was this lack of robustness against model reduction which motivated Herrmann and Imkeller [**50**] to look for different measures of quality of periodic tuning for diffusion trajectories. These notions are designed to depend only on the rough inter well motion of the diffusion. The measure treated in the setting of one-dimensional diffusion processes subject to periodic forcing of small frequency is related to the transition probability during a fixed time window of exponential length $T(\varepsilon) = \exp(\frac{\mu}{\varepsilon})$ parametrized by a free energy parameter μ according to the

Kramers–Eyring formula. The corresponding exit rate is maximized in μ to account for optimal tuning. The methods of investigation of stochastic resonance in [**50**] are heavily based on comparison arguments which are not an appropriate tool from dimension 2 on. Time inhomogeneous diffusion processes such as the ones under consideration are compared to piecewise homogeneous diffusions by freezing the potential's time dependence on small intervals.

In Herrmann et al. [**51**] this approach is extended to the general setting of *finite dimensional diffusion processes* with two meta-stable states. Since the stochastic resonance criterion considered in [**50**] is based on transition times between them, our analysis relies on a suitable notion of transition or exit time parametrized again by the free energy parameter μ from $T(\varepsilon) = \exp(\frac{\mu}{\varepsilon})$ as a natural measure of scale. Assume now that the depths of the two equilibria of the potential in analogy to the scenarios considered before are smooth periodic functions of time of period 1 given for one of them by $\frac{v(t)}{2}$, and for the other one by the same function with some phase delay (for instance by $\frac{1}{2}$). Therefore, at time s the system needs energy $v(s)$ to leave the domain of attraction of the equilibrium. Hence an exit from this set should occur at time

$$a_\mu = \inf\{t \geq 0 \colon v(t) \leq \mu\}$$

in the diffusion's natural time scale, in the time re-scaled by $T(\varepsilon)$ thus at time $a_\mu \cdot T(\varepsilon)$. To find a quality measure of periodic tuning depending only on the transition dynamics, we look at the probabilities of transitions to the other domain within a time window $[(a_\mu - h)T(\varepsilon), (a_\mu + h)T(\varepsilon)]$ centered at $a_\mu \cdot T(\varepsilon)$ for small $h > 0$. If τ is the random time at which the diffusion roughly reaches the other domain of attraction (to be precise, one has to look at first entrance times of small neighborhoods of the corresponding equilibrium), we use the quantity (again, to be precise, we use the worst case probability for the diffusion starting in a point of a small neighborhood of the equilibrium of the starting domain)

$$\mathcal{M}^h(\varepsilon, \mu) = \mathbb{P}\Big(\tau \in [(a_\mu - h)T(\varepsilon), (a_\mu + h)T(\varepsilon)]\Big).$$

To symmetrize this quality measure with respect to switching of the equilibria, we refine it by taking its minimum with the analogous expression for interchanged equilibria. In order to exclude trivial or chaotic transition behavior, the scale parameter μ has to be restricted to an interval I_R of reasonable values which we call *resonance interval*. With this measure of quality, the stochastic resonance point may be determined as follows. We first fix ε and the window width parameter $h > 0$, and maximize $\mathcal{M}^h(\varepsilon, \mu)$ in μ, eventually reached for the time scale $\mu_0(h)$. Then the eventually existing limit $\lim_{h\to 0} \mu_0(h)$ will be the *resonance point*.

To calculate $\mu_0(h)$ for fixed positive h we use large deviations techniques. In fact, our main result consists in an extension of the Freidlin–Wentzell large deviations result to weakly time inhomogeneous dynamical systems perturbed by small Gaussian noise which states that

$$\lim_{\varepsilon\to 0} \varepsilon \ln\Big(1 - \mathcal{M}^h(\varepsilon, \mu)\Big) = \mu - v(a_\mu - h),$$

again in a form which is symmetric for switched equilibria. We show that this asymptotic relation holds uniformly w.r.t. μ on compact subsets of I_R, a fact which enables us to perform a maximization and find $\mu_0(h)$. The resulting notion of stochastic resonance is strongly related to the notions of periodic tuning based on *interspike intervals* (see [**49**]), which describe the probability distribution for transitions as functions of time with exponentially decaying spikes near the integer multiples of the forcing periods. It has the big advantage of being robust for model reduction, i.e. the passage from the diffusion to the two state Markov chain describing its reduced dynamics.

The techniques needed to prove this main result feature non-trivial extensions and refinements of the fundamental large deviations theory for time homogeneous diffusions by Freidlin–Wentzell [**40**]. We prove a large deviations principle for the inhomogeneous diffusion (0.1) and further strengthen this result to get uniformity in system parameters. Similarly to the time homogeneous case, where large deviations theory is applied to the problem of *diffusion exit* culminating in a mathematically rigorous proof of the Kramers–Eyring law, we study the problem of diffusion exit from a domain which is carefully chosen in order to allow for a detailed analysis of transition times. The main idea behind our analysis is that the natural time scale is so large that re-scaling in these units essentially leads to an asymptotic freezing of the time inhomogeneity, which has to be carefully controlled, to hook up to the theory of large deviations of time homogeneous diffusions.

The material in the book is organized as follows. In Chapter 1 we give a detailed treatment of the heuristics behind our mathematical approach, mostly in space dimension 1. We start by giving a fairly thorough account of the paradigm of glacial cycles which was the historical root of physical models exhibiting stochastic resonance. It gives rise to the model equation of a weakly periodically forced dynamical system with noise that can be interpreted as the motion of an overdamped physical particle in a weakly periodically forced potential landscape subject to noise. The heuristics of exit and transition behavior between domains of attraction (potential wells) of such systems based on the classical large deviations theory is explained in two steps: first for time independent potential landscapes, then for potentials switching discontinuously between two anti-symmetric states every half period. Freidlin's quasi-deterministic motion is seen to not cover the concept of optimal periodic tuning between weak periodic input and randomly amplified output. They determine stochastic resonance through measures of quality of periodic tuning such as the spectral power amplification or the signal-to-noise ratio. The latter concepts are studied first for finite state Markov chains capturing the dynamics of the underlying diffusions reduced to the meta-stable states, and then for the diffusions with time continuous periodic potential functions. The robustness defect of the classical notions of resonance in passing from Markov chain to diffusion is pointed out. Then alternative notions of resonance are proposed which are based purely on the asymptotic behavior of transition times. Finally, examples of systems exhibiting stochastic resonance features from different areas of science are presented and briefly discussed. They document the ubiquity of the phenomenon of stochastic resonance.

Our approach is based on concepts of large deviations. Therefore Chapter 2 is devoted to a self-contained treatment of the theory of large deviations for randomly perturbed dynamical systems in finite dimensions. Following a direct and elegant approach of Baldi and Roynette [**3**], we describe Brownian motion in its Schauder decomposition. It not only allows a direct approach to its regularity properties in terms of Hölder norms on spaces of continuous functions. It also allows a derivation of Schilder's large deviation principle (LDP) for Brownian motion from the elementary LDP for one-dimensional Gaussian random variables. The key to this elegant and direct approach is Ciesielski's isomorphism of normed spaces of continuous functions with sequence spaces via Fourier representation. The proof of the LDP for Brownian motion using these arguments is given after recalling general notions and basic concepts about large deviations, especially addressing their construction from exponential decay rates of probabilities of basis sets of topologies, and their transport between different topological spaces via continuous mappings (contraction principle). Since we only consider diffusion processes with additive noise for which Itô's map is continuous, an appeal to the contraction principle provides the LDP for the homogeneous diffusion processes we study. Finally, we follow Dembo and Zeitouni [**25**] to derive the exit time laws due to Freidlin and Wentzell [**40**] for time homogeneous diffusions from domains of attraction of underlying dynamical systems in the small noise limit.

Chapter 3 deals with an approach to stochastic resonance for diffusions with weakly time periodic drift and additive noise in the spirit of the associated Markovian semigroups and their spectral theory. This approach, presented in space dimension 1, is clearly motivated by the physical notions of periodic tuning, in particular the spectral power amplification coefficient. It describes the average spectral component of the diffusion trajectories corresponding to the frequency of the periodic input signal given by the drift term. We first give a rigorous account of Freidlin's quasi-deterministic limiting motion for potential double well diffusions of this type. We then follow the paradigm of the physics literature, in particular NcNamara and Wiesenfeld [**74**], and introduce the effective dynamics of our weakly periodically forced double well diffusions given by reduced continuous time Markov chains jumping between their two meta-stable equilibria. In this setting, different notions of periodic tuning can easily be investigated. We not only consider the physicists' favorites, spectral power amplification and signal-to-noise ratio, but also other reasonable concepts in which the energy carried by the Markov chain trajectories or the entropy of their invariant measures are used. Turning to diffusions with weakly time periodic double well potentials and additive noise again, we then develop an asymptotic analysis of their spectral power amplification coefficient based on the spectral theory of their infinitesimal generators. It is based on the crucial observation that in the case of double well potentials its spectrum has a gap between the second and third eigenvalue. Therefore we have to give the corresponding eigenvalues and eigenfunctions a more detailed study, in particular with respect to their asymptotic behavior in the small noise limit. Its results then enable us to give a related small noise asymptotic expansion both of the densities of the associated invariant measures as for the spectral power amplification coefficients. We finally compare spectral power amplification coefficients of the Markov chains describing the reduced dynamics and the associated diffusions, to find that in the small noise limit they may be essentially different, caused by the *small oscillations catastrophe* near the potential wells' bottoms.

This motivates us in Chapter 4 to look for notions of periodic tuning for the solution trajectories of diffusions in spaces of arbitrary finite dimension with weakly periodic drifts and additive small noise which do not exhibit this robustness defect. We aim at notions related to the maximal probabilities that the random exit or transition times between different domains of attraction of the underlying dynamical systems happen in time windows parametrized by free energy parameters on an exponential scale. For the two-state Markov chains describing the effective dynamics of the diffusions with slow and weak time inhomogeneity this optimal transition rate is readily calculated. This concept moreover has the advantage that their related transition times, as well as the corresponding ones for diffusions with a weak noise dependent time inhomogeneity, allow a treatment by methods of large deviations in the small noise limit. We therefore start with a careful extension of large deviations theory to diffusions with slow time inhomogeneity. The central result for the subsequent analysis of their exit times is contained in a large deviations principle, uniform with respect to the energy parameter. It allows us in the sequel to derive upper and lower bounds for the asymptotic exponential exit rate from domains of attraction for slowly time dependent diffusions. They combine to the main large deviations result describing the exact asymptotic exponential exit rates for slowly and weakly time inhomogeneous diffusions in the small noise limit. This central result is tailor made for providing the optimal tuning rate related to maximal probability of transition during an exponential time window. We finally compare the resulting stochastic resonance point to the ones obtained for the Markov chains of the reduced dynamics, and conclude that they agree in the small noise limit, thus establishing robustness.

In two appendices — for easy reference in the text — we collect some standard results about Gronwall's lemma and Laplace's method for integrals with exponential singularities of the integrand.

CHAPTER 1

Heuristics of noise induced transitions

1.1. Energy balance models of climate dynamics

The simple concept of energy balance models stimulated research not only in the area of conceptual climate models, but was at the cradle of a research direction in physics which subsequently took important examples from various domains of biology, chemistry and neurology: it was one of the first examples for which the phenomenon of *stochastic resonance* was used to explain the transition dynamics between different stable states of physical systems. For a good overview see Gammaitoni et al. [**43**] or Jung [**62**].

In the end of the 70's, Nicolis [**83**] and Benzi et al. [**5**] almost simultaneously tried *stochastic resonance* as a rough and qualitative explanation for the glaciation cycles in earth's history. They were looking for a simple mathematical model appropriate to explain experimental findings from deep sea core measurements according to which the earth has seen ten glacial periods during the last million years, alternating with warm ages rather regularly in periods of about 100 000 years. Mean temperature shifts between warm age and glacial period are reported to be of the order of 10 K, and relaxation times, i.e. transition times between two relatively stable mean temperatures as rather short, of the order of only 100 years. Mathematically, their explanation was based on an equation stating the global energy balance in terms of the average temperature $T(t)$, where the global average is taken meridionally (i.e. over all latitudes), zonally (i.e. over all longitudes), and annually around time t. The global radiative power change at time t is equated to the difference between incoming solar (short wave) radiative power R_{in} and outgoing (long wave) radiative power R_{out}.

The power R_{in} is proportional to the global average of the solar constant $Q(t)$ at time t. To model the periodicity in the glaciation cycles, one assumes that Q undergoes periodic variations due to one of the so-called *Milankovich cycles*, based on periodic perturbations of the earth's orbit around the sun. Two of the most prominent cycles are due to a small periodic variation between 22.1 and 24.5 degrees of the angle of inclination (*obliquity*) of the earth's rotation axis with respect to its plane of rotation, and a very small periodic change of only about 0.1 percent of the *eccentricity*, i.e. the deviation from a circular shape, of the earth's trajectory around the sun. The obliquity cycle has a duration of about 41 000 years, while the eccentricity cycle corresponds to the 100 000 years observed in the temperature proxies from deep sea core measurements mentioned above. They are caused by gravitational influences of other planets of our solar system. In formulas, Q was assumed to be of the form

$$Q(t) = Q_0 + b \sin \omega t,$$

with some constants Q_0, b and a frequency $\omega = 10^{-5}[\frac{1}{y}]$.

The other component determining the absorbed radiative power R_{in} is a rough and difficult to model averaged *surface albedo* of the earth, i.e. the proportion of the solar power absorbed. It is supposed to be just (average) temperature dependent. For temperatures below $\underline{T}$, for which the surface water on earth is supposed to have turned into ice, and the surface is thus constantly bright, the albedo is assumed to be constantly equal to $\overline{a}$, for temperatures above $\overline{T}$, for which all ice has melted, and the surface constantly brown, it is assumed to be given by a constant $\underline{a} < \overline{a}$. For temperatures between $\underline{T}$ and $\overline{T}$, the two constant values $\underline{a}$ and $\overline{a}$ are simply linearly interpolated in the most basic model. The rough albedo function has therefore the *ramp function* shape depicted in Figure 1.1.

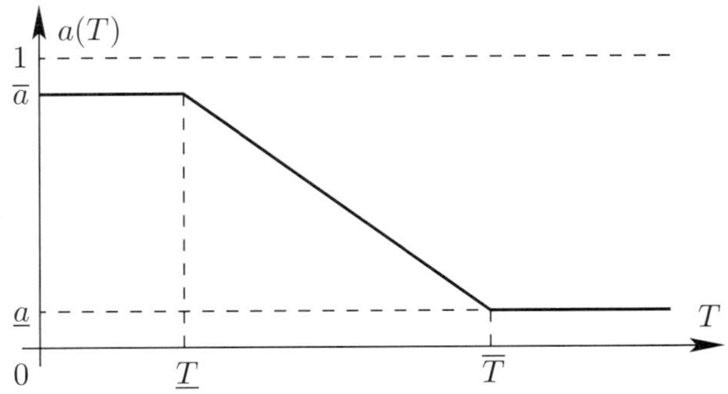

FIGURE 1.1. The albedo function $a = a(T)$.

To have a simple model of R_{out}, the earth is assumed to behave approximately as a *black body radiator*, for which the emitted power is described by the Stefan–Boltzmann law. It is proportional to the fourth power of the body's temperature and is given by $\gamma T^4(t)$, with a constant γ proportional to the Stefan constant.

Hence the simple energy balance equation with periodic input Q on which the model is built is given by

$$c\frac{d}{dt}T(t) = Q(t)\Big(1 - a(T(t))\Big) - \gamma T(t)^4, \tag{1.1}$$

where the constant c describes a global thermal inertia. According to (1.1), (quasi-) stationary states of average temperature should be given by the solutions of the equation $\frac{dT(t)}{dt} = 0$. If the model is good, they should reasonably well interpret glacial period and warm age temperatures. Graphically, they are given by the intersections of the curves of absorbed and emitted radiative power, see Figures 1.2 and 1.3.

As we shall more carefully explain below, the lower $(T_1(t))$ and upper $(T_3(t))$ quasi-equilibria are stable, while the middle one $(T_2(t))$ is unstable. The equilibrium $T_1(t)$ should represent an ice age temperature, $T_3(t)$ a warm age, while $T_2(t)$ is not observed over noticeably long periods. In their dependence on t they should describe small fluctuations due to the variations in the solar constant. But here one encounters a serious problem with this purely deterministic model. If the fluctuation amplitude of Q is small, then we will observe the two disjoint branches of stable solutions T_1 and T_3 (Figure 1.4).

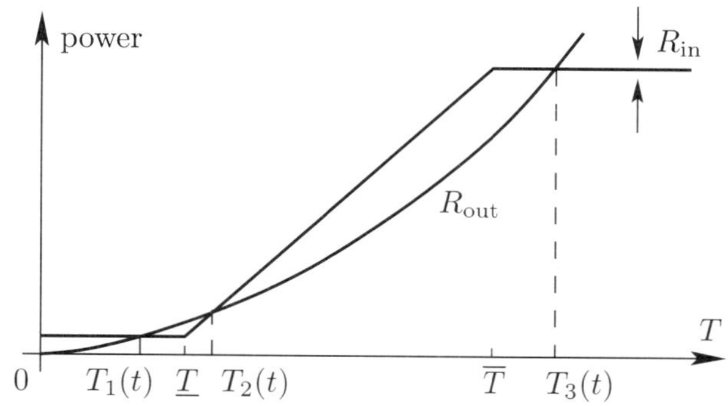

FIGURE 1.2. Incoming vs. outgoing power.

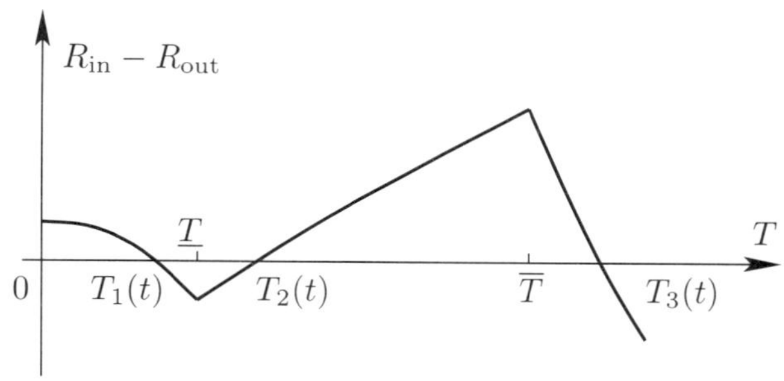

FIGURE 1.3. Difference of the powers of incoming and outgoing radiation.

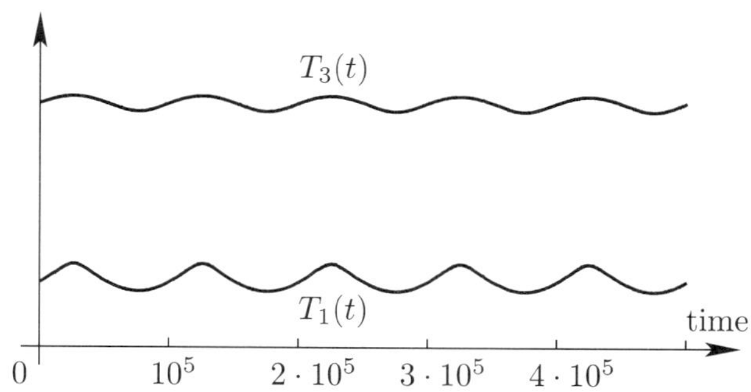

FIGURE 1.4. Equilibrium temperatures $T_1(t)$ and $T_3(t)$ for small fluctuation amplitude b.

However for both branches alone — besides being unrealistically low or high — the difference between minimal and maximal temperature can by no means account for the observed shift of about 10 K, and also the relaxation times are much too long. But the most important shortcoming of the model is the lacking possibility of transitions between the two branches.

If we allow the fluctuation amplitude b to be large, the picture is still very unrealistic: There are intervals during which one of the two branches T_1 or T_3

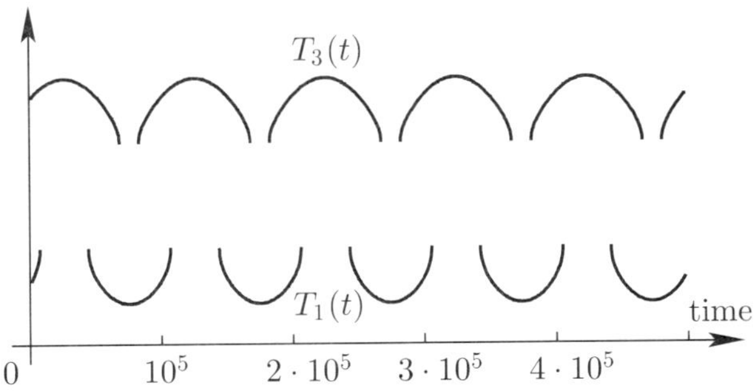

FIGURE 1.5. Unrealistic equilibrium temperatures $T_1(t)$ and $T_3(t)$ for large fluctuation amplitude b.

vanishes completely, and transitions are still impossible, unless one is willing to accept discontinuous behavior (Figure 1.5).

For this reason, Nicolis [**83**] and Benzi et al. [**5**] proposed to add a noise term in (1.1). Despite the fact that then negative temperatures become possible, they worked with the equation

$$cT^{\varepsilon}{}'(t) = Q(t)\Big(1 - a(T^{\varepsilon}(t))\Big) - \gamma T^{\varepsilon}(t)^4 + \sqrt{\varepsilon}\,\dot{W}_t, \tag{1.2}$$

$\varepsilon > 0$, where $\dot{W}$ is a white noise. In passing to (1.2), stable equilibria of the deterministic system become — approximately at least — meta-stable states of the stochastic system. And more importantly, the unbounded noise process W makes spontaneous transitions (*tunneling*) between the meta-stable states $T_1(t)$ and $T_3(t)$ possible. In fact, the *random hopping* between the meta-stable states immediately exhibits two features which make the model based on (1.2) much more attractive for a qualitative explanation of glaciation cycles: a) the transitions between T_1 and T_3 allow for far more realistic temperature shifts, b) relaxation times are random, but very short compared to the periods the process solving (1.2) spends in the stable states themselves.

But now a new problem arises, which actually provided the name *stochastic resonance*.

If, seen on the scale of the period of Q, ε is too small, the solution may be trapped in one of the states T_1 or T_3. By the periodic variation of Q, there are well defined periodically returning time intervals during which $T_1(t)$ is the *more probable* state, while $T_3(t)$ takes this role for the rest of the time. So if ε is small, the process, initially in T_1, may for example fail to leave this state during a whole period while the other one is more probable. The solution trajectory may then look as in Figure 1.6.

If, on the other hand, ε is too large, the big random fluctuation may lead to eventual excursions from the actually more probable equilibrium during its domination period to the other one. The trajectory then typically looks like in Figure 1.7.

In both cases it will be hard to speak of a random periodic curve. Good tuning with the periodic forcing by Q is destroyed by a random mechanism being too slow or too fast to follow. It turned out in numerous simulations in a number of similar systems that there is, however, an *optimal* parameter value ε for which the solution

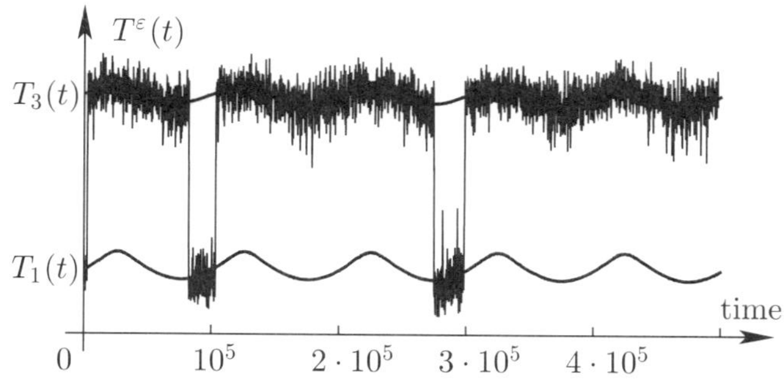

FIGURE 1.6. A typical solution trajectory of equation (1.2) for the small noise amplitude.

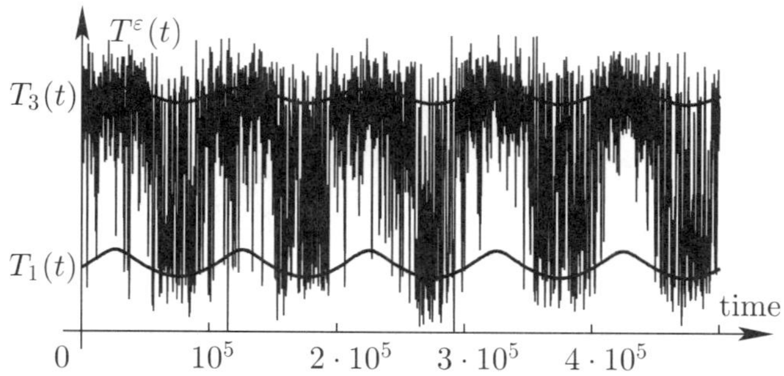

FIGURE 1.7. A typical solution trajectory of equation (1.2) for the large noise amplitude.

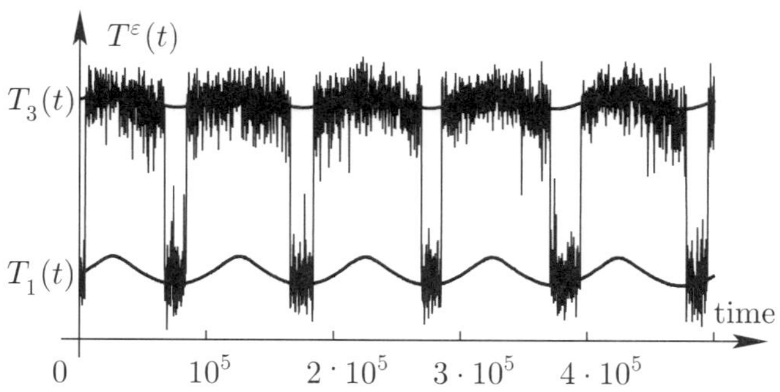

FIGURE 1.8. A typical solution trajectory of equation (1.2), the noise amplitude well tuned.

curves are well tuned with the periodic input. A typical well tuned curve is shown in Figure 1.8.

The optimally tuned system is then said to be in *stochastic resonance.* Nicolis [**83**] and Benzi et al. [**5**], by tuning the noise parameter ε to appropriate values, were able to give qualitative explanations for glaciation cycles based on this phenomenon.

Stochastic resonance proved to be relevant in other elementary climate models than the energy balance models considered so far. In Penland et al. [**87**], Wang et al. [**107, 106**], a two-dimensional stochastic model for a qualitative explanation of the ENSO (El Niño Southern Oscillation) phenomenon also leads to stochastic resonance effects: for certain parameter ranges the model exhibits random tuned transitions between two stable sea surface temperatures. New evidence for the presence of stochastic resonance phenomena in paleo-climatic time series was added by Ganopolski and Rahmstorf [**45**]. Their paper interprets the GRIP ice core record representing temperature proxies from the Greenland glacier that extend over a period of roughly 90 000 years, and showing the fine structure of the temperature record of the last glacial period. The time series shows about 20 intermediate warmings during the last glacial period commonly known under the name of Dansgaard–Oeschger events. These events are clearly marked by rapid spontaneous increases of temperature by about $6K$ followed by slower coolings to return to the initial basic cold age temperature. It was noted in [**45**] that a histogram of the number of Dansgaard–Oeschger events with a duration of $k \cdot 1480$ years, with $k = 1, 2, 3, \ldots$ exhibits the typical shape of a stochastic resonance spike train consistent for instance with the results of Herrmann and Imkeller [**49**] for Markov chains describing the effective diffusion dynamics, or Berglund and Gentz [**13**] for diffusion processes with periodic forcing.

1.2. Heuristics of our mathematical approach

The rigorous mathematical elaboration of the concept of *stochastic resonance* is the main objective of this book. We start its mathematically sound treatment by giving a heuristical outline of the main stream of ideas and arguments based on the methods of *large deviations* for random dynamical systems in the framework of the Freidlin–Wentzell theory. Freidlin [**39**] is able to formulate Kramers' [**65**] very old seminal approach mathematically rigorously in a very general setting, and this way provides a lower estimate for the good tuning (see also the numerical results by Milstein and Tretyakov [**77**]). To obtain an upper estimate, we finally argue by embedding time discrete Markov chains into the diffusion processes that describe the effective dynamics of noise induced transitions. Optimal tuning results obtained for the Markov chains will then be transferred to the original diffusion processes.

To describe the idea of our approach, let us briefly return to our favorite example explained in the preceding section. Recall that the function

$$f(t,T) = Q(t)\Big(1 - a(T)\Big) - \gamma T^4, \quad T, t \geq 0,$$

describes a multiple of $R_{\text{in}} - R_{\text{out}}$, and its very slow periodicity in t is initiated by the assumption on the solar constant $Q(t) = Q_0 + b\sin(\omega t)$. Let us compare this quantity, sketched in Figure 1.9 schematically for two times, say t_1, t_2 such that Q takes its minimum at t_1 and its maximum at t_2. The graph of f moves periodically between the two extreme positions. Note that in the one-dimensional situation considered, $f(t,\cdot)$ can be seen as the negative gradient of a potential function $U(t,\cdot)$ which depends periodically on time t.

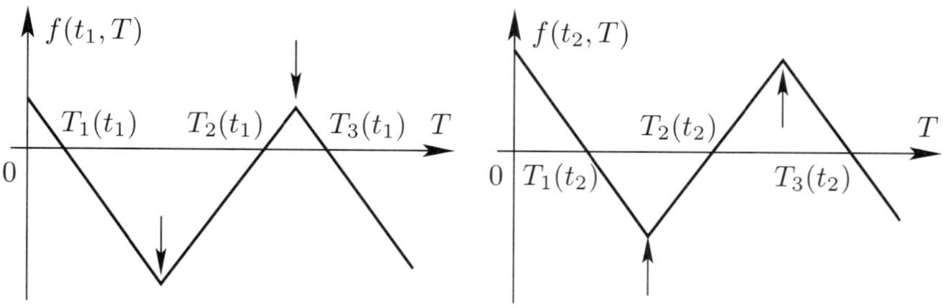

FIGURE 1.9. Schematical form of radiation power difference at times t_1 and t_2

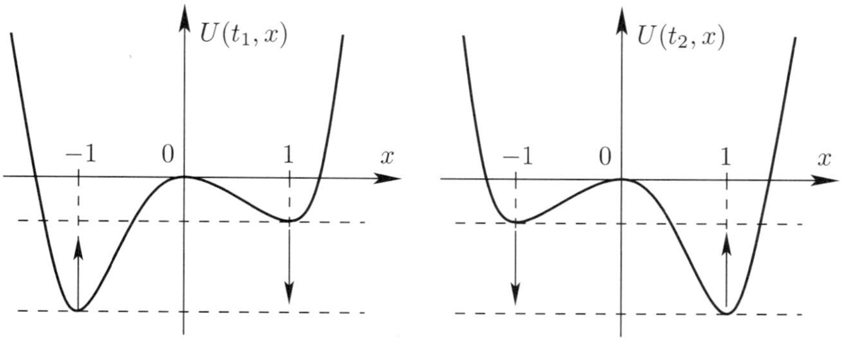

FIGURE 1.10. Potential function U at time instants t_1 and t_2.

We now turn to a more general context. For simplicity of the heuristical exposition still sticking to a one-dimensional scenario, we start by considering a temporally varying *potential function* U and set

$$f(t, \cdot) = -\frac{\partial}{\partial x} U(t, \cdot), \quad t \geq 0.$$

We assume that U oscillates in time between the two extreme positions depicted schematically in Figure 1.10.

In Figure 1.10 (l.), the potential well on left hand side is deeper than on the right hand side, in Figure 1.10 (r.) the role of the deeper well has changed. As t varies, we will observe a smoothly time dependent potential with two wells of periodically and smoothly fluctuating relative depth. Just the function describing the position of the deepest well will in general be discontinuous. It will play a crucial role in the analysis now sketched.

We assume in the sequel for simplicity that $U(t, x)$, $t \geq 0$, $x \in \mathbb{R}$, is a smooth function such that for all $t \geq 0$, $U(t, \cdot)$ has exactly two minima, one at -1, the other at 1, and that the two wells at -1 and 1 are separated by the saddle 0, where $U(t, 0)$ is assumed to take the value 0. Two moment pictures of the potential may look as in Figure 1.10.

We further assume time periodicity for U, more formally that

$$U(t, \cdot) = U(t + 1, \cdot).$$

The variable period of the input will be denoted by some positive number T. We therefore consider the stochastic differential equation

$$\frac{d}{dt}X_t^\varepsilon = f\Big(\frac{t}{T}, X_t^\varepsilon\Big) + \sqrt{\varepsilon}\,\dot{W}_t, \tag{1.3}$$

with a one-dimensional Wiener process W (white noise $\dot{W}$). We may circumscribe a more mathematical concept of stochastic resonance like this: given T ($\omega = \frac{1}{T}$), find the parameter $\varepsilon = \varepsilon(T)$ such that X^ε is *optimally tuned* with the periodic input $f(\frac{t}{T}, \cdot)$. We pose the problem in the following (almost equivalent) way: given $\varepsilon > 0$, find the good scale $T = T(\varepsilon)$ such that optimal tuning of X^ε with the periodic input is given, at least in the limit $\varepsilon \to 0$.

1.2.1. Random motion of a strongly damped Brownian particle. The analogy with the motion of a physical particle in a periodically changing double well potential landscape alluded to in (1.3) (see also Mazo [**72**] and Schweitzer [**97**]) motivates us to pause for a moment and give it a little more thought. As in the previous section, let us concentrate on a one-dimensional setting, remarking that our treatment easily generalizes to a finite-dimensional setting. Due to Newton's law, the motion of a particle is governed by the impact of all forces acting on it. Let us denote F the sum of these forces, m the mass, x the space coordinate and v the velocity of the particle. Then

$$m\dot{v} = F.$$

Let us first assume the potential to be turned off. In their pioneering work at the turn of the twentieth century, Marian Smoluchowski and Paul Langevin introduced stochastic concepts to describe the Brownian particle motion by claiming that at time t

$$F(t) = -\gamma v(t) + \sqrt{2kT\gamma}\dot{W}_t.$$

The first term results from friction γ and is velocity dependent. An additional stochastic force represents random interactions between Brownian particles and their simple molecular random environment. The white noise $\dot{W}$ (the formal derivative of a Wiener process) plays the crucial role. The diffusion coefficient (standard deviation of the random impact) is composed of Boltzmann's constant k, friction and environmental temperature T. It satisfies the condition of the fluctuation-dissipation theorem expressing the balance of energy loss due to friction and energy gain resulting from noise. The equation of motion becomes

$$\begin{cases} \dot{x}(t) = v(t), \\ \dot{v}(t) = -\dfrac{\gamma}{m}\,v(t) + \dfrac{\sqrt{2kT\gamma}}{m}\dot{W}_t. \end{cases}$$

In equilibrium, the stationary Ornstein–Uhlenbeck process provides its solution:

$$v(t) = v(0)\,e^{-\frac{\gamma}{m}t} + \frac{\sqrt{2kT\gamma}}{m}\int_0^t e^{-\frac{\gamma}{m}(t-s)}\,dW_s.$$

The ratio $\beta := \frac{\gamma}{m}$ determines the dynamic behavior. Let us focus on the over-damped situation with large friction and very small mass. Then for $t \gg \frac{1}{\beta} = \tau$ (relaxation time), the first term in the expression for velocity can be neglected, while the stochastic integral represents a Gaussian process. By integrating, we obtain in the over-damped limit ($\beta \to \infty$) that v and thus x is Gaussian with almost constant mean

$$m(t) = x(0) + \frac{1 - e^{-\beta t}}{\beta} v(0) \approx x(0)$$

and covariance close to the covariance of white noise, see Nelson [**82**]:

$$\begin{aligned} K(s,t) &= \frac{2kT}{\gamma} \min(s,t) + \frac{kT}{\gamma\beta}\Big(-2 + 2e^{-\beta t} + 2e^{-\beta s} - e^{-\beta|t-s|} - e^{-\beta(t+s)}\Big) \\ &\approx \frac{2kT}{\gamma} \min(s,t), \quad s,t \geq 0. \end{aligned}$$

Hence the time-dependent change of the velocity of the Brownian particle can be neglected, the velocity rapidly converges to thermal equilibrium ($\dot{v} \approx 0$), while the spatial coordinate remains far from it. In the so-called adiabatic transformation, the evolution of the particle's position is thus given by the transformed Langevin equation

$$\dot{x}(t) = \frac{\sqrt{2kT}}{\gamma} \dot{W}_t.$$

Let us next suppose that we face a Brownian particle in an external field of force, associated with a potential $U(t,x)$, $t \geq 0$, $x \in \mathbb{R}$. This then leads to the Langevin equation

$$\begin{cases} \dot{x}(t) = v(t), \\ m\dot{v}(t) = -\gamma\, v(t) - \dfrac{\partial U}{\partial x}(t, x(t)) + \sqrt{2kT\gamma}\, \dot{W}_t. \end{cases}$$

In the over-damped limit, after relaxation time, the adiabatic elimination of the fast variables (see Gardiner [**46**]) then leads to an equation similar to the one encountered in the previous section, namely

$$\dot{x}(t) = -\frac{1}{\gamma}\frac{\partial U}{\partial x}(t, x(t)) + \frac{\sqrt{2kT}}{\gamma}\dot{W}_t.$$

1.2.2. Time independent potential. We now continue discussing the heuristics of stochastic resonance for systems described by equations of the type encountered in the previous two sections. To motivate the link to the theory of large deviations, we first study the case in which $U(t,\cdot)$ is given by some time independent potential function U for all t. Following Freidlin and Wentzell [**40**], the description of the asymptotics contained in the *large deviations principle* requires the crucial notion of *action functional*. It is defined for $T > 0$ and absolutely continuous functions $\varphi\colon [0,T] \to \mathbb{R}$ with derivative $\dot{\varphi}$ by

$$S_{0T}(\varphi) = \frac{1}{2}\int_0^T \Big[\dot{\varphi}_s - \Big(-\frac{\partial}{\partial x} U\Big)(\varphi_s)\Big]^2 ds.$$

By means of the action functional we can define the *quasipotential function*

$$V(x,y) = \inf\{S_{0T}(\varphi)\colon \varphi_0 = x, \varphi_T = y, T > 0\},$$

for x, $y \in \mathbb{R}$. Intuitively, $V(x,y)$ describes the minimal work to be done in the potential landscape given by U to pass from x to y. Keeping this in mind, the relationship between U and V is easy to understand (for a more formal argument see Chapter 3). If x and y are in the same potential well, we have

$$V(x,y) = 2(U(y) - U(x))^+, \tag{1.4}$$

where $a^+ = a \vee 0 = \max\{a, 0\}$ denotes the positive part of a real number a. In particular, if $U(y) < U(x)$, then $V(x, y) = 0$, i.e. going downhill in the landscape does not require work. If, however, x and y are in different potential wells, we have (recall $U(0) = 0$)

$$V(x, y) = -2U(x). \tag{1.5}$$

This equation reflects the fact that the minimal work to do to pass to y consists in reaching the saddle 0, since then one can just go downhill.

Rudiments of the following arguments can also be found in the explanation of stochastic resonance by McNamara and Wiesenfeld [**74**]. The main ingredient is the *exit time law* by Freidlin and Wentzell [**40**] (see also Eyring [**37**], Kramers [**65**] and Bovier et al. [**14**]). For $y \in \mathbb{R}$, $\varepsilon > 0$ the first time y is visited is defined by

$$\tau_y^\varepsilon = \inf\{t \geq 0 \colon X_t^\varepsilon = y\}.$$

If $\mathbb{P}_x$ denotes the law of the diffusion $(X_t^\varepsilon)_{t \geq 0}$ started at $x \in \mathbb{R}$, the exit time law states that for any $\delta > 0$, $x \in \mathbb{R}$ we have

$$\mathbb{P}_x\left(e^{\frac{V(x,y)-\delta}{\varepsilon}} \leq \tau_y^\varepsilon \leq e^{\frac{V(x,y)+\delta}{\varepsilon}}\right) \to 1 \tag{1.6}$$

as $\varepsilon \to 0$.

In other words, in the limit $\varepsilon \to 0$, the process started at x takes approximately time $\exp(\frac{V(x,y)}{\varepsilon})$ to reach y, or more roughly

$$\varepsilon \ln \tau_y^\varepsilon \cong V(x, y)$$

as $\varepsilon \to 0$. As a consequence, one finds that as $\varepsilon \to 0$, on time scales $T(\varepsilon)$ at least as long as $\exp(\frac{V(x,y)}{\varepsilon})$ or such that

$$\varepsilon \ln T(\varepsilon) > V(x, y),$$

we may expect with $\mathbb{P}_x$-probability close to 1 that the process $X_{tT(\varepsilon)}^\varepsilon$ has reached y by time 1. Remembering (1.4) and (1.5) one obtains the following statement formulated much more generally by Freidlin. Suppose

$$\lim_{\varepsilon \to 0} \varepsilon \ln T(\varepsilon) > 2 \max\{-U(-1), -U(1)\}, \tag{1.7}$$

and $U(-1) < U(1)$. Then the Lebesgue measure of the set

$$\left\{t \in [0, 1] \colon |X_{tT(\varepsilon)}^\varepsilon - (-1)| > \delta\right\} \tag{1.8}$$

tends to 0 in $\mathbb{P}_x$-probability as $\varepsilon \to 0$, for any $\delta > 0$.

In other words, the process X^ε, run in a time scale $T(\varepsilon)$ large enough, will spend most of the time in the deeper potential well. Excursions to the other well are *exponentially negligible* on this scale, as $\varepsilon \to 0$. The picture is roughly as deployed in Figure 1.11.

1.2.3. Periodic step potentials and quasi-deterministic motion.

As a rough approximation of temporally continuously varying potential functions we may consider periodic *step function* potentials such as

$$U(t, \cdot) = \begin{cases} U_1(\cdot), & t \in [k, k + \frac{1}{2}), \\ U_2(\cdot), & t \in [k + \frac{1}{2}, k + 1), \quad k \in \mathbb{N}_0. \end{cases} \tag{1.9}$$

We assume that both U_1 and U_2 are of the type described above, that $U_1(x) = U_2(-x)$, $x \in \mathbb{R}$, and that U_1 has a well of depth $\frac{V}{2}$ at -1, and a well of depth $\frac{v}{2}$ at

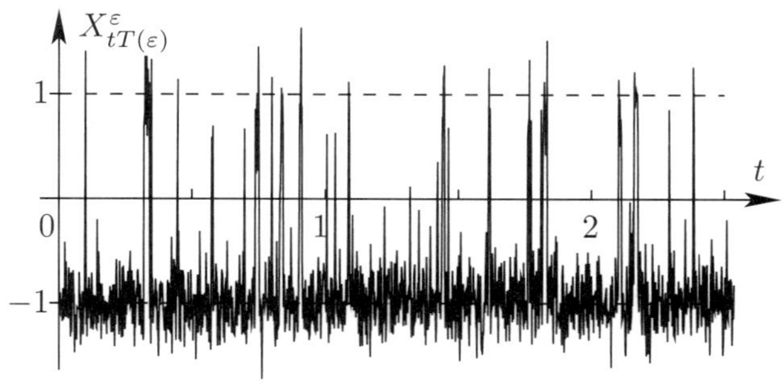

FIGURE 1.11. Solution trajectory of the diffusion $X^{\varepsilon}_{tT(\varepsilon)}$ in the time independent double-well potential U.

1, with $V > v$ (and U_2 wells with respectively opposite roles). Let us briefly point out the main features of the transition times for periodic step potentials described in (1.6). According to (1.6) the exponential rate of the transition time from -1 to 1 in U_1 in the small noise limit is asymptotically given by $\exp(\frac{V}{\varepsilon})$, as long as the time scale T of the diffusion allows no switching of the potential states before, i.e. as long as $T = T(\varepsilon) > \exp(\frac{V}{\varepsilon})$. Accordingly, the transition time from 1 to -1 in U_1 is given by $\exp(\frac{v}{\varepsilon})$, as long as $T = T(\varepsilon) > \exp(\frac{v}{\varepsilon})$. Similar statements hold for transitions between states of U_2. It is therefore also plausible that (1.8) generalizes to the following statement due to Freidlin [**39**, Theorem 2].

Suppose

$$\lim_{\varepsilon\to 0} \varepsilon \ln T(\varepsilon) > V. \tag{1.10}$$

Define

$$\phi(t) = \begin{cases} -1, & t \in [k, k+\frac{1}{2}), \\ 1, & t \in [k+\frac{1}{2}, k+1), \quad k \in \mathbb{N}_0. \end{cases}$$

Then the Lebesgue measure of the set

$$\left\{t \in [0,1] \colon |X^{\varepsilon}_{t\,T(\varepsilon)} - \phi(t)| > \delta\right\} \tag{1.11}$$

tends to 0 as $\varepsilon \to 0$ in $\mathbb{P}_x$-probability, for any $\delta > 0$, $x \in \mathbb{R}$.

Again, this just means that the process X^{ε}, run in a time scale $T(\varepsilon)$ large enough, will spend most of the time in the minimum of the deepest potential well which is given by the time periodic function ϕ. Excursions to the other well are *exponentially negligible* on this scale, as $\varepsilon \to 0$. The picture is typically the one depicted in Figure 1.12.

1.2.4. Periodic potentials and quasi-deterministic motion. Since the function ϕ appearing in the previous theorem is already discontinuous, it is plausible that the step function potential is in fact a reasonable approximation of the general case of continuously and (slowly) periodically changing potential functions. It is intuitively clear how the result has to be generalized to this situation. We just have to replace the periodic step potentials by potentials *frozen* along a partition of the period interval on the potential state taken at its starting point, and finally let the mesh of the partition tend to 0. To continue the discussion in the spirit of the previous section and with the idea of instantaneously frozen potential states, we

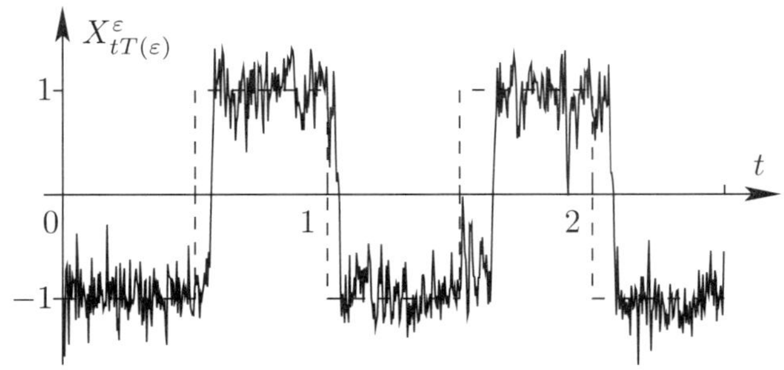

FIGURE 1.12. Solution trajectory of the diffusion $X^{\varepsilon}_{tT(\varepsilon)}$ in the double-well periodic step potential.

have to explain the asymptotics of the minimal time a Brownian particle needs to exit from the (frozen) starting well, say the left one. Freezing the potential at some time s, the asymptotics of its exit time is derived from the classical large deviation theory of randomly perturbed dynamical systems, see Freidlin and Wentzell [**40**]. Let us assume that U is locally Lipschitz continuous. We recall that for any $t \geq 0$ the potential $U(t, \cdot)$ has its minima at -1 and 1, separated by the saddle point 0. The law of the first exit time $\tau_1^{\varepsilon} = \inf\{t \geq 0\colon X_t^{\varepsilon} > 0\}$ is described by some particular functional related to large deviation. For $t > 0$, we introduce the *action functional* on the space of real valued continuous functions $C([0,t], \mathbb{R})$ on $[0,t]$ by

$$S_t^s(\varphi) = \begin{cases} \dfrac{1}{2}\displaystyle\int_0^t \left(\dot{\varphi}_u + \dfrac{\partial}{\partial x} U(s, \varphi_u)\right)^2 du \text{ if } \varphi \text{ is absolutely continuous,} \\ +\infty \text{ otherwise,} \end{cases}$$

which is non-negative and vanishes on the set of solutions of the ordinary differential equation $\dot{\varphi} = -\frac{\partial}{\partial x} U(s, \varphi)$. Let x and y be real numbers. With respect to the (frozen) action functional, we define the (frozen) quasipotential

$$V_s(x, y) = \inf\{S_t^s(\varphi)\colon \varphi \in C([0,t], \mathbb{R}),\ \varphi_0 = x,\ \varphi_t = y,\ t \geq 0\}$$

which represents the minimal work the diffusion with a potential frozen at time s and starting in x has to do in order to reach y. To switch wells, the Brownian particle starting in the left well's bottom -1 has to overcome the barrier. So we let

$$\overline{V}_s = V_s(-1, 0).$$

This minimal work needed to exit from the left well can be computed explicitly, and is equal to twice its depth at time s. The asymptotic behavior of the exit time is expressed by

$$\lim_{\varepsilon \to 0} \varepsilon \ln \mathbb{E}_x \tau_1^{\varepsilon} = \overline{V}_s$$

or in generalization of (1.6)

$$\lim_{\varepsilon \to 0} \mathbb{P}_x\left(e^{\frac{\overline{V}_s - \delta}{\varepsilon}} < \tau_1^{\varepsilon} < e^{\frac{\overline{V}_s + \delta}{\varepsilon}}\right) = 1 \text{ for any } \delta > 0 \text{ and } x < 0.$$

Let us now assume that the left well is the deeper one at time s. If the Brownian particle has enough time to cross the barrier, i.e. if $T(\varepsilon) > e^{\frac{\overline{V}_s}{\varepsilon}}$, then, generalizing

(1.8), Freidlin in [**39**, Theorem 1] proves that independently of the starting point x it should stay near -1 in the following sense. The Lebesgue measure of the set

$$\left\{t \in [0,1]\colon |X^{\varepsilon}_{tT(\varepsilon)} - (-1)| > \delta\right\}$$

converges to 0 in probability as $\varepsilon \to 0$. If $T(\varepsilon) < e^{\frac{\overline{V}_s}{\varepsilon}}$, the time left is not long enough for any crossing: the particle, starting at x, stays in the starting well, near the stable equilibrium point. In other words, the Lebesgue measure of the set

$$\left\{t \in [0,1]\colon |X^{\varepsilon}_{tT(\varepsilon)} - (-\mathbb{I}_{(-\infty,0)}(X^{\varepsilon}_{tT(\varepsilon)}) + \mathbb{I}_{[0,\infty)}(X^{\varepsilon}_{tT(\varepsilon)}))| > \delta\right\}$$

converges to 0 in the small noise limit. This observation is at the basis of Freidlin's law of quasi-deterministic periodic motion discussed in the subsequent section. The lesson it teaches is this: to observe switching of the position to the energetically most favorable well, $T(\varepsilon)$ should be larger than some critical level $e^{\frac{\lambda}{\varepsilon}}$, where $\lambda = \inf_{s\geq 0} \overline{V}_s$. Measuring time in exponential scales by μ through the equation $T(\varepsilon) = e^{\frac{\mu}{\varepsilon}}$, the condition translates into $\mu > \lambda$. Continuing the reasoning of the preceding subsection, if this condition is satisfied, we may define a periodic function ϕ denoting the deepest well position in dependence on t. Then, in generalization of (1.11), the process X^{ε} will spend most of the time, measured by Lebesgue's measure, near ϕ for small ε.

1.2.5. Quality of periodic tuning and reduced motion. Do the manifestations of quasi-deterministic motion in instantaneously frozen potentials just discussed explain *stochastic resonance*? The problem is obvious. They just give lower bounds for the scale $T(\varepsilon) = e^{\frac{\mu}{\varepsilon}}$ for which noise strength ε leads to random switches between the *most probable* potential wells near the (periodic) deterministic times when the role of the deepest well switches. But if μ is too big, occasional excursions into the higher well will destroy a truly periodic tuning with the potential (see Figure 1.12). Just the duration of the excursions, being exponentially smaller than the periods of dwelling in the deeper well, will not be noticed by the residence time criteria discussed. We therefore also need an upper bound for possible scales. In order to find this optimal tuning scale $\mu_R > \lambda$, we first have to measure *goodness* of periodic tuning of the trajectories of the solution. In the huge physics literature on stochastic resonance, two families of criteria can be distinguished. The first one is based on invariant measures and spectral properties of the infinitesimal generator associated with the diffusion X^{ε}. Now, X^{ε} is not time autonomous and consequently does not admit invariant measures. By taking into account deterministic motion of time in the interval of periodicity and considering the time autonomous process $Z^{\varepsilon}_t = (t \bmod T(\varepsilon), X^{\varepsilon}_t)$, $t \geq 0$, we obtain a Markov process with an invariant measure $\nu^{\varepsilon}_t(x)\, dt\, dx$. In particular, for $t \geq 0$ the law of $X^{\varepsilon}_t \sim \nu^{\varepsilon}_t(x)\, dx$ and the law of $X^{\varepsilon}_{t+T(\varepsilon)} \sim \nu^{\varepsilon}_{t+T(\varepsilon)}(x)\, dx$, under this measure are the same for all $t \geq 0$. Let us present the most important notions of quality of tuning (see Jung [**62**], or Gammaitoni et al. [**43**]):

- the *spectral power amplification* (SPA) which plays an eminent role in the physics literature and describes the energy carried by the spectral component of the averaged trajectories of X^{ε} corresponding to the period of the signal:

$$\eta^X(\varepsilon, T) = \left| \int_0^1 \mathbb{E}_{\nu} X^{\varepsilon}_{sT} \cdot e^{2\pi i s}\, ds \right|^2, \quad \varepsilon > 0,\ T > 0.$$

- the total energy of the averaged trajectories

$$\mathrm{En}^X(\varepsilon,T)=\int_0^1\left|\mathbb{E}_\nu X_{sT}\right|^2 ds,\quad \varepsilon>0,\ T>0.$$

The second family of criteria is more probabilistic. It refers to quality measures purely based on the location of transition times between domains of attraction of the local minima, and residence time distributions measuring the time spent in one well between two transitions, or interspike times. This family, to be discussed in more detail in Section 1.4 below is certainly less popular in the physics community.

As will turn out later, these physical notions of quality of periodic tuning of random trajectories exhibit one important drawback: they are not robust with respect to model resolution. It is here that an important concept of model reduction enters the stage. It is based on the conjecture that the effective dynamical properties of periodically forced diffusion processes as given by (1.3) can be traced back to finite state Markov chains periodically hopping between the stable equilibria of the potential function underlying the diffusion, for which the smallness parameter of the noise intensity is simply reflected in the transition matrix. These Markov chains should be designed to capture the essential information about the *inter-well* dynamics of the diffusion, while *intra-well* small fluctuations of the diffusion near the potential minima are neglected. Investigating goodness of tuning according to the different physical measures of quality makes sense both for the Markov chains as for the diffusions. This idea of model reduction was captured and followed in the physics literature in Eckmann and Thomas [**32**], McNamara and Wiesenfeld [**74**], and Nicolis [**83**]. In fact, theoretical work on the concept of stochastic resonance in the physics literature is based on the model reduction approach, see the surveys Anishchenko et al. [**1**], Gammaitoni et al. [**43, 44**], Moss et al. [**79**], and Wellens et al. [**108**].

As we shall see in Chapter 3, the optimal tuning relations between ε and T do not necessarily agree for Markov chains and diffusions. Even in the small noise limit discrepancies may persist that are caused by very subtle geometric properties of the potential function. It is our goal to present a notion of quality of periodic tuning which possesses this robustness property when passing from the Markov chains capturing the effective dynamics to the original diffusions. For this reason we shall study the different physical notions of quality of tuning first in the context of typical finite state Markov chains with periodically forced transition matrices.

1.3. Markov chains for the effective dynamics and the physical paradigm of spectral power amplification

To keep this heuristic exposition of the main ideas of our mathematical approach as simple as possible, besides allowing only two states for our Markov chain that play the role of the stable equilibria of the potential -1 and 1, let us also discretize time. We continue to assume as in the discussion of periodically switching potential states above that $U_1(-1)=U_2(1)=-\frac{V}{2}$, and $U_1(1)=U_2(-1)=-\frac{v}{2}$. In a setting better adapted to our continuous time diffusion processes, in Chapter 3 time continuous Markov chains switching between two states will capture the effective diffusion dynamics. Hence, we follow here Pavlyukevich [**86**] and Imkeller and Pavlyukevich [**58**] and shall assume in this section that the parameter T in our model describing the period length, is an even integer. So for $T\in 2\mathbb{N}$, $\varepsilon>0$, consider a Markov chain $Y^\varepsilon = (Y^\varepsilon(k))_{k\ge 0}$ on the state space $\mathcal{S} = \{-1, 1\}$. Let $P_T(k)$ be the matrix of one-step transition probabilities at time k. If we denote $p_T^-(k) = \mathbb{P}(Y^\varepsilon(k) = -1)$, $p_T^+(k) = \mathbb{P}(Y^\varepsilon(k) = 1)$, and write P^* for the transposed matrix, we have

$$\begin{pmatrix} p_T^-(k+1) \\ p_T^+(k+1) \end{pmatrix} = P_T^*(k) \begin{pmatrix} p_T^-(k) \\ p_T^+(k) \end{pmatrix}.$$

In order to model the periodic switching of the double-well potential in our Markov chains, we define the transition matrix P_T to be periodic in time with period T. More precisely,

$$P_T(k) = \begin{cases} Q_1, & 0 \le k \bmod T \le \frac{T}{2} - 1, \\ Q_2, & \frac{T}{2} \le k \bmod T \le T - 1, \end{cases}$$

with

$$Q_1 = \begin{pmatrix} 1-\varphi & \varphi \\ \psi & 1-\psi \end{pmatrix}, \quad Q_2 = \begin{pmatrix} 1-\psi & \psi \\ \varphi & 1-\varphi \end{pmatrix}, \tag{1.12}$$
$$\varphi = pe^{-V/\varepsilon}, \quad \psi = qe^{-v/\varepsilon},$$

where $0 \le p, q \le 1$, $0 < v < V < +\infty$, $0 < \varepsilon < \infty$.

The entries of the transition matrices clearly are designed to mimic transition rates between -1 and 1 or vice versa that correspond to the transition times of the diffusion processes between the meta-stable equilibria, given according to the preceding section by $\exp(\frac{V}{\varepsilon})$ resp. $\exp(\frac{v}{\varepsilon})$. The exponential factors in the one-step transition probabilities are just chosen to be the inverses of those mean transition times. This is exactly what elementary Markov chain theory requires *in equilibrium*. The phenomenological *prefactors* p and q, chosen between 0 and 1, add asymmetry to the picture.

It is well known that for a time-homogeneous Markov chain on $\{-1, 1\}$ with transition matrix P_T one can talk about *equilibrium*, given by the stationary distribution, to which the law of the chain converges exponentially fast. The stationary distribution can be found by solving the matrix equation $\pi = P_T^*\pi$ with norming condition $\pi^- + \pi^+ = 1$.

For time non homogeneous Markov chains with time periodic transition matrix, the situation is quite similar. Enlarging the state space $\mathcal{S}$ to $\mathcal{S}_T = \{-1, 1\} \times \{0, 1, \ldots, T-1\}$, we recover a time homogeneous chain by setting

$$Z^\varepsilon(k) = (Y^\varepsilon(k), k \bmod T), \quad k \ge 0,$$

to which the previous remarks apply. For convenience of notation, we assume $\mathcal{S}_T$ to be ordered in the following way:

$$\mathcal{S}_T = \Big((-1,0), (1,0), (-1,1), (1,1), \ldots, (-1, T-1), (1, T-1)\Big).$$

Writing A_T for the matrix of one-step transition probabilities of Z^ε, the stationary distribution $R = (r(i,j))^*$ is obtained as a normalized solution of the matrix equation $(A_T^* - E)R = 0$, E being the identity matrix. We shall be dealing with the following variant of stationary measure, which is not normalized in time. Let $\pi_T(k) = (\pi_T^-(k), \pi_T^+(k))^* = (r(-1,k), r(1,k))^*$, $0 \le k \le T-1$. We call the family $\pi_T = (\pi_T(k))_{0\le k\le T-1}$ the *stationary distribution* of the Markov chain Y^ε.

The matrix A_T of one-step transition probabilities of Z^ε is explicitly given by

$$A_T = \begin{pmatrix} 0 & Q_1 & 0 & 0 & \cdots & 0 & 0 & 0 \\ 0 & 0 & Q_1 & 0 & \cdots & 0 & 0 & 0 \\ \vdots & \vdots & \vdots & \vdots & \ddots & \vdots & \vdots & \vdots \\ 0 & 0 & 0 & 0 & \cdots & 0 & Q_2 & 0 \\ 0 & 0 & 0 & 0 & \cdots & 0 & 0 & Q_2 \\ Q_2 & 0 & 0 & 0 & \cdots & 0 & 0 & 0 \end{pmatrix}.$$

The matrix A_T has block structure. In this notation 0 means a 2×2-matrix with all entries equal to zero, Q_1, and Q_2 are the 2-dimensional matrices defined in (1.12).

Applying some algebra we see that the equation $(A_T^* - E)R = 0$ is equivalent to $A'_T R = 0$, where

$$A'_T = \begin{pmatrix} \widehat{Q} - E & 0 & 0 & 0 & \cdots & 0 & 0 & 0 \\ Q_1^* & -E & 0 & 0 & \cdots & 0 & 0 & 0 \\ \vdots & \vdots & \vdots & \vdots & \ddots & \vdots & \vdots & \vdots \\ 0 & 0 & 0 & 0 & \cdots & -E & 0 & 0 \\ 0 & 0 & 0 & 0 & \cdots & P_2^* & -E & 0 \\ 0 & 0 & 0 & 0 & \cdots & 0 & Q_2^* & -E \end{pmatrix}$$

and $\widehat{Q} = Q_2^* Q_2^* \cdots Q_1^* Q_1^* = (Q_2^*)^{\frac{T}{2}}(Q_1^*)^{\frac{T}{2}}$. But A'_T is a block-wise lower diagonal matrix, and so $A'_T R = 0$ can be solved in the usual way resulting in the following formulas.

For every $T \in 2\mathbb{N}$, the stationary distribution π_T of Y^ε with matrices of one-step probabilities defined in (1.12) is given by

$$\begin{cases} \pi_T^-(l) = \dfrac{\psi}{\varphi+\psi} + \dfrac{\varphi-\psi}{\varphi+\psi}\cdot\dfrac{(1-\varphi-\psi)^l}{1+(1-\varphi-\psi)^{\frac{T}{2}}}, \\ \pi_T^+(l) = \dfrac{\varphi}{\varphi+\psi} - \dfrac{\varphi-\psi}{\varphi+\psi}\cdot\dfrac{(1-\varphi-\psi)^l}{1+(1-\varphi-\psi)^{\frac{T}{2}}}, \\ \pi_T^-(l+\frac{T}{2}) = \pi_T^+(l), \\ \pi_T^+(l+\frac{T}{2}) = \pi_T^-(l), \qquad 0 \le l \le \frac{T}{2}-1. \end{cases} \tag{1.13}$$

The proof of (1.13) is easy and instructive, and will be contained in the following arguments. Note that $\pi_T(0)$ satisfies the matrix equation

$$\Big((Q_2^*)^{\frac{T}{2}}(Q_1^*)^{\frac{T}{2}} - E\Big)\pi_T(0) = 0$$

with additional condition $\pi_T^-(0) + \pi_T^+(0) = 1$. To calculate $(Q_2^*)^{\frac{T}{2}}(Q_1^*)^{\frac{T}{2}}$, we use a formula for the k-th power of 2×2-matrices $Q = \begin{pmatrix} 1-a & a \\ b & 1-b \end{pmatrix}$, $a, b \in \mathbb{R}$, proved in a straightforward way by induction on k which reads

$$\begin{pmatrix} 1-a & a \\ b & 1-b \end{pmatrix}^k = \frac{1}{a+b}\begin{pmatrix} b & a \\ b & a \end{pmatrix} + \frac{(1-a-b)^k}{a+b}\begin{pmatrix} a & -a \\ -b & b \end{pmatrix}.$$

Using some more elementary algebra we find

$$\begin{aligned}(Q_2^*)^{\frac{T}{2}}(Q_1^*)^{\frac{T}{2}} &= \left((Q_1)^{\frac{T}{2}}(Q_2)^{\frac{T}{2}}\right)^* = \begin{pmatrix}1-\psi & \psi \\ \varphi & 1-\varphi\end{pmatrix}^{\frac{T}{2}}\begin{pmatrix}1-\varphi & \varphi \\ \psi & 1-\psi\end{pmatrix}^{\frac{T}{2}} \\ &= \frac{1}{\varphi+\psi}\begin{pmatrix}\varphi & \varphi \\ \psi & \psi\end{pmatrix} + (1-\varphi-\psi)^{\frac{T}{2}}\frac{\varphi-\psi}{\varphi+\psi}\begin{pmatrix}-1 & -1 \\ 1 & 1\end{pmatrix} \\ &+ \frac{(1-\varphi-\psi)^T}{\varphi+\psi}\begin{pmatrix}\varphi & -\psi \\ -\varphi & \psi\end{pmatrix},\end{aligned}$$

from which another straightforward calculation yields

$$\begin{cases}\pi_T^-(0) = \dfrac{\varphi}{\varphi+\psi} + \dfrac{\psi}{\varphi+\psi}\cdot\dfrac{(1-\varphi-\psi)^{\frac{T}{2}}}{1+(1-\varphi-\psi)^{\frac{T}{2}}}, \\ \pi_T^+(0) = \dfrac{\psi}{\varphi+\psi} + \dfrac{\varphi}{\varphi+\psi}\cdot\dfrac{(1-\varphi-\psi)^{\frac{T}{2}}}{1+(1-\varphi-\psi)^{\frac{T}{2}}}.\end{cases}$$

To compute the remaining entries, we use $\pi_T(l) = (Q_1^*)^l\pi_T(0)$ for $0 \le l \le \frac{T}{2}-1$, and $\pi_T(l) = (Q_2^*)^l(Q_1^*)^{\frac{T}{2}}\pi_T(0)$ for $\frac{T}{2} \le l \le T-1$ to obtain (1.13). Note also the symmetry $\pi_T^-(l+\frac{T}{2}) = \pi_T^+(l)$ and $\pi_T^+(l+\frac{T}{2}) = \pi_T^-(l)$, $0 \le l \le \frac{T}{2}-1$.

To motivate the physical quality of tuning concept of *spectral power amplification*, we first remark that our Markov chain Y^ε can be interpreted as amplifier of the periodic input signal of period T. In the stationary regime, i.e. if the law of Y^ε is given by the measure π_T, the power carried by the output Markov chain at frequency a/T is a random variable

$$\xi_T(a) = \frac{1}{T}\sum_{l=0}^{T-1} Y^\varepsilon(l)e^{\frac{2\pi i a}{T}l}.$$

We define the *spectral power amplification (SPA)* as the relative expected power carried by the component of the output with (input) frequency $\frac{1}{T}$. It is given by

$$\eta^Y(\varepsilon,T) = \left|\mathbb{E}_{\pi_T}\xi_T(1)\right|^2, \quad \varepsilon > 0, T \in 2\mathbb{N}.$$

Here $\mathbb{E}_{\pi_T}$ denotes expectation w.r.t. the stationary law π_T.

The explicit description of the invariant measure now readily yields an explicit formula for the spectral power amplification. In fact, using (1.13) one immediately gets

$$\begin{aligned}\mathbb{E}_{\pi_T}\xi_T(1) &= \frac{1}{T}\sum_{k=0}^{T-1}\mathbb{E}_{\pi_T}Y^\varepsilon(k)e^{\frac{2\pi i}{T}k} = \frac{1-e^{\pi i}}{T}\sum_{k=0}^{\frac{T}{2}-1}(\pi_T^+(k)-\pi_T^-(k))e^{\frac{2\pi i}{T}k} \\ &= \frac{4}{T}\frac{\varphi-\psi}{\varphi+\psi}\Big(\frac{1}{1-e^{\frac{2\pi i}{T}}} - \frac{1}{1-(1-\varphi-\psi)e^{\frac{2\pi i}{T}}}\Big).\end{aligned}$$

Elementary algebra then leads to the following description of the spectral power amplification coefficient of the Markov chain Y^ε for $\varepsilon > 0$, $T \in 2\mathbb{N}$:

$$\eta^Y(\varepsilon,T) = \frac{4}{T^2\sin^2(\frac{\pi}{T})}\cdot\frac{(\varphi-\psi)^2}{(\varphi+\psi)^2+4(1-\varphi-\psi)\sin^2(\frac{\pi}{T})}. \tag{1.14}$$

Note now that the one-step probabilities Q_1 and Q_2 depend on the parameters noise level ε. Our next goal is to *tune* this parameter to a value which maximizes the amplification coefficient $\eta^Y(\varepsilon, T)$ as a function of ε. So the *stochastic resonance* point is marked by the maximum of the spectral power amplification coefficient as a function of ε. To calculate it, substitute $e^{-1/\varepsilon} = x$, and differentiate the explicit formula (1.14). The resulting relationship between period length $T(\varepsilon)$ and *noise intensity* ε marking the stochastic resonance point can be recast in the formula

$$T(\varepsilon) \cong \frac{1}{2\pi}\sqrt{pq\frac{V-v}{v}}\exp\left(\frac{V+v}{2\varepsilon}\right).$$

The maximal value of spectral power amplification is given by

$$\lim_{\varepsilon \to 0} \eta^Y(\varepsilon, T(\varepsilon)) = \frac{4}{\pi^2}$$

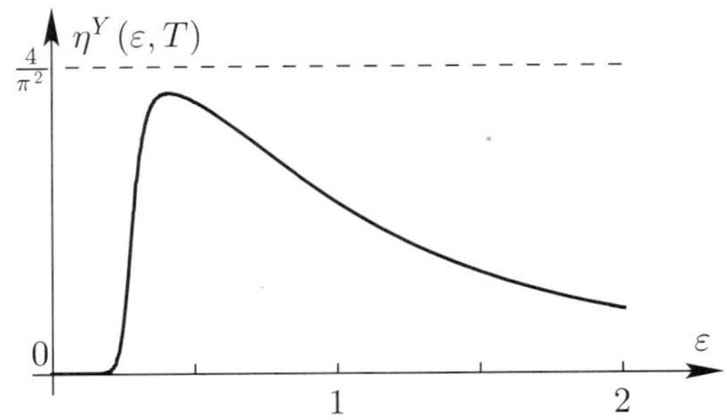

FIGURE 1.13. The coefficient of the spectral power amplification $\varepsilon \mapsto \eta^Y(\varepsilon, T)$ for $p = q = 0.5$, $V = 2$, $v = 1$, $T = 10\,000$.

We also see that the spectral power amplification as a measure of quality of stochastic resonance allows to distinguish a unique time scale, and find its exponential rate $\frac{V+v}{2}$ together with the pre-exponential factor. The optimal exponential rate is therefore given by the arithmetic mean of the two potential barriers marked by the deep and shallow well of our double well potential. This basic relationship will appear repeatedly at different stages of our mathematical elaboration of concepts of optimal tuning.

We may summarize our findings so far for discrete Markov chains that capture the effective dynamics of the potential diffusions which are our main subject of interest. Following the physics literature (e.g. Gammaitoni et al. [**43**] and McNamara and Wiesenfeld [**74**]) we understand *stochastic resonance* as *optimal spectral power amplification*. The closely related notion of *signal-to-noise ratio* and other reasonable concepts based on quality measures such as the relative entropy of invariant laws are discussed for Markov chains in Chapter 3 (see Section 3.2). The spectral power amplification coefficient measures the power carried by the expected Fourier coefficient in equilibrium of the Markov chain switching between the stable equilibria of the potential landscape of the diffusion which corresponds to the frequency of the underlying periodic deterministic signal.

1.4. Diffusions with continuously varying potentials

The concept of spectral power amplification is readily extended to Markov chains in continuous time, still designed to capture the effective diffusion dynamics in higher dimensions, as well as to potential diffusions themselves. This will be done in detail in Chapter 3. However, it will turn out that diffusions and their reduced dynamics Markov chains are not as similar as expected. Indeed, in a reasonably large time window around the resonance point for Y^ε, the tuning picture of the spectral power amplification for the diffusion is different. Under weak regularity conditions on the potential, it exhibits strict monotonicity in the window. Hence optimal tuning points for diffusion and Markov chain differ essentially. In other words, the diffusion's SPA tuning behavior is not robust for passage to the reduced model (see Chapter 3, subsection 3.4.4). This strange deficiency is difficult to explain. The main reason of this subtle effect appears to be that the diffusive nature of the Brownian particle is neglected in the reduced model. In order to point out this feature, we may compute the SPA coefficient of $g(X^\varepsilon)$ where g is a particular function designed to cut out the small fluctuations of the diffusion in the neighborhood of the bottoms of the wells, by identifying all states there. So $g(x) = -1$ (resp. 1) in some neighborhood of -1 (resp. 1) and otherwise g is the identity. This results in

$$\tilde{\eta}^X(\varepsilon, T) = \left| \int_0^1 \mathbb{E}_\nu g(X^\varepsilon_{sT})\, e^{2\pi i s}\, ds \right|^2, \quad \varepsilon > 0,\ T \geq 0.$$

In the small noise limit this quality function admits a local maximum close to the resonance point of the reduced model: the growth rate of $T_{\mathrm{opt}}(\varepsilon)$ is also given by the arithmetic mean of the wells' depths. So the lack of robustness seems to be due to the small fluctuations of the particle in the wells' bottoms.

In any case, this clearly calls for other quality measures to be used to transfer properties of the reduced model to the original one. Our discussion indicates that due to their emphasis on the pure transition dynamics, a second more probabilistic family of quality measures should be used. This will be made mathematically rigorous in Chapter 4. The family is composed of quality measures based on transition times between the domains of attraction of the local minima, residence times distributions measuring the time spent in one well between two transitions, or interspike times. To explain its main features there is no need to restrict to landscapes frozen in time independent potential states on half period intervals. So from now on the potential $U(t, x)$ is a continuous function in (t, x). For simplicity — remaining in the one-dimensional case — we further suppose that its local minima are given by ± 1, and its only saddle point by 0, independently of time. So the only meta-stable states on the time axis are ± 1. Let us denote by $\frac{v_-(t)}{2}$ (resp. $\frac{v_+(t)}{2}$ the depth of the left (resp. right) well. These function are continuous and 1-periodic. We shall assume that they are strictly monotonous between their global extrema. Let us now consider the motion of the Brownian particle in this landscape. As in the preceding case, according to Freidlin's law of quasi-deterministic motion its trajectory gets close to the global minimum, if the period is large enough. The exponential rate of the period should be large enough to permit transitions: if $T(\varepsilon) = e^{\mu/\varepsilon}$ with $\mu \geq \max_{i=\pm} \sup_{t\geq 0} v_i(t)$ meaning that μ is larger than the maximal work needed to cross the barrier, then the particle often switches between the two wells and should stay close to the deepest position in the landscape. By defining $\phi(t) = 2\mathbb{I}_{\{v_+(t) > v_-(t)\}} - 1$, in the small noise limit the Lebesgue measure of the set

$$\{t \in [0, 1] \colon |X_{tT} - \phi(t)| > \delta\}$$

converges to 0 in probability for any $\delta > 0$. But in this case many transitions occur in practice, and the trajectory looks chaotical instead of periodic. So we have to choose smaller periods even if we cannot assure that the particle stays close to the global minimum since it needs some time to cross the barrier. Let us study the transition times. For this we assume that the starting point is -1 corresponding to the bottom of the deepest well. If the depth of the well is always larger than $\mu = \varepsilon \ln T(\varepsilon)$, then the particle does not have enough time during one period to climb the barrier and should therefore stay in the starting well. On the contrary if the depth of the starting well becomes smaller than μ, the transition can and will happen. More precisely, for $\mu \in (\inf_{t\geq 0} v_-(t), \sup_{t\geq 0} v_-(t))$ we define

$$a_\mu^-(s) = \inf\{t \geq s \colon v_-(t) \leq \mu\}.$$

The first transition time from -1 to 1 denoted τ_+ has the following asymptotic behavior in the small noise limit: $\tau_+/T(\varepsilon) \to a_\mu^-(0)$. The second transition which lets the particle return to the starting well will appear near the deterministic time $a_\mu^+(a_\mu^-(s))T(\varepsilon)$. The definitions of the coefficients a_μ^- and a_μ^+ are similar, the depth of the left well just being replaced by that of the right well. In order to observe periodic behavior of the trajectory, the particle has to stay a little time in the right well before going back. This will happen under the assumption $v_+(a_\mu(0)) > \mu$, that is, the right well is the deepest one at the transition time. In fact we can then define the *resonance interval* I_R, the set of all values μ such that the trajectories look periodic in the small noise limit:

$$I_R = \Big(\max_{i=\pm} \inf_{t\geq 0} v_i(t), \inf_{t\geq 0} \max_{i=\pm} v_i(t) \Big).$$

On this interval trajectories approach some deterministic periodic limit. We now outline the construction of a quality measure that is based on these observations, to be optimized in order to obtain *stochastic resonance* as the best possible response to periodic forcing. The measure we consider is based on the probability that a random transition of the diffusion happens during a small time window around the limiting deterministic transition time. Recall the transition times $\tau_{\pm 1}^\varepsilon$ of X^ε to ± 1. For $h > 0, \varepsilon > 0, T \geq 0$ let

$$\mathcal{M}^h(\varepsilon, T) = \min_{i=\pm} \mathbb{P}_i\Big(\frac{\tau_{\mp 1}^\varepsilon}{T(\varepsilon)} \in [a_\mu^i - h, a_\mu^i + h] \Big).$$

In the small noise limit, this quality measure tends to 1 and optimal tuning can be obtained due to its asymptotic behavior described by the formula

$$\lim_{\varepsilon\to 0} \varepsilon \ln(1 - \mathcal{M}^h(\varepsilon, T)) = \max_{i=\pm}\{\mu - v_i(a_\mu^i - h)\}$$

for $\mu \in I_R$, uniformly on each compact subset. This property results from classical large deviation techniques applied to an approximation of the diffusion which is supposed to be locally time homogeneous, and will be derived in Chapter 4. Now we minimize the term on the left hand side in the preceding equality. In fact, if the window length $2h$ is small then $\mu - v_i(a_\mu^i - h) \approx 2hv_i'(a_\mu^i)$ since $v_i(a_\mu^i) = \mu$ by definition. The value $v_i'(a_\mu^i)$ is of course negative. Thus the position in which its absolute value is maximal should be identified. At this position the depth of the starting well drops most rapidly below the level μ.

It is clear that for h small the eventually existing global minimizer $\mu_R(h)$ is a good candidate for the resonance point. To get rid of the dependence on h, we shall consider the limit of $\mu_R(h)$ as $h \to 0$ denote by μ_R. This limit, if it exists, is called the *resonance point* of the diffusion with time periodic landscape U. Let us note that for $v_-(t) = \frac{V+v}{4} + \frac{V-v}{4}\cos(2\pi t)$ and $v_+(t) = v_-(t+\pi)$, which corresponds to the case of periodically switching wells' depths between $\frac{v}{2}$ to $\frac{V}{2}$ as in the *frozen* landscape case described above. Then the optimal tuning is $T(\varepsilon) = \exp(\frac{\mu_R}{\varepsilon})$ with $\mu_R = \frac{v+V}{2}$. This optimal rate is equivalent to the optimal rate given by the SPA coefficient.

The big advantage of the quality measure based on the transition times is its robustness. Let us therefore consider the reduced model consisting in a two-state Markov chain with the infinitesimal generator

$$Q(t) = \begin{pmatrix} -\varphi(t) & \varphi(t) \\ \psi(t) & -\psi(t) \end{pmatrix},$$

where $\varphi(t) = \exp(-\frac{v_-(t/T)}{\varepsilon})$ and $\psi(t) = \exp(-\frac{v_+(t/T)}{\varepsilon})$. The distribution of transition times of this Markov chain is well known (see Chapter 4) and, divided by the period length, converges to a^i_μ. The reduced dynamics of the diffusion is captured by the Markov chain, and the optimization of the quality measure $\mathcal{M}^h(\varepsilon, T)$ for the Markov chain and the diffusion leads to the same resonance points.

Our investigation focuses essentially on two criteria: one concerning the family of spectral measures, especially the spectral power amplification coefficient, and the other one dealing with transitions between the local minima of the potential. Many other criteria for optimal tuning between weak periodic signals in dynamical systems and stochastic response can be employed (see Chapter 3). The relation between long deterministic periods and noise intensity usually is expressed in exponential form $T(\varepsilon) = \exp(\frac{\mu}{\varepsilon})$, since the particle needs exponentially large times to cross the barrier separating the wells. This approach relies on the basic assumption that the barrier height is bounded below uniformly in time. This assumption which seems natural in the simple energy balance model of climate dynamics may be questionable in other situations. If the barrier height becomes small periodically on a scale related to the noise intensity, the Brownian particle does not need to wait an exponentially long time to climb it. In this scaling trajectories may appear periodic in the small noise limit. The modulation is assumed to be slow, but the time dependence does not have to be assumed exponentially slow in the noise intensity. In a series of papers [**8, 9, 10, 11, 12**] and in a monograph [**13**], Berglund and Gentz study the case in which the barrier between the wells becomes low twice per period: at time zero the right-hand well becomes almost flat and at the same time the bottom of the well and the saddle approach each other; half a period later, the scenario with the roles of the wells switched occurs. Even in this situation, there is a threshold value for the noise intensity under which transitions are unlikely and, above this threshold, trajectories typically exhibit two transitions per period. In this particular situation, optimal tuning can be described in terms of the concentration of sample paths in small space-time sets.

1.5. Stochastic resonance in models from electronics to biology

As described in the preceding sections, the paradigm of stochastic resonance can quite generally and roughly be seen as the optimal amplification of a weak periodic signal in a dynamical system triggered by random forcing. In this section, we shall briefly deviate from the presentation of our mathematical approach of optimal tuning by large deviations methods, illustrate the ubiquity of the phenomenon of stochastic resonance. We will briefly discuss some prominent examples of dynamical systems arising in different areas of natural sciences in which it occurs, following several big reviews on stochastic resonance from the point of view of natural sciences such as [**1, 43, 44, 79, 108**]. We refer the reader to these references for ample further information on a huge number of examples where stochastic resonance appears. Finally we will briefly comment on computational aspects of stochastic resonance that are important in particular in high dimensional applications.

1.5.1. Resonant activation and Brownian ratchets. The two popular examples we mention here are elementary realizations of transition phenomena corresponding roughly to our paradigm of an overdamped Brownian particle in a potential landscape subject to weak periodic variation of some parameters. Here we face the examples of one-well potentials resp. asymmetric periodic multi-well potentials.

The effect of the so-called *resonant activation* arises in the simple situation in which an overdamped Brownian particle exits from a single potential well with randomly fluctuating potential barrier. In the case we consider the potential barrier can be considered to undergo weak periodic deterministic fluctuations in contrast. Even in the simplest situation, in which the height of the potential barrier is given by a Markov chain switching between two states, one can observe a non-linear dependence of the mean first exit time from the potential well and the intensity of the switching (see e.g. Doering and Elston [**28**]).

Noise induced transport in Brownian ratchets addresses the directed motion of the Brownian particle in a spatially asymmetric periodic potential having the shape of a long chain of downward directed sawtooths of equal length. It arises as another exit time phenomenon, since random exits over the lower potential barrier on the right hand side of the particle's actual position are highly favored. For instance in the context of an electric conductor, this effect creates a current in the downward direction indicated, see Doering et al. [**28, 27**] and Reimann [**91**]. An important application of this effect is the biomolecular cargo transport, see e.g. Elston and Peskin [**35**] and Vanden–Eijnden et al. [**80, 26**].

1.5.2. Threshold models and the Schmitt trigger. Models of stochastic resonance based on a bistable weakly periodic dynamical system of the type (0.1) are often referred to as dynamical models in contrast to the so-called non-dynamical or *threshold models*. These are models usually consisting of a biased deterministic input which may be periodic or not, and a multi-state output. In the simplest situation, the output takes a certain value as the input crosses a critical threshold. The simplest model of this type is the Schmitt trigger, an *electronic device* studied first by Fauve and Heslot [**38**] and Melnikov [**76**] (see also [**1, 43, 69, 70, 74**]). It is given by a well-known electronic circuit, characterized by a two-state output and a hysteretic loop. The circuit is supplied with the input voltage $w = w_t$, which is an arbitrary function of time. In the ideal Schmitt trigger the output voltage $Y = Y_t$ has only two possible values, say $-V$ and V. Let w increase from $-\infty$. Then $Y = V$ until w reaches the critical voltage level V_+. As this happens, the output jumps instantaneously to the level $-V$. Decreasing w does not affect the output Y until w reaches the critical voltage V_-. Then Y jumps back. Therefore, the Schmitt trigger is a bistable system with hysteresis, see Figure 1.14. The width of the hysteresis loop is $V_+ - V_-$. Applying a periodic voltage of small amplitude a

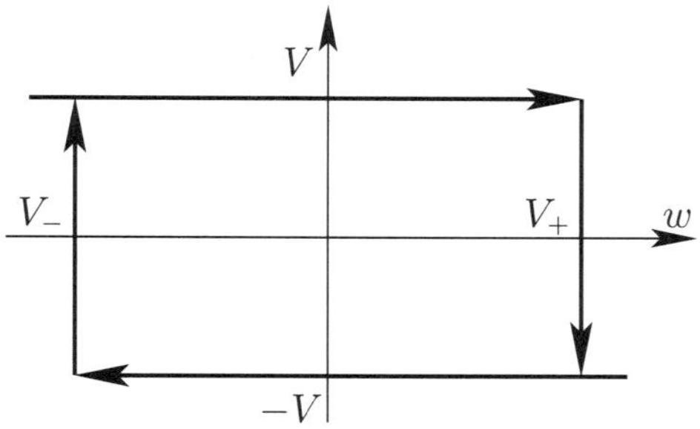

FIGURE 1.14. The input-output characteristic (hysteresis loop) of the Schmitt trigger.

and period $T > 0$, for example, to V_+, we periodically modulate the critical level. After adding a random noise at the input, the system is able to jump between the two states $\pm V$. As in the example of glacial cycles we can consider a discontinuous modulation, for instance given by $V_+(t) = a\,\mathrm{sign}(\sin(\frac{2\pi t}{T}))$. The whole picture is now similar to the one in (1.3). Here the periodic modulation of the reference voltage corresponds to the tilting of the potential wells.

Fauve and Heslot [**38**] studied the power spectrum of the system and, as in the glacial cycle example, established that the energy carried by the spectral component of Y at a given driving frequency has a local maximum for a certain intensity of the input noise.

The Schmitt trigger provides another interpretation to the phenomenon of stochastic resonance. A system displaying stochastic resonance can be considered as a random amplifier. The weak periodic signal which cannot be detected in the absence of noise, can be successfully recovered if the system (the Schmitt trigger or (1.3)) is appropriately tuned. In other words, the weak underlying periodicity is exhibited at appropriately chosen non-zero levels of noise, and gets lost if noise is either too small or too large.

To date, the most important application area of threshold models is neural dynamics (see Bulsara et al. [**17**], Douglass et al. [**29**], Patel and Kosko [**85**]) and transmission of information (see Neiman et al. [**81**], Simonotto et al. [**99**], Stocks [**101**], Moss et al. [**79**]). The recent book [**73**] by McDonnell et al. gives a very complete account on the theory of non-dynamic or threshold stochastic resonance.

1.5.3. The paddlefish. In this well known and frequently discussed example stochastic resonance appears in the noise-enhanced feeding behavior of the paddlefish *Polyodon spathula* (see Greenwood et al. [**47**], Russel et al. [**95**], Freund et al. [**42**]). This species of fish lives in the Midwest of the United States and in the Yangtze River in China, and feeds on the zooplankton Daphnia. To detect its prey animals under limited visibility conditions at river bottoms, the paddlefish uses the long rostrum in front of its mouth as an electrosensory antenna. The frequency range of sensitivity of the rostrum's electroreceptors well overlaps with the range of frequencies produced by the prey. Roughly, the capture probability is observed as a function of the position of the prey relative to the rostrum. In experiments, external noise was generated by electrodes connected to an electric noise generator. It was observed that the spatial distribution and number of strike locations is a function of the external noise intensity, with a maximum of captures of more distant plankton at some optimal external noise intensity. If experimentally noisy electric signals improve the sensitivity of the electroreceptors, nature itself should also provide sources of noise. In [**95**] it was conjectured that, besides the signal, such a noise might be produced by the populations of prey animals themselves. In [**42**] this conjecture was confirmed by measurements of the noise strength produced by single Daphnia in the vicinity of a swarm. In the simple quantitative approaches, quality of tuning is measured by Fisher information, a concept that may be comparable to the entropy notions in Chapter 3, Section 3.2.

1.5.4. The FitzHugh–Nagumo system. A more detailed modeling of neural activities of living systems underlies this well known and studied example. It deals with action potentials and electric currents transmitted through systems of ion channels provided by the axons in neural networks, triggered by their mutual interaction and the interaction of the system with the biological environment. Neurons communicate with each other or with muscle cells by means of electric signals. Each single neuron can be modeled as an excitable dynamical system: in the rest state characterized by a negative potential gap with respect to the extracellular environment, no current flows through the membrane of the neuronal cell. If this threshold potential barrier disappears due to noisy perturbations created by the environment (neighboring cells, external field), ion channels through the membrane are opened and currents appear in form of a spike or firing, followed by a deterministic recovery to the rest state. During a finite (refractory) time interval, the membrane potential is hyperpolarized by the current flow, and any firing impossible. The theory that captures the above-mentioned features of neuronal dynamics, including the finite refractory time, is described by the FitzHugh–Nagumo (FHN) equations (see Kanamaru et al. [**63**]). In the diffusively coupled form, a system of N coupled neurons is described by the system of equations (see [**63**])

$$\tau\dot{u}_i(t) = \Big(-v_i + \Big(u_i - \frac{u_i^3}{3}\Big) + S(t)\Big) + \sqrt{\varepsilon}\,\dot{W}_i(t) + \frac{1}{N}\sum_{j=1}^{N}(u_i - u_j),$$
$$\dot{v}_i(t) = u_i - \beta v_i + \gamma.$$

Here u_i describes the membrane potential of neuron i, v_i a variable describing whether and to which degree neuron i is in the refractory interval of time after firing, S describes an external periodic pulse acting on the potential levels, while $W_1, \ldots, W_n$ is a vector of independent Brownian motions. Finally, τ, β, ε and γ are system parameters. In the infinite particle limit, the system becomes a stochastic partial differential equation. Roughly, total throughput current will be a function of the model parameters, and stochastic resonance appears as its optimal value for a suitable parameter choice (synchronization). The paper by Wiesenfeld et al. [**109**] reports about a much simplified form of this system, in which action potentials of single mechanoreceptor cells of the crayfish Procambarus clarkii are concerned. The mechanosensory system of the crayfish consists of hairs located on its tailfan, connected to mechanoreceptor cells. Streaming water moves the hair and so provides the external excitation that causes the mechanoreceptor cell to fire. Experimentally (see [**109**]), a piece of tailfan containing the hair and sensory neuron was extracted and put into a saline solution environment. Then, periodic pressure modulations and random noise were imposed on the environment. The firings produced by the mechanoreceptor cell were recorded for different noise levels, and show clear stochastic resonance peaks as functions of noise intensity. Similar phenomena are encountered on a much more general basis in the exchange of substances or information through ionic channels on cell membranes in living organisms.

1.5.5. Physiological systems. The fact that sensory neurons are excitable systems leading to the FitzHugh–Nagumo equations in the preceding subsection, is also basic for many suggestions of how to make use of the phenomenon of stochastic resonance in medicine. Disfunctions arising in sensory organs responsible for hearing, tactile or visual sensations or for balance control could result from relatively higher sensitivity thresholds compared with those of healthy organs. To raise the sensitivity level, a natural idea seems to be to apply the right amount of external noise to these dysfunctional organs, in order to let stochastic resonance effects amplify weak signal responses. In experiments reported in Collins et al. [**21**], local indentations were applied to the tips of digits of test persons who had to correctly identify whether a stimulus was presented. Stimuli generated by subthreshold signals garnished with noise led to improvements in correct identification, with some optimal noise level indicating a stochastic resonance point. Results like this may be used for designing practical devices, as for instance gloves, for individuals with elevated cutaneous sensory thresholds. Similarly, randomly vibrating shoe inserts may help restoring balance control (see Priplata et al. [**89**]). Stochastic resonance effects may be used for treating disfunctions of the human blood pressure system (baroreflex system) featuring a negative feedback between blood pressure and heart rate resp. width of blood vessels. Blood pressure is monitored by two types of receptors, for arteries and veins. In Hidaka [**54**], a weak periodic input was introduced at the venous blood pressure receptor, whereas noise was added to the arterial receptor. It was shown that the power of the output signal of the heart rate (measured by an electrocardiogram) as a function of noise intensity exhibits a bell-shape form, typical for a curve with a stochastic resonance point. Another group of possible medical applications of the amplification effects of stochastic resonance is related to the human brains information processing activity (see Mori and Kai [**78**]). In an experiment in Usher and Feingold [**103**] the effect of stochastic resonance in the speed of memory retrieval was exhibited. Test persons were proven to perform single digit calculations (e.g. $7 \times 8 =?$) significantly faster when exposed to an optimal level of acoustic noise (via headphones).

1.5.6. Optical systems. In optical systems, stochastic resonance was first observed in McNamara et al. [**75**] and Vemuri and Roy [**104**] in a bidirectional ring laser, i.e. a ring resonator with a dye as lasing medium. This laser system supports two meta-stable states realized as modes of the same frequency that travel in opposite directions. They are strongly coupled to each other by the lasing medium, thus permitting a bistable operation. When the pumping exceeds the lasing threshold, either clockwise or counterclockwise modes propagate in the laser, with switchings between those two modes initiated by spontaneous emission in the active medium, and fluctuations of the pump laser. The net gains of the two propagating modes in opposite directions can be controlled by an acousto-optical modulator inside the cavity, which thus can be used both to impose a periodic switching rate between the modes and to inject noise. Therefore the resulting semiclassical laser equations are equivalent to those describing overdamped motion of a particle in a periodically modulated double well potential, as described in the prototypical example of Section 1.2.

The choice of examples we discussed in more detail is rather selective. The effects of stochastic resonance have been found in a big number of dynamical systems in various further areas of the sciences, and studied by a variety of physical measures of quality of tuning. We just mention a big field of applications in microscopic systems underlying the laws of quantum mechanics in which intrinsic quantum tunneling effects interfere with the interpretation of potential barrier tunneling that can be seen as causing noise induced transitions in diffusion dynamics. See [**108**] for a comprehensive survey. Stochastic resonance has further been observed in passive optical bistable systems [**30**], in experiments with magnetoelastic ribbons [**100**], in chemical systems [**67**], as well as in further biological ones [**94, 60, 41**]. Stochastic resonance may even be observed in more general systems in which the role of periodic deterministic signals is taken by some other physical mechanisms (see [**108**]).

1.5.7. Computational aspects of large deviation theory related to stochastic resonance. Our theoretical approach aimed at explaining stochastic resonance conceptually by means of space-time large deviations of weakly periodic dynamical systems does not touch at all the field of numerical algorithms and scientific computing for stochastic resonance related quantities which become very important for applications especially in high dimensions. In the framework of the classical Freidlin–Wentzell theory, first exit time estimates as well as large deviations rates are analytically expressed by the *quasi-potential* (see Chapter 2) which can be calculated more or less explicitly for gradient systems. To determine and minimize the quasi-potential in high-dimensional scenarios is an analytically hardly accessible task. In Vanden–Eijnden et al. [**31, 53**] practically relevant algorithms with numerous applications for this task have been developed. They have been applied to various problems in different areas of application of stochastic resonance.

CHAPTER 2

Transitions for time homogeneous dynamical systems with small noise

The trigger of stochastic resonance are diffusion exits from domains, described by the exit times of randomly perturbed dynamical systems from domains of attraction of their stable fixed points. These exits can be rigorously understood by means of large deviations for dynamical systems with small random perturbations, as has been shown in Freidlin and Wentzell [**40**]. The core ingredient of our approach to stochastic resonance are therefore concepts of large deviations. In this chapter we present a self-contained treatment of the theory of large deviations for randomly perturbed dynamical systems with additive Gaussian noise. They are typically given by solutions of stochastic differential equations of the type

$$dX_t^\varepsilon = b(X_t^\varepsilon)\,dt + \sqrt{\varepsilon}\,dW_t, \quad X_0^\varepsilon = x, \tag{2.1}$$

with a smooth mapping b of Euclidean space $\mathbb{R}^d$, and a d-dimensional Brownian motion W. To fix ideas in the framework of the most important paradigm for our treatment, think of $b = -\nabla_x U$ for a typically double well potential U on $\mathbb{R}^d$. By the action of the Brownian noise, the two well bottoms become meta-stable, and we are primarily interested in describing the asymptotic exit and transition behavior of the trajectories of the solution of (2.1) between their domains of attraction, in the limit $\varepsilon \to 0$. Asymptotic exit rates for these random times are derived via large deviations principles (LDP) for the laws $\mu_\varepsilon = \mathbb{P} \circ X^\varepsilon$ of the diffusion process as $\varepsilon \to 0$. In this case, μ_ε lives on the space of $\mathbb{R}^d$-valued continuous functions defined on $\mathbb{R}_+$, endowed with the topology of uniform convergence on compact subintervals of $\mathbb{R}_+$. To derive these LDP, we start with the LDP for the Wiener process given by Schilder's theorem. Since by our choice of noise, the solution of (2.1) depends continuously on Brownian motion, the LDP for W may be transferred to X^ε via a contraction principle. To establish Schilder's theorem from an elementary perspective, we use the Schauder representation of Brownian motion allowing an elegant approach according to Baldi and Roynette [**3**]. The derivation of the asymptotic properties of exit times in the small noise limit follows the classical treatment by Dembo and Zeitouni [**25**]. Here the famous Kramers–Eyring formula for the expected exit time is underpinned with mathematical rigor. In terms of the pseudopotential V associated with a dissipative drift b it states that the average time it takes to leave the domain of attraction of one of the two meta-stable equilibria, say x_+, is given to exponential order by $\exp(\frac{\overline{V}}{\varepsilon})$. Here $\overline{V} = \inf_{y\in\chi} V(x_+, y)$ is the critical cost to provide for leaving the domain, and χ is the manifold separating the basins of attraction.

In Section 2.1 we start by representing one-dimensional Brownian motion in its Schauder decomposition (Theorem 2.5). It allows a direct approach to its regularity properties in terms of Hölder norms on spaces of continuous functions (Theorem 2.6). The key to this elegant and direct approach is Ciesielski's isomorphism of normed spaces of continuous functions with sequence spaces via Fourier representation (Theorems 2.2 and 2.3). In Section 2.2 we recall general notions and basic concepts about large deviations, especially addressing their construction from exponential decay rates of probabilities of basis sets of topologies (Theorem 2.15), and their transport between different topological spaces via continuous mappings (contraction principle) (Theorems 2.17, 2.21). In Section 2.3 the elegant Schauder representation of Brownian motion then allows a derivation of Schilder's large deviation principle (LDP) for Brownian motion (Theorem 2.28) from the elementary LDP for one-dimensional Gaussian random variables (Theorem 2.22). The Freidlin–Wentzell theory, extended to locally Lipschitz diffusion coefficients, and culminating in the LDP for diffusion processes (Theorem 2.36) is presented in Section 2.4. Finally, in Section 2.5 we follow Dembo and Zeitouni [**25**] to derive the exit time laws due to Freidlin and Wentzell [**40**] for time homogeneous diffusions from domains of attraction of underlying dynamical systems in the small noise limit (Theorem 2.42).

2.1. Brownian motion via Fourier series

In this section, we shall present Brownian motion in an approach based on Fourier series with respect to the orthonormal system of Haar functions. This approach will be seen to open an easy and fast route to large deviations principles for Brownian motion, the basic noise process added to deterministic dynamical systems to provide the time homogeneous randomly perturbed dynamical systems that are the main objects of interest for this section. In fact, we shall present a direct proof of Schilder's Theorem which only uses this Fourier series representation and the large deviation principle for one-dimensional Gaussian variables. The basic idea of this approach for large deviations on function spaces is triggered by an observation by Ciesielski according to which smoothness properties of functions in Hölder spaces can be studied via a universal Banach space isomorphism through convergence properties of sequences. We first present Ciesielski's isomorphism.

2.1.1. The Ciesielski isomorphism of Hölder and sequence spaces. For $0 < \alpha \leq 1$, let $\mathcal{C}^\alpha([0,1],\mathbb{R})$ be the space of all α-Hölder continuous functions $f\colon [0,1] \to \mathbb{R}$ starting at zero, $f(0) = 0$. This space is a Banach space endowed with the Hölder norm

$$\|f\|_\alpha = \sup_{0\leq t<s\leq 1} \frac{|f(t) - f(s)|}{|t-s|^\alpha}.$$

Denote, moreover, by $l^\infty(\mathbb{N}_0,\mathbb{R})$ the Banach space of all bounded real valued sequences $\eta = (\eta_n)_{n\geq 0}$ endowed with the norm $\|\eta\|_\infty = \sup_{n\geq 0} |\eta_n|$. We call two Banach spaces *isomorphic* if there exists a one-to-one linear map between the spaces. By means of Fourier decomposition of Hölder continuous functions with the Schauder basis we first prove that $\mathcal{C}^\alpha([0,1],\mathbb{R})$ and $l^\infty(\mathbb{N}_0,\mathbb{R})$ are isomorphic.

For this purpose we introduce the *Haar functions*. For $t \in [0,1]$ let $\chi_0(t) \equiv 1$ and

$$\chi_{2^k+l}(t) = \begin{cases} \sqrt{2^k} & \text{if } t \in \left[\frac{2l}{2^{k+1}}, \frac{2l+1}{2^{k+1}}\right), \\ -\sqrt{2^k} & \text{if } t \in \left[\frac{2l+1}{2^{k+1}}, \frac{2l+2}{2^{k+1}}\right), \quad k \geq 0, \quad l = 0, 1, \ldots, 2^k - 1, \\ 0 & \text{otherwise.} \end{cases}$$

The family $(\chi_n)_{n \geq 0}$ is a complete orthonormal system of $L^2([0,T], \mathbb{R}^d)$. The *Schauder functions* are just the primitives of the Haar system, given for $t \in [0,1]$ by

$$\phi_n(t) = \int_0^t \chi_n(s)\, ds.$$

If $f \in \mathcal{C}^\alpha([0,1], \mathbb{R})$ possesses a square integrable density $\dot{f} \in L^2([0,1], \mathbb{R})$ so that for $t \in [0,1]$ we have

$$f(t) = \int_0^t \dot{f}(s)\, ds,$$

we can write

$$\dot{f} = \sum_{n=0}^{\infty} \langle \chi_n, \dot{f} \rangle \chi_n,$$

and therefore

$$f = \sum_{n=0}^{\infty} \langle \chi_n, \dot{f} \rangle \phi_n.$$

Indeed, due to the fact that for $k \geq 0$ fixed, and $0 \leq l_1,\ l_2 \leq 2^k - 1$, $l_1 \neq l_2$, the supports of the functions $\phi_{2^k+l_1}$ and $\phi_{2^k+l_2}$ are disjoint and the functions are uniformly bounded by $2^{-\frac{k}{2}-1}$, we may estimate for $K \in \mathbb{N}$, $q \geq p \geq 2^K$ by means of Cauchy–Schwarz' inequality

$$\begin{aligned} \Big\| \sum_{n=p}^{q} \langle \chi_n, \dot{f} \rangle \phi_n \Big\| &\leq \sum_{k=K}^{\infty} \Big\| \sum_{l=0}^{2^k-1} \langle \chi_{2^k+l}, \dot{f} \rangle \phi_{2^k+l} \Big\| \\ &\leq \sum_{k=K}^{\infty} \sup_{0 \leq l \leq 2^k-1} |\langle \chi_{2^k+l}, \dot{f} \rangle| \cdot 2^{-\frac{k}{2}-1} \qquad (2.2) \\ &\leq \sum_{k=K}^{\infty} \Big(\int_0^1 \dot{f}^2(s)\, ds \Big)^{\frac{1}{2}} \cdot 2^{-\frac{k}{2}-1}. \end{aligned}$$

This clearly implies the convergence of the series in the uniform norm. We shall now see by following Ciesielski [**20**] that this representation may be extended to Hölder spaces. For this purpose denote for $n = 2^k + l$

$$\langle \chi_n, \dot{f} \rangle = \sqrt{2^k} \Big[2f\Big(\frac{2l+1}{2^{k+1}}\Big) - f\Big(\frac{2l+2}{2^{k+1}}\Big) - f\Big(\frac{2l}{2^{k+1}}\Big) \Big].$$

This just gives the integral of the function χ_n with respect to f as an integrator.

LEMMA 2.1. *Let $0 < \alpha \leq 1$, and let $f \in \mathcal{C}^\alpha([0,1], \mathbb{R})$. Then*

$$f = \sum_{n=0}^{\infty} \langle \chi_n, \dot{f} \rangle \phi_n,$$

with convergence with respect to the uniform norm.

PROOF. It can be seen easily that f may be approximated in the uniform norm by a sequence $(f_m)_{m\in\mathbb{N}}$ of functions possessing square integrable densities $(\dot f_m)_{m\in\mathbb{N}}$, and with α-Hölder norms bounded by the one of f. Take for instance a sequence obtained from f by smoothing with a sequence of smooth approximations of the unit. More precisely, let $\phi\colon[-1,1]\to\mathbb{R}_+$ be a C^∞ function such that $\int_{-1}^1\phi(x)\,dx = 1$. For $m\in\mathbb{N}$, let $\phi_m(\cdot)=m\phi(m\,\cdot)$, and $f_m(t)=\int_{-1}^1 f(t-x)\phi_m(x)\,dx$, $t\in[0,1]$ (here we assume f to be trivially extended continuously to $[-1,2]$ by constant branches). Obviously, $(f_m)_{m\in\mathbb{N}}$ converges to f in the uniform norm, and for each m, f_m possesses a square integrable density $\dot f_m$. Moreover, for $s,t\in[0,1]$, $m\in\mathbb{N}$, $\alpha\in(0,1]$,

$$\frac{|f_m(t)-f_m(s)|}{|t-s|^\alpha}\le\int_{-1}^1\frac{|f(t-x)-f(s-x)|}{|t-s|^\alpha}\phi_m(x)\,dx,$$

hence $\|f_m\|_\alpha\le\|f\|_\alpha$.

Since we know from the above discussion that the desired representations hold for f_m for all $m\in\mathbb{N}$, a dominated convergence argument shows that we have to prove

$$\sup_{m\in\mathbb{N}}\Big\|\sum_{n=p}^q\langle\chi_n,\dot f_m\rangle\phi_n\Big\|\to 0\quad\text{as}\quad q\ge p\to\infty. \tag{2.3}$$

To do this, we have to modify the estimate (2.2) a bit. In fact, for any $m\in\mathbb{N}$, and $K\in\mathbb{N}$, $q\ge p\ge 2^K$ we have

$$\begin{aligned}\Big\|\sum_{n=p}^q\langle\chi_n,\dot f_m\rangle\phi_n\Big\| &\le\sum_{k=K}^\infty\Big\|\sum_{l=0}^{2^k-1}\langle\chi_{2^k+l},\dot f_m\rangle\phi_{2^k+l}\Big\|\\ &\le\sum_{k=K}^\infty\sup_{0\le l\le 2^k-1}|\langle\chi_{2^k+l},\dot f_m\rangle|2^{-\frac{k}{2}-1}\\ &\le\sum_{k=K}^\infty\|f_m\|_\alpha 2^{-\alpha k}\\ &\le\|f\|_\alpha\cdot\sum_{k=K}^\infty 2^{-\alpha k}.\end{aligned}\tag{2.4}$$

The latter expression obviously converges to 0 as $K\to\infty$. This completes the proof. □

The following Theorem states that $\mathcal{C}^\alpha([0,1],\mathbb{R})$ and $l^\infty(\mathbb{N}_0,\mathbb{R})$ are isomorphic.

THEOREM 2.2. *Let $0<\alpha<1$. For $n\in\mathbb{N}_0$ let*

$$c_n(\alpha)=\begin{cases}1, & n=0,\\ 2^{k(\alpha-\frac12)+\alpha-1}, & n=2^k+l,\ k\ge 0,\ 0\le l\le 2^k-1.\end{cases}$$

Define

$$T_\alpha\colon\mathcal{C}^\alpha([0,1],\mathbb{R})\to l^\infty(\mathbb{N}_0,\mathbb{R}),\quad f\mapsto(c_n(\alpha)\,\langle\chi_n,\dot f\rangle)_{n\ge0}.$$

Then

$$T_\alpha^{-1}\colon l^\infty(\mathbb{N}_0,\mathbb{R})\to\mathcal{C}^\alpha([0,1],\mathbb{R}),\quad(\eta_n)_{n\ge0}\mapsto\sum_{n=0}^\infty\frac{\eta_n}{c_n(\alpha)}\phi_n,$$

T_α is an isomorphism, and for the operator norms we have the following inequalities:

$$\|T_\alpha\| = 1, \quad \|T_\alpha^{-1}\| \le \frac{2}{(2^\alpha - 1)(2^{1-\alpha} - 1)}.$$

PROOF. By definition, for $n = 2^k + l$, $k \ge 0$, $0 \le l \le 2^k - 1$ we have

$$|\langle \chi_n, \dot{f} \rangle| \le 2^{-(k+1)\alpha + \frac{k}{2} + 1} \|f\|_\alpha = \frac{1}{c_n(\alpha)} \|f\|_\alpha. \tag{2.5}$$

Therefore, T_α is well defined, and we have

$$\|T_\alpha\| \le 1.$$

Moreover, for $f(t) = t$, $0 \le t \le 1$, we have $\langle \chi_0, \dot{f} \rangle = 1$, while for $n \in \mathbb{N}$ we have $\langle \chi_n, \dot{f} \rangle = 0$. Hence $\|T_\alpha f\|_\infty = \|f\|_\alpha$. This implies $\|T_\alpha\| = 1$. Lemma 2.1 shows that T_α is one-to-one and that T_α^{-1} is its inverse.

We next prove the inequality for the operator norm of T_α^{-1}. For $\eta = (\eta_n)_{n \ge 0} \in l^\infty(\mathbb{N}_0, \mathbb{R})$ given, choose $0 \le s < t \le 1$, and write $f = T_\alpha^{-1}(\eta)$. Then we have

$$|f(t) - f(s)| \le \|\eta\|_\infty \cdot \Big(|t - s| + \sum_{k=0}^\infty \sum_{l=0}^{2^k - 1} \frac{1}{c_{2^k + l}(\alpha)} |\phi_{2^k + l}(t) - \phi_{2^k + l}(s)| \Big). \tag{2.6}$$

Now choose $k_0 \ge 0$ such that

$$2^{-k_0 - 1} < |t - s| \le 2^{-k_0}.$$

Then for $0 \le k < k_0$ due to the fact that the supports of $\phi_{2^k + l}$, $0 \le l \le 2^k - 1$, are disjoint

$$\begin{aligned} \sum_{l=0}^{2^k - 1} \frac{1}{c_{2^k + l}(\alpha)} |\phi_{2^k + l}(t) - \phi_{2^k + l}(s)| &= \sup_{0 \le l \le 2^k - 1} \frac{1}{c_{2^k + l}(\alpha)} |\phi_{2^k + l}(t) - \phi_{2^k + l}(s)| \\ &\le 2^{-k(\alpha - \frac{1}{2}) - \alpha + 1} \cdot 2^{\frac{k}{2}} \cdot |t - s| \\ &\le 2^{k(1-\alpha) - \alpha + 1 - k_0(1-\alpha)} \cdot |t - s|^\alpha \\ &= 2^{(1-\alpha)(1 + k - k_0)} \cdot |t - s|^\alpha, \end{aligned} \tag{2.7}$$

while for $k \ge k_0$

$$\begin{aligned} \sum_{l=0}^{2^k - 1} \frac{1}{c_{2^k + l}(\alpha)} |\phi_{2^k + l}(t) - \phi_{2^k + l}(s)| &\le 2^{-k(\alpha - \frac{1}{2}) - \alpha + 1} \cdot 2^{-\frac{k}{2}} \\ &\le 2^{-k\alpha - \alpha + 1 + (k_0 + 1)\alpha} \cdot |t - s|^\alpha \\ &= 2^{\alpha(k_0 - k)} \cdot |t - s|^\alpha. \end{aligned} \tag{2.8}$$

Combining (2.6), (2.7) and (2.8), we obtain the estimate

$$\frac{|f(t) - f(s)|}{|t - s|^\alpha} \le \frac{2}{(2^\alpha - 1)(2^{1-\alpha} - 1)} \|\eta\|_\infty,$$

and therefore

$$\|T_\alpha^{-1}\| \le \frac{2}{(2^\alpha - 1)(2^{1-\alpha} - 1)}.$$

□

The spaces we are ultimately interested in are those in which almost all sample paths of the Brownian motion are living. We therefore have to extend the isomorphism of Theorem 2.2 to the following subspaces of Hölder continuous functions. For $0 < \alpha \leq 1$, let $\mathcal{C}_0^\alpha([0,1],\mathbb{R})$ be the subspace of $\mathcal{C}^\alpha([0,1],\mathbb{R})$ composed of all functions f for which $f(0) = 0$ and

$$\lim_{\delta\to 0} \sup_{\substack{0\leq s<t\leq 1,\\ |t-s|\leq\delta}} \frac{|f(t)-f(s)|}{|t-s|^\alpha} = 0.$$

The isomorphism of Theorem 2.2 will then be restricted to the subspace $c_0(\mathbb{N}_0,\mathbb{R})$ of all sequences $\eta = (\eta_n)_{n\geq 0}$ in $l^\infty(\mathbb{N}_0,\mathbb{R})$ which converge to 0 as $n\to\infty$ as a target space. The following Theorem holds.

THEOREM 2.3. *Let $0 < \alpha < 1$. Let $(c_n(\alpha))_{n\geq 0}$ be defined as in Theorem 2.2. Define*

$$T_{\alpha,0}\colon \mathcal{C}_0^\alpha([0,1],\mathbb{R}) \to c_0(\mathbb{N}_0,\mathbb{R}), \quad f \mapsto (c_n(\alpha)\,\langle\chi_n,\dot{f}\rangle)_{n\geq 0}.$$

Then

$$T_{\alpha,0}^{-1}\colon c_0(\mathbb{N}_0,\mathbb{R}) \to \mathcal{C}_0^\alpha([0,1],\mathbb{R}), \quad (\eta_n)_{n\geq 0} \mapsto \sum_{n=0}^\infty \frac{\eta_n}{c_n(\alpha)}\phi_n.$$

The mapping $T_{\alpha,0}$ is an isomorphism, and for the operator norms we have the following inequalities

$$\|T_{\alpha,0}\| = 1, \quad \|T_{\alpha,0}^{-1}\| \leq \frac{2}{(2^\alpha-1)(2^{1-\alpha}-1)}.$$

PROOF. Note first that (2.5) can be strengthened by definition to read

$$|\langle\chi_n,\dot{f}\rangle| \leq \frac{1}{c_n(\alpha)} \sup_{\substack{0\leq s<t\leq 1,\\ |t-s|\leq 2^{-k-1}}} \frac{|f(t)-f(s)|}{|t-s|^\alpha}. \tag{2.9}$$

Hence, by definition of $\mathcal{C}_0^\alpha([0,1],\mathbb{R})$, we obviously obtain that $T_{\alpha,0}$ is well defined. To prove that also $T_{\alpha,0}^{-1}$ is well defined, we just have a closer inspection of the arguments that led to the operator norm inequality in the proof of Theorem 2.2. First, note that for $\eta = (\eta_n)_{n\geq 0} \in c_0(\mathbb{N}_0,\mathbb{R})$, with $f = T_{\alpha,0}^{-1}(\eta)$

$$|f(t)-f(s)| \leq |\eta_0(\alpha)|\cdot|t-s| + \sum_{k=0}^\infty \sum_{l=0}^{2^k-1} \frac{1}{c_{2^k+l}(\alpha)}|\eta_{2^k+l}|\cdot|\phi_{2^k+l}(t)-\phi_{2^k+l}(s)|. \tag{2.10}$$

Now choose again $k_0 \geq 0$ such that

$$2^{-k_0-1} < |t-s| \leq 2^{-k_0},$$

and denote $\tau_n = \sup_{k\geq n}|\eta_k|$. Then for $0 \leq k < k_0$

$$\begin{aligned}
&\sum_{l=0}^{2^k-1} \frac{1}{c_{2^k+l}(\alpha)}|\eta_{2^k+l}|\cdot|\phi_{2^k+l}(t)-\phi_{2^k+l}(s)| \\
&\qquad = \sup_{0\leq l\leq 2^k-1} \frac{1}{c_{2^k+l}(\alpha)}|\eta_{2^k+l}|\cdot|\phi_{2^k+l}(t)-\phi_{2^k+l}(s)| \\
&\qquad \leq \tau_{2^k}\cdot 2^{-k(\alpha-\frac{1}{2})-\alpha+1}\cdot 2^{\frac{k}{2}}|t-s| \\
&\qquad \leq \tau_{2^k}\cdot 2^{k(1-\alpha)-\alpha+1-k_0(1-\alpha)}\cdot|t-s|^\alpha \\
&\qquad = \tau_{2^k}\cdot(2^{1-\alpha})^{(1+k-k_0)}\cdot|t-s|^\alpha,
\end{aligned} \tag{2.11}$$

while for $k \geq k_0$

$$\begin{aligned}
\sum_{l=0}^{2^k-1} \frac{1}{c_{2^k+l}(\alpha)} |\eta_{2^k+l}| \cdot |\phi_{2^k+l}(t) - \phi_{2^k+l}(s)| \\
\leq \tau_{2^{k_0}} \cdot 2^{-k(\alpha - \frac{1}{2}) - \alpha + 1} \cdot 2^{-\frac{k}{2}} \\
\leq \tau_{2^{k_0}} \cdot 2^{-k\alpha - \alpha + 1 + (k_0+1)\alpha} \cdot |t-s|^\alpha \\
= \tau_{2^{k_0}} \cdot (2^\alpha)^{(k_0-k)} \cdot |t-s|^\alpha.
\end{aligned} \tag{2.12}$$

Hence (2.10), (2.11) and (2.12) imply

$$\frac{|f(t)-f(s)|}{|t-s|^\alpha} \leq \|\eta\|_\infty \cdot |t-s|^{1-\alpha} + \sum_{k \leq k_0} \tau_{2^k} \cdot (2^{1-\alpha})^{(1+k-k_0)} + \frac{1}{2^\alpha - 1} \tau_{2^{k_0}}.$$

Now $k_0 \to \infty$ as $|t-s| \to 0$. This, however, entails that $f \in \mathcal{C}_0^\alpha([0,1], \mathbb{R})$. All arguments used in the proof of Theorem 2.2 to show the inequalities for the operator norms are valid here. Just note that the function $f(t) = t$, $0 \leq t \leq 1$, belongs to $\mathcal{C}_0^\alpha([0,1], \mathbb{R})$. □

2.1.2. The Schauder representation of Brownian motion. We shall now present an approach to the study of one-dimensional Brownian motion which is close to Wiener's representation of Brownian motion by Fourier series with trigonometric functions as a basis. Our basis will be given by the Haar functions and their primitives. In fact, the trajectories of Brownian motion will be described just as in the preceding section continuous functions were isomorphically described by sequences. Given a Brownian motion W indexed by the unit interval, with the same notation as in the preceding section we write it sample by sample as a series with coefficients $\langle \chi_n, \dot{W} \rangle$, $n \in \mathbb{N}_0$. Due to the scaling properties and the structure of Haar functions, these random coefficients are i.i.d. standard normal random variables. This, in turn, allows us to construct Brownian motion indexed by the unit interval by taking any sequence of i.i.d. standard normal variables $(Z_n)_{n \in \mathbb{N}_0}$ on a probability space $(\Omega, \mathcal{F}, \mathbb{P})$, and defining the stochastic process

$$W_t = \sum_{n=0}^\infty Z_n \phi_n(t), \quad t \in [0,1]. \tag{2.13}$$

To get information about the quality of convergence of this Fourier series, we need to control the size of the random sequence $(Z_n)_{n \in \mathbb{N}_0}$ in the following Lemma.

LEMMA 2.4. *Let $(Z_n)_{n \geq 2}$ be a sequence of standard gaussian random variables. There exists a real valued random variable C such that the inequality*

$$\sup_{n \geq 2} |Z_n(\omega)| \leq C(\omega) \sqrt{\ln n} \tag{2.14}$$

holds true $\mathbb{P}$-a.s.

PROOF. For $x \geq 1$ and $n \geq 2$ we have

$$\mathbb{P}(|Z_n| \geq x) = \sqrt{\frac{2}{\pi}} \int_x^\infty e^{-\frac{u^2}{2}} \, du \leq \sqrt{\frac{2}{\pi}} \int_x^\infty u e^{-\frac{u^2}{2}} \, du = \sqrt{\frac{2}{\pi}} e^{-\frac{x^2}{2}}.$$

Hence for any $\beta > 1$

$$\mathbb{P}\Big(|Z_n| \geq \sqrt{2\beta \ln n}\Big) \leq \sqrt{\frac{2}{\pi}} e^{-\beta \ln n} = \sqrt{\frac{2}{\pi}} \frac{1}{n^\beta}.$$

Therefore, the lemma of Borel–Cantelli yields that $|Z_n(\omega)| \leq \sqrt{2\beta \ln n}$ for almost all $n \geq 2$ with probability 1. Hence the random variable $C(\omega) = \sup_{n\geq 2} \frac{|Z_n(\omega)|}{\sqrt{2\beta \ln n}}$ is almost surely finite, and the desired inequality follows. □

The preceding Lemma enables us to state that the convergence in (2.13) is absolute and therefore the process continuous. Its law has the characteristics of the law of a Brownian motion, as the following Theorem shows.

THEOREM 2.5. *The series in* (2.13) *converges with probability one absolutely in the space of continuous functions* $C([0,1],\mathbb{R})$ *w.r.t. the uniform norm to a continuous process* W *which is a Brownian motion on* $[0,1]$.

PROOF. Let us first prove the absolute convergence of the series in the uniform norm. This will evidently imply that W is continuous. Let k, p, $q \in \mathbb{N}$ be such that $q \geq p \geq 2^k$. Then with probability 1 we have with the random variable $C(\omega)$ of the preceding Lemma

$$\sup_{t\in[0,1]} \sum_{n=p}^{q} |Z_n(\omega)|\phi_n(t) \leq C(\omega) \sup_{t\in[0,1]} \sum_{n=p}^{q} \sqrt{\ln n} \cdot \phi_n(t)$$

$$\leq C(\omega) \sup_{t\in[0,1]} \sum_{j=k}^{\infty} \sum_{l=0}^{2^j-1} \sqrt{j+1}\sqrt{\ln 2} \cdot \phi_{2^j+l}(t) \leq C(\omega) \sum_{j=k}^{\infty} \sqrt{j+1} \cdot 2^{-\frac{j}{2}-1},$$

which converges to 0 as q, p tend to ∞.

To prove that W is a Gaussian process with $\mathbb{E}W_t = 0$ and $\mathrm{cov}(W_t, W_s) = s \wedge t$, for $0 \leq s, t \leq 1$, we first note that the series also converges in $L^2(\Omega, \mathcal{F}, \mathbb{P})$. Indeed, we have for $t \in [0,1]$, k, p, $q \in \mathbb{N}$ such that $q \geq p \geq 2^k$ by the law properties of Z_n, $n \geq 0$,

$$\mathbb{E}\Big(\sum_{n=p}^{q} Z_n \phi_n(t)\Big)^2 = \sum_{n=p}^{q} \phi_n(t)^2 \leq \sum_{j=k}^{\infty} 2^{-j-2},$$

which converges to 0 as $p, q \to \infty$. Next, let $m \in \mathbb{N}$, $0 \leq t_1 < \cdots < t_m \leq 1$, and $\theta = (\theta_1, \cdots, \theta_m) \in \mathbb{R}^d$ be given. We compute the characteristic function $\varphi(\theta)$ of the vector $(W_{t_1}, \ldots, W_{t_m})$ at θ. We have by dominated convergence and the law properties of Z_n, $n \geq 0$, again

$$\varphi(\theta) = \mathbb{E}\exp\Big(i\sum_{j=1}^{m} \theta_j W_{t_j}\Big) = \mathbb{E}\exp\Big(i\sum_{j=1}^{m} \theta_j \sum_{n=0}^{\infty} Z_n \phi_n(t_j)\Big)$$

$$= \prod_{n=0}^{\infty} \mathbb{E}\exp\Big(iZ_n \sum_{j=1}^{m} \theta_j \phi_n(t_j)\Big) = \prod_{n=0}^{\infty} \exp\Big(-\frac{1}{2}\Big(\sum_{j=1}^{m} \theta_j \phi_n(t_j)\Big)^2\Big)$$

$$= \exp\Big(-\frac{1}{2}\sum_{n=0}^{\infty}\Big(\sum_{j=1}^{m} \theta_j \phi_n(t_j)\Big)^2\Big) = \exp\Big(-\frac{1}{2}\sum_{j,k=1}^{m} \theta_j\theta_k \sum_{n=0}^{\infty} \phi_n(t_j)\phi_n(t_k)\Big).$$

Now observe that Parseval's equation implies for $1 \leq j, k \leq m$

$$t_j \wedge t_k = \langle \mathbb{I}_{[0,t_j]}, \mathbb{I}_{[0,t_k]} \rangle = \sum_{n=0}^{\infty} \langle \mathbb{I}_{[0,t_j]}, \chi_n \rangle \, \langle \mathbb{I}_{[0,t_k]}, \chi_n \rangle = \sum_{n=0}^{\infty} \phi_n(t_j)\phi_n(t_k).$$

Therefore we finally obtain

$$\varphi(\theta) = \exp\Big(-\frac{1}{2} \sum_{j,k=1}^{m} \theta_j \theta_k \cdot t_j \wedge t_k \Big).$$

But this means that $(W_{t_1}, \ldots, W_{t_m})$ is Gaussian with the mean value 0 and covariance matrix $(t_j \wedge t_k)_{1 \leq j,k \leq m}$. It is easy to see that these properties imply that the process W possesses independent increments which are Gaussian with mean 0 and variance corresponding to the length of the increment intervals. This, however, characterizes a Brownian motion. □

The usual one-dimensional Brownian motion $W = (W_t)_{t \geq 0}$ can now be introduced by starting with a sequence of independent Brownian motions $(W^n)_{n \in \mathbb{N}}$ indexed by the unit interval and given by the construction discussed above on the basis of Schauder representations. Then we can evidently set

$$W_t = \sum_{k=1}^{[t]} W_1^k + W_{t-[t]}^{[t]+1}, \quad t \geq 0.$$

Finally, the d-dimensional Brownian motion $W = (W^1, \ldots, W^d)$ is just a d-tuple of independent one-dimensional Brownian motions indexed by $\mathbb{R}_+$.

We now use the Schauder representation of Brownian motion to show its Hölder continuity properties.

THEOREM 2.6. *With probability one, the Brownian motion* $W = (W_t)_{0 \leq t \leq 1}$ *is Hölder continuous of order* $\alpha < 1/2$. *Its trajectories are a.s. nowhere Hölder continuous of order* $\alpha > 1/2$. *Moreover we have (Lévy's modulus of continuity)*

$$\mathbb{P}\Big(\sup_{0 \leq s < t \leq 1} \frac{|W_t - W_s|}{h(|t-s|)} < \infty \Big) = 1, \tag{2.15}$$

where $h(u) = \sqrt{u \log(1/u)}$, $u > 0$. *In particular, for* $\alpha < \frac{1}{2}$, *the trajectories of* W *are* $\mathbb{P}$*-a.s. contained in the space* $C_0^\alpha([0,1], \mathbb{R})$.

PROOF. Let first $0 < \alpha < 1$, $(c_n)_{n \geq 0}$ be a sequence of real numbers for which there exists $c \in \mathbb{R}$ such that for $n = 2^k + l$, $0 \leq l \leq 2^k - 1$ we have

$$|c_n| \leq c\sqrt{k}.$$

Let

$$f(t) = \sum_{n=0}^{\infty} c_n \, \phi_n(t), \quad t \in [0,1].$$

We shall prove that $\sup_{0 \leq s < t \leq 1} \frac{|f(t) - f(s)|}{h(|t-s|)} < \infty$. Due to Lemma 2.4, this will imply the claimed formula for the Brownian motion. Indeed since $\phi_0(t) = t$ is infinitely differentiable on $[0,1]$, we may assume that $c_0 = 0$. Then for $0 \leq s < t \leq 1$

$$|f(t) - f(s)| \leq \sum_{k=0}^{\infty} \sum_{l=0}^{2^k - 1} |c_{2^k+l}| \cdot |\phi_{2^k+l}(t) - \phi_{2^k+l}(s)|. \tag{2.16}$$

Now choose $k_0 \geq 0$ such that

$$2^{-k_0 - 1} < |t - s| \leq 2^{-k_0}.$$

Then for $0 \le k < k_0$

$$\begin{aligned}\sum_{l=0}^{2^k-1} |c_{2^k+l}|\cdot|\phi_{2^k+l}(t) - \phi_{2^k+l}(s)| &\le \sup_{0\le l\le 2^k-1} |c_{2^k+l}| \cdot |\phi_{2^k+l}(t) - \phi_{2^k+l}(s)| \\ &\le c\sqrt{k}\cdot 2^{\frac{k}{2}} \cdot |t-s| \le c\sqrt{k}\cdot 2^{\frac{k-k_0}{2}} \cdot |t-s|^{\frac{1}{2}} \\ &\le \frac{c}{\sqrt{\ln 2}} \cdot \sqrt{\frac{k}{k_0}} \cdot 2^{\frac{k-k_0}{2}} \cdot \sqrt{|t-s| \ln \frac{1}{|t-s|}},\end{aligned} \tag{2.17}$$

while for $k \ge k_0$

$$\begin{aligned}\sum_{l=0}^{2^k-1} |c_{2^k+l}|\cdot|\phi_{2^k+l}(t) - \phi_{2^k+l}(s)| &\le c\sqrt{k}\cdot 2^{-\frac{k}{2}} \\ &\le \frac{c}{\sqrt{\ln 2}} \sqrt{\frac{k}{k_0}}\, 2^{\frac{k_0+1-k}{2}} \cdot \sqrt{|t-s| \ln \frac{1}{|t-s|}}.\end{aligned} \tag{2.18}$$

It is easy to see, for instance by estimating $\int_a^\infty \sqrt{x}2^{-x}\,dx$ and $\int_1^a \sqrt{x}2^x\,dx$ for $a \ge 1$ using integration by parts that the sum in k of the two estimates can be taken and yields a finite upper bound which does not depend on k_0. Hence (2.16), (2.17) and (2.18) imply

$$\frac{|f(t) - f(s)|}{\sqrt{|t-s| \ln \frac{1}{|t-s|}}} \le c',$$

for some constant c' independent of s and t. This implies the desired inequality, and all claims about Hölder continuity for $\alpha < \frac{1}{2}$.

Let us next fix $\alpha > \frac{1}{2}$. For $c > 0$ and $\varepsilon > 0$ let

$$G(\alpha, c, \varepsilon) = \Big\{\omega \in \Omega\colon \exists\, s \in [0,1] \text{ such that } \sup_{\substack{t\in[0,1]\\ 0<|t-s|\le\varepsilon}} \frac{W_t(\omega) - W_s(\omega)}{|s-t|^\alpha} \le c\Big\}$$

We will show that $\mathbb{P}(G(\alpha, c, \varepsilon)) = 0$ for all $c, \varepsilon > 0$, and thus that W is a.s. nowhere Hölder continuous of order α. To this end, for all $m, n \in \mathbb{N}$, $m \le n$, and $0 \le k < n$ let

$$X_{m,k} = \max_{j=k,\dots,k+m} \Big|W_{\frac{j}{n}} - W_{\frac{j+1}{n}}\Big|.$$

Let $\omega \in G(\alpha, c, \varepsilon)$. Choose $n \in \mathbb{N}$ such that $\frac{m}{n} \le \varepsilon$. Let $s \in [0,1]$ be given such that for all $t \in [0,1]$ satisfying $0 < |s-t| \le \varepsilon$ we have $|W_t(\omega) - W_s(\omega)| \le c|s-t|^\alpha$. Choose $0 \le k \le n-m$ such that $\frac{k}{n} \le s < \frac{k+m}{n}$. Then for $k \le j < k+m$

$$\begin{aligned}\Big|W_{\frac{j}{n}}(\omega) - W_{\frac{j+1}{n}}(\omega)\Big| &\le \Big|W_{\frac{j}{n}}(\omega) - W_s(\omega)\Big| + \Big|W_s(\omega) - W_{\frac{j+1}{n}}(\omega)\Big| \\ &\le c\Big|\frac{j}{n} - s\Big|^\alpha + c\Big|s - \frac{j+1}{n}\Big|^\alpha \le 2c\Big(\frac{m}{n}\Big)^\alpha.\end{aligned}$$

This proves that

$$G(\alpha, c, \varepsilon) \subseteq \Big\{\min_{0\le k\le n-m} X_{m,k} \le 2c\Big(\frac{m}{n}\Big)^\alpha\Big\}. \tag{2.19}$$

Let us now estimate the probability of the latter set. Indeed, we use independence and stationarity of the laws of the increments of W, and its scaling properties to obtain

$$\begin{aligned}\mathbb{P}\Big(\min_{0\le k\le n-m} X_{m,k} \le 2c\Big(\frac{m}{n}\Big)^\alpha\Big) &\le n\mathbb{P}\Big(X_{m,1} \le 2c\Big(\frac{m}{n}\Big)^\alpha\Big)\\ &\le n\mathbb{P}\Big(\Big|W_{\frac{1}{n}}\Big| \le 2c\Big(\frac{m}{n}\Big)^\alpha\Big)^m = n\mathbb{P}\Big(\Big|W_1\Big| \le 2c\sqrt{n}\Big(\frac{m}{n}\Big)^\alpha\Big)^m\\ &\le n\Big[2c\sqrt{\frac{2n}{\pi}}\Big(\frac{m}{n}\Big)^\alpha\Big]^m = n^{1+(\frac{1}{2}-\alpha)m}\Big[2c\sqrt{\frac{2}{\pi}}m^\alpha\Big]^m.\end{aligned}$$

Now choose m such that $1 + (\frac{1}{2} - \alpha)m < 0$ and let $n \to \infty$ to obtain that

$$\mathbb{P}(G(\alpha, c, \varepsilon)) = 0,$$

as desired. □

2.2. The large deviation principle

Throughout this book, we shall mainly be concerned with the calculation of large deviation rates for diffusion processes X^ε derived from dynamical systems perturbed by additive Brownian noise of small intensity ε. The rates will be calculated in two steps: first we shall establish the large deviations principle for small Brownian motion $\sqrt{\varepsilon}W$. In a second step diffusions will be considered as continuous maps of Brownian motion, and large deviations principles transferred via the contraction principle. In this section, we shall prepare these steps by discussing the general framework of large deviations theory.

2.2.1. Concept and basic properties. To state the large deviation principle, and investigate its basic properties, let $(\mu_\varepsilon)_{\varepsilon>0}$ be a family of probability measures on a topological (Hausdorff) space $(\mathbf{X}, \mathcal{B})$, $\mathcal{B}$ being the Borel σ-algebra. Think of μ_ε as the law of $\sqrt{\varepsilon}W$ or X^ε, $\varepsilon > 0$. In particular, we consider a space $\mathbf{X} = C([0, 1], \mathbb{R})$ or a Hölder space $\mathbf{X} = \mathcal{C}_0^\alpha([0, 1], \mathbb{R})$. The large deviation principle concerns the limiting behavior of exponential rates of $(\mu_\varepsilon)_{\varepsilon>0}$ as $\varepsilon \to 0$ in terms of a *rate function.*

DEFINITION 2.7. A *rate function* is a lower semicontinuous function $I\colon \mathbf{X} \to [0, \infty]$, i.e. for all $\alpha \in [0, \infty)$, the level sets

$$\Psi_I(\alpha) = \{x \in \mathbf{X}\colon I(x) \le \alpha\}$$

are closed. I is called *good rate function,* if all level sets are compact.

REMARK 2.8. If the topology of $\mathbf{X}$ has a countable basis, lower semicontinuity of I is equivalent to the property

$$\liminf_{n\to\infty} I(x_n) \ge I(x)$$

for all sequences $(x_n)_{n\in\mathbb{N}} \subseteq \mathbf{X}$ converging to $x \in \mathbf{X}$.

DEFINITION 2.9. Let I be a rate function. A family of probability measures $(\mu_\varepsilon)_{\varepsilon>0}$ on $(\mathbf{X}, \mathcal{B})$ satisfies *the large deviation principle (LDP) with rate function* I if for all $\Gamma \in \mathcal{B}$ we have

$$-\inf_{x\in\Gamma^o} I(x) \le \liminf_{\varepsilon\to0} \varepsilon \ln \mu_\varepsilon(\Gamma) \le \limsup_{\varepsilon\to0} \varepsilon \ln \mu_\varepsilon(\Gamma) \le -\inf_{x\in\overline{\Gamma}} I(x).$$

Here Γ^o resp. $\overline{\Gamma}$ denote the open kernel resp. the closed hull of the set Γ.

The following equivalent characterization is evident, but often more practical to prove.

REMARK 2.10. $(\mu_\varepsilon)_{\varepsilon>0}$ satisfies a LDP with rate function I iff the following conditions are satisfied.

(a) For every $\alpha < \infty$ and every $\Gamma \in \mathcal{B}$ such that $\inf_{x\in\overline{\Gamma}} I(y) \geq \alpha$ we have

$$\limsup_{\varepsilon\to 0} \varepsilon \ln \mu_\varepsilon(\Gamma) \leq -\alpha.$$

(b) For $x \in \mathbf{X}$ with $I(x) < \infty$ and any $\Gamma \in \mathcal{B}$ with $x \in \Gamma^o$ we have

$$\liminf_{\varepsilon\to 0} \varepsilon \ln \mu_\varepsilon(\Gamma) \geq -I(x).$$

REMARK 2.11. Let $(\mu_\varepsilon)_{\varepsilon>0}$ be a family of probability measures, I a rate function. Then the LDP is equivalent to the following statements:

(a) for any closed set $F \subset \mathbf{X}$ we have

$$\limsup_{\varepsilon\to 0} \varepsilon \ln \mu_\varepsilon(F) \leq -\inf_{x\in F} I(x);$$

(b) for any open set $G \subset \mathbf{X}$ we have

$$\liminf_{\varepsilon\to 0} \varepsilon \ln \mu_\varepsilon(G) \geq -\inf_{x\in G} I(x).$$

PROOF. 1. The LDP evidently implies (a) and (b).

2. Assume that (a) and (b) are satisfied, and let $\Gamma \in \mathcal{B}$. By (a) we have

$$\limsup_{\varepsilon\to 0} \varepsilon \ln \mu_\varepsilon(\Gamma) \leq \limsup_{\varepsilon\to 0} \varepsilon \ln \mu_\varepsilon(\overline{\Gamma}) \leq -\inf_{x\in\overline{\Gamma}} I(x).$$

By (b) we have

$$\liminf_{\varepsilon\to 0} \varepsilon \ln \mu_\varepsilon(\Gamma) \geq \liminf_{\varepsilon\to 0} \varepsilon \ln \mu_\varepsilon(\Gamma^o) \geq -\inf_{x\in\Gamma^o} I(x).$$

Combining the two inequalities gives the defining property. □

DEFINITION 2.12. We say that $(\mu_\varepsilon)_{\varepsilon>0}$ satisfies a *weak LDP* with rate function I, if (a) for compact instead of closed sets and (b) of the preceding remark are satisfied.

In practice, one often has the validity of the weak LDP. To conclude from this the validity of the LDP, some appropriate tightness condition is needed.

DEFINITION 2.13. A family $(\mu_\varepsilon)_{\varepsilon>0}$ is said to be *exponentially tight* if for every $\alpha < \infty$ there exists a compact set $K_\alpha \subseteq \mathbf{X}$ such that

$$\limsup_{\varepsilon\to 0} \varepsilon \ln \mu_\varepsilon(K_\alpha^c) < -\alpha.$$

We show how exponential tightness can be used to deduce a LD principle.

LEMMA 2.14. *Let $(\mu_\varepsilon)_{\varepsilon>0}$ be an exponentially tight family of probability measures. Then we have:*

*(a) The condition (*lower bound*)*

$$\limsup_{\varepsilon\to 0} \varepsilon \ln \mu_\varepsilon(K) \leq -\inf_{x\in K} I(x), \quad K \subseteq \mathbf{X} \textit{ compact}$$

implies the lower bound for closed sets $F \subseteq \mathbf{X}$.

*(b) The condition (*upper bound*)*

$$\limsup_{\varepsilon\to 0} \varepsilon \ln \mu_\varepsilon(G) \geq -\inf_{x\in G} I(x), \quad G \subseteq \mathbf{X} \textit{ open}$$

implies that I is a good rate function.

PROOF. 1. Let $F \subseteq \mathbf{X}$ be closed, and $\alpha < \infty$ such that $\inf_{x\in F} I(x) \geq \alpha$. Choose K_α according to the definition of exponential tightness. Then for any $\varepsilon > 0$

$$\mu_\varepsilon(F) \leq \mu_\varepsilon(F \cap K_\alpha) + \mu_\varepsilon(K_\alpha^c).$$

Now, for $a, b\colon \mathbb{R}_+ \to \mathbb{R}_+$, due to the inequality

$$\ln\Big(a(\varepsilon) + b(\varepsilon)\Big) \leq \ln(2a(\varepsilon)) \vee \ln(2b(\varepsilon)) = \ln(a(\varepsilon)) \vee \ln(b(\varepsilon)) + \ln 2$$

we have

$$\limsup_{\varepsilon\to 0} \varepsilon \ln\Big(a(\varepsilon) + b(\varepsilon)\Big) \leq \limsup_{\varepsilon\to 0} \varepsilon \ln(a(\varepsilon)) \vee \ln(b(\varepsilon))$$

and therefore by hypothesis applied to $F \cap K_\alpha$

$$\begin{aligned}
\limsup_{\varepsilon\to 0} \varepsilon \ln \mu_\varepsilon(F) &\leq \limsup_{\varepsilon\to 0} \varepsilon \ln\Big(\mu_\varepsilon(F\cap K_\alpha) + \mu_\varepsilon(K_\alpha^c)\Big) \\
&\leq \limsup_{\varepsilon\to 0} \varepsilon \ln\Big(\mu_\varepsilon(F\cap K_\alpha)\Big) \vee \limsup_{\varepsilon\to 0} \varepsilon \ln\Big(\mu_\varepsilon(K_\alpha^c)\Big) \\
&= \limsup_{\varepsilon\to 0} \varepsilon \ln\Big(\mu_\varepsilon(F\cap K_\alpha)\Big) \\
&\leq - \inf_{x\in F\cap K_\alpha} I(x) \leq - \inf_{x\in F} I(x).
\end{aligned}$$

2. For $\alpha < \infty$ let K_α be chosen according to the definition of exponential tightness. We have to show that the level set $\Psi_I(\alpha)$ is compact. Apply the lower bound to the open set K_α^c. Then we have

$$- \inf_{x\in K_\alpha^c} I(x) \leq \liminf_{\varepsilon\to 0} \varepsilon \log \mu_\varepsilon(K_\alpha^c) < -\alpha,$$

or equivalently

$$\inf_{x\in K_\alpha^c} I(x) > \alpha,$$

which means that the inequality $I(x) \leq \alpha$ implies $x \in K_\alpha$. Hence $\Psi_I(\alpha) \subset K_\alpha$ is compact. □

2.2.2. Construction of LDP from exponential rates of elementary sets. Large deviations principles state exponential rates for all open and closed sets of a topological space. Suppose that originally the exponential rates are only known for some simple sets for instance belonging to a basis of the topology. We shall now give a sufficient criterion under which from those rates one can obtain an LDP. In fact, we start with discussing a weak LDP.

THEOREM 2.15. *Let $\mathcal{G}_0$ be a collection of open sets in the topology of $(\mathbf{X}, \mathcal{B})$ such that for each open set $G \in \mathcal{B}$ and each $x \in G$ there is $G_0 \in \mathcal{G}_0$ such that $x \in G_0 \subseteq G$. Let I be a rate function and $(\mu_\varepsilon)_{\varepsilon>0}$ a family of probability measures. Assume that for every $G \in \mathcal{G}_0$ we have*

$$- \inf_{x\in G} I(x) = \lim_{\varepsilon\to 0} \varepsilon \ln \mu_\varepsilon(G).$$

Then $(\mu_\varepsilon)_{\varepsilon>0}$ satisfies a weak LDP with rate function I.

PROOF. Let us first establish the lower bound. For an open set G choose $x \in G$, and a basis set G_0 such that $x \in G_0 \subseteq G$. Then evidently

$$\liminf_{\varepsilon\to 0} \varepsilon \ln \mu_\varepsilon(G) \geq \liminf_{\varepsilon\to 0} \varepsilon \ln \mu_\varepsilon(G_0) = - \inf_{y\in G_0} I(y) \geq -I(x).$$

Now the lower bound follows readily by taking the sup of $-I(x)$, $x \in G$, on the right hand side, the left hand side not depending on x.

For the upper bound, fix a compact subset K of $\mathbf{X}$. For $\delta > 0$ denote

$$I^\delta(x) = (I(x) - \delta) \wedge \frac{1}{\delta}, \quad x \in \mathbf{X}.$$

For any $x \in K$, use the lower semicontinuity of I, more precisely that the set $\{y \in \mathbf{X} : I(y) > I^\delta(x)\}$ is open, to choose a set $G_x \in \mathcal{G}_0$ such that

$$-I^\delta(x) \geq \limsup_{\varepsilon \to 0} \varepsilon \ln \mu_\varepsilon(G_x).$$

Use compactness of K to extract from the open cover $K \subset \cup_{x \in K} G_x$ a finite subcover $K \subset \cup_{i=1}^n G_{x_i}$. Then with an argument as in the proof of Lemma 2.14 we obtain

$$\limsup_{\varepsilon \to 0} \varepsilon \ln \mu_\varepsilon(K) \leq \max_{1 \leq i \leq n} \limsup_{\varepsilon \to 0} \varepsilon \ln \mu_\varepsilon(G_{x_i}) \leq -\min_{1 \leq i \leq n} I^\delta(x_i) \leq -\inf_{x \in K} I^\delta(x).$$

Now let $\delta \to 0$ to complete the proof. □

COROLLARY 2.16. Let $\mathcal{G}_0$ be a collection of open sets in the topology of $(\mathbf{X}, \mathcal{B})$ such that for each open set G and each $x \in G$ there is $G_0 \in \mathcal{G}_0$ such that $x \in G_0 \subseteq G$. Let I be a rate function and $(\mu_\varepsilon)_{\varepsilon > 0}$ an exponentially tight family of probability measures. Assume that for every $G \in \mathcal{G}_0$ we have

$$-\inf_{x \in G} I(x) = \lim_{\varepsilon \to 0} \varepsilon \ln \mu_\varepsilon(G).$$

Then $(\mu_\varepsilon)_{\varepsilon > 0}$ satisfies an LDP with good rate function I.

PROOF. Apply Lemma 2.14 to Theorem 2.15. □

2.2.3. Transformations of LDP. Assume we have established an LDP for a family of probability measures $(\mu_\varepsilon)_{\varepsilon > 0}$ on a topological space $(\mathbf{X}, \mathcal{B})$, $f : \mathbf{X} \to \mathbf{Y}$ is a continuous map to a topological (Hausdorff) space $(\mathbf{Y}, \mathcal{C})$, and $\nu_\varepsilon := \mu_\varepsilon \circ f^{-1}$, $\varepsilon > 0$. Then we will show that the family $(\nu_\varepsilon)_{\varepsilon > 0}$ also satisfies an LDP.

THEOREM 2.17 (contraction principle). *Let $(\mathbf{X}, \mathcal{B}), (\mathbf{Y}, \mathcal{C})$ be topological spaces, and $f : \mathbf{X} \to \mathbf{Y}$ a continuous mapping. Let $I : \mathbf{X} \to [0, \infty]$ be a good rate function.*

(a) For $y \in \mathbf{Y}$ let

$$I'(y) = \inf\{I(x) : x \in \mathbf{X}, y = f(x)\}.$$

Then I' is a good rate function on $\mathbf{Y}$.

(b) Suppose $(\mu_\varepsilon)_{\varepsilon > 0}$ satisfies an LDP with rate function I, and $\nu_\varepsilon = \mu_\varepsilon \circ f^{-1}$, $\varepsilon > 0$. Then the family $(\nu_\varepsilon)_{\varepsilon > 0}$ satisfies an LDP with respect to the rate function I'.

PROOF. (a) We have to show that for $\alpha < \infty$ the level set

$$\Psi_{I'}(\alpha) = \{y \in \mathbf{Y} : I'(y) \leq \alpha\}$$

is compact. Indeed, by continuity of f and by definition we have $f^{-1}(\{y\}) \cap \Psi_I(\alpha + \varepsilon) \neq \emptyset$ for any $\varepsilon > 0$. Thus by compactness

$$\Psi_{I'}(\alpha) = \{f(x) \in \mathbf{Y} : I(x) \leq \alpha\} \subseteq f(\Psi_I(\alpha)),$$

while the opposite inclusion is trivial. Since f is continuous and $\Psi_I(\alpha)$ compact, the compactness of $\Psi_{I'}(\alpha)$ follows.

(b) Let $H \subseteq \mathbf{Y}$ be open. Then $f^{-1}(H) \subseteq \mathbf{X}$ is open, and we have

$$\begin{aligned}\liminf_{\varepsilon\to 0} \varepsilon \ln \nu_\varepsilon(H) &= \liminf_{\varepsilon\to 0} \varepsilon \ln \mu_\varepsilon(f^{-1}(H)) \\ &\geq - \inf_{x\in f^{-1}(H)} I(x) = - \inf_{y\in H} I'(y).\end{aligned}$$

An analogous statement holds for closed sets. □

Can we get a similar transfer of large deviation principles in the other direction? This question becomes relevant for example in the context of different topologies on the same topological space. Under which additional assumptions can the LDP be transferred from the coarser to a finer topology?

THEOREM 2.18 (inverse contraction principle). *Let $(\mathbf{X}, \mathcal{B})$, $(\mathbf{Y}, \mathcal{C})$ be topological spaces, and $g\colon \mathbf{Y} \to \mathbf{X}$ a continuous bijection. Let $(\nu_\varepsilon)_{\varepsilon>0}$ be a family of probability measures on $(\mathbf{Y}, \mathcal{C})$ which is exponentially tight. Let $\mu_\varepsilon = \nu_\varepsilon \circ g^{-1}$, $\varepsilon > 0$. If $(\mu_\varepsilon)_{\varepsilon>0}$ satisfies a LDP with rate function I, then $(\nu_\varepsilon)_{\varepsilon>0}$ satisfies an LDP with rate function $I' = I \circ g$.*

PROOF. 1. We prove that I' is a rate function. Indeed, let $\alpha < \infty$ be given. Then

$$\Psi_{I'}(\alpha) = \{y \in \mathbf{Y}\colon I'(y) \leq \alpha\} = \{y \in \mathbf{Y}\colon I(g(y)) \leq \alpha\} = g^{-1}(\Psi_I(\alpha)).$$

Since $\Psi_I(\alpha)$ is closed and g continuous, $\Psi_{I'}(\alpha)$ is closed. Hence I' is a rate function.

2. To prove a LDP for $(\nu_\varepsilon)_{\varepsilon>0}$, according to the exponential tightness of the family and Lemma 2.14 all we have to establish is the lower bound and the upper bound for compact sets.

(a) We show that for any compact $K \subseteq \mathbf{Y}$

$$\limsup_{\varepsilon\to 0} \varepsilon \log \nu_\varepsilon(K) \leq - \inf_{y\in K} I'(y).$$

Indeed since g is a continuous bijection, we have

$$\begin{aligned}\limsup_{\varepsilon\to 0} \varepsilon \ln \nu_\varepsilon(K) &= \limsup_{\varepsilon\to 0} \varepsilon \ln \nu_\varepsilon \circ g^{-1}(g((K))) \\ &= \limsup_{\varepsilon\to 0} \varepsilon \ln \mu_\varepsilon(g(K)) \leq - \inf_{x\in g(K)} I(x) \\ &= - \inf_{y\in K} I'(y).\end{aligned}$$

(b) We show that for any open set G and $y \in G$ we have

$$\liminf_{\varepsilon\to 0} \varepsilon \ln \nu_\varepsilon(G) \geq -I'(y).$$

Once this is established, the lower bound follows readily by taking the sup of $-I'(y)$ over $y \in G$ on the right hand side, the left hand side not depending on y.

So let $y \in \mathbf{Y}$, and G open with $y \in G$. Let $\alpha = I'(y) = I(g(y))$. Choose $K_\alpha \subseteq \mathbf{Y}$ compact such that

$$\limsup_{\varepsilon\to 0} \varepsilon \nu_\varepsilon(K_\alpha^c) < -\alpha.$$

Since by the continuous bijection property of g we have $g(K_\alpha)^c \subset \mathbf{X}$ open, we may infer from the LDP for $(\mu_\varepsilon)_{\varepsilon>0}$

$$- \inf_{x\in g(K_\alpha^c)} I(x) \leq \limsup_{\varepsilon\to 0} \varepsilon \nu_\varepsilon(K_\alpha^c) < -\alpha.$$

Hence in particular $g(y) \in g(K_\alpha)$. Moreover, g is a homeomorphism on K_α. Now the identity is a continuous map from $g(K_\alpha)$ to $\mathbf{X}$ with the induced topology. Hence by Theorem 2.17 the family $(\mu_\varepsilon)_{\varepsilon>0}$, restricted to $g(K_\alpha)$ satisfies an LDP. And so

$$\liminf_{\varepsilon\to 0} \varepsilon \log \nu_\varepsilon(G \cap K_\alpha) = \liminf_{\varepsilon\to 0} \varepsilon \log \mu_\varepsilon(g(G \cap K_\alpha) \geq -I'(y).$$

Hence

$$\liminf_{\varepsilon\to 0} \varepsilon \ln \nu_\varepsilon(G) \geq \liminf_{\varepsilon\to 0} \varepsilon \ln \nu_\varepsilon(G \cap K_\alpha) \geq -I'(y).$$

This completes the proof. □

As a corollary, we note that exponential tightness helps to transfer an LDP from a coarser to a finer topology.

COROLLARY 2.19. Let $\mathbf{X}$ be a set with two topologies τ_1 and τ_2 such that $\tau_2 \subseteq \tau_1$. Denote the Borel sets with respect to τ_1 by $\mathcal{B}$, those with respect to τ_2 by $\mathcal{C}$. Let $(\mu_\varepsilon)_{\varepsilon>0}$ be an exponentially tight family of probability measures on $(\mathbf{X}, \mathcal{C})$. If $(\mu_\varepsilon)_{\varepsilon>0}$ satisfies an LDP with respect to τ_2, it satisfies an LDP with respect to τ_1.

PROOF. Let $g\colon \mathbf{X} \to \mathbf{X}$ be the identity mapping. It is a continuous bijection from the finer to the coarser topology. Now apply the Theorem 2.18. □

We next consider the situation in which two families of measures are given which describe the laws of families of processes. If the processes are asymptotically close in the sense of the following definition, we can prove that large deviations principles are transferred from one family to the other.

DEFINITION 2.20. Let $(\Omega, \mathcal{F}, \mathbb{P})$ be a probability space, $(\mathbf{X}, \mathcal{B})$ a metric measure space with metric d. For any $\varepsilon > 0$ let Z_ε, $\tilde{Z}_\varepsilon$, be $\mathbf{X}$-valued random variables with joint law $P_\varepsilon = \mathbb{P} \circ (Z_\varepsilon, \tilde{Z}_\varepsilon)^{-1}$, and laws $\mu_\varepsilon = \mathbb{P} \circ Z_\varepsilon^{-1}$, $\tilde{\mu}_\varepsilon = \mathbb{P} \circ \tilde{Z}_\varepsilon^{-1}$. We call $(Z_\varepsilon)_{\varepsilon>0}$ and $(\tilde{Z}_\varepsilon)_{\varepsilon>0}$ *exponentially equivalent* if for every $\delta > 0$, setting

$$\Gamma_\delta = \{(y, z) \in \mathbf{X} \times \mathbf{X} : d(y, z) > \delta\},$$

and supposing that $d(Z_\varepsilon, \tilde{Z}_\varepsilon)$ be measurable, we have

$$\limsup_{\varepsilon\to 0} \varepsilon \ln P_\varepsilon(\Gamma_\delta) = \limsup_{\varepsilon\to 0} \varepsilon \ln \mathbb{P}(d(Z_\varepsilon, \tilde{Z}_\varepsilon) > \delta) = -\infty.$$

In this case the families $(\mu_\varepsilon)_{\varepsilon>0}$ and $(\tilde{\mu}_\varepsilon)_{\varepsilon>0}$ are also called *exponentially equivalent.*

Let us now show that exponentially equivalent families are not distinguished by the LDP.

THEOREM 2.21. *Let $(\mathbf{X}, \mathcal{B})$ be a metric measurable space with metric d, $(\mu_\varepsilon)_{\varepsilon>0}$ a family of probability measures which satisfies an LDP with good rate function I on $(\mathbf{X}, \mathcal{B})$. Let $(\tilde{\mu}_\varepsilon)_{\varepsilon>0}$ be exponentially equivalent to $(\mu_\varepsilon)_{\varepsilon>0}$. Then also $(\tilde{\mu}_\varepsilon)_{\varepsilon>0}$ satisfies an LDP with good rate function I.*

PROOF. The proof will be divided into 4 steps.
1. We show that for any $y \in \mathbf{X}$

$$I(y) = -\inf_{\delta>0} \limsup_{\varepsilon\to 0} \varepsilon \ln \tilde{\mu}_\varepsilon(B_\delta(y)) = -\inf_{\delta>0} \liminf_{\varepsilon\to 0} \varepsilon \ln \tilde{\mu}_\varepsilon(B_\delta(y)).$$

Let $\delta > 0$, $y \in \mathbf{X}$ be given. For any $\varepsilon > 0$ we have

$$\mu_\varepsilon(B_\delta(y)) \leq \tilde{\mu}_\varepsilon(B_{2\delta}(y)) + P_\varepsilon(\Gamma_\delta).$$

The lower bounds in the LDP for μ_ε further reveal

$$\begin{aligned}
-\inf_{z\in B_\delta(y)} I(z) &\le \liminf_{\varepsilon\to 0} \varepsilon \ln \mu_\varepsilon(B_\delta(y)) \\
&\le \liminf_{\varepsilon\to 0} \varepsilon \ln \Big(\tilde{\mu}_\varepsilon(B_{2\delta}(y)) + P_\varepsilon(\Gamma_\delta)\Big) \\
&\le \Big(\liminf_{\varepsilon\to 0} \varepsilon \ln \tilde{\mu}_\varepsilon(B_{2\delta}(y))\Big) \vee \Big(\liminf_{\varepsilon\to 0} \varepsilon \ln P_\varepsilon(\Gamma_\delta)\Big).
\end{aligned}$$

Since $(\mu_\varepsilon)_{\varepsilon>0}$ is exponentially equivalent to $(\tilde{\mu}_\varepsilon)_{\varepsilon>0}$, we further obtain

$$-\inf_{z\in B_\delta(y)} I(z) \le \liminf_{\varepsilon\to 0} \varepsilon \ln \tilde{\mu}_\varepsilon(B_{2\delta}(y)).$$

Repeating the arguments leading to this estimate with the roles of $\tilde{Z}_\varepsilon$ and Z_ε reversed gives

$$-\inf_{z\in \overline{B_{2\delta}(y)}} I(z) \ge \limsup_{\varepsilon\to 0} \varepsilon \ln \tilde{\mu}_\varepsilon(B_\delta(y)).$$

Now, noting that $\overline{B_{2\delta}(y)} \subset B_{3\delta}(y)$, we may take the $\inf_{\delta>0}$ on both sides of the preceding inequalities, and recall the lower semicontinuity of I, to get for $\rho > 0$

$$I(y) \ge \inf_{\delta>0} \inf_{z\in B_\delta(y)} I(z) \ge I(y) - \rho,$$

and hence

$$-I(y) \le \inf_{\delta>0} \liminf_{\varepsilon\to 0} \varepsilon \ln \tilde{\mu}_\varepsilon(B_\delta(y)) \le \inf_{\delta>0} \limsup_{\varepsilon\to 0} \varepsilon \ln \tilde{\mu}_\varepsilon(B_\delta(y)) \le -I(y).$$

This proves the claim.

2. We show that for $y \in \mathbf{X}$, and $G \subseteq \mathbf{X}$ open with $y \in G$ we have

$$-I(y) \le \liminf_{\varepsilon\to 0} \varepsilon \ln \tilde{\mu}_\varepsilon(G),$$

remarking that this statement implies the LDP lower bound for the family $(\tilde{\mu}_\varepsilon)_{\varepsilon>0}$.

To see this, use part 1. and the fact that due to $y \in G$ there exists $\delta > 0$ such that $B_\delta(y) \subseteq G$ to deduce

$$-I(y) = \inf_{\delta>0} \limsup_{\varepsilon\to 0} \varepsilon \ln \tilde{\mu}_\varepsilon(B_\delta(y)) \le \liminf_{\varepsilon\to 0} \varepsilon \ln \tilde{\mu}_\varepsilon(G).$$

This is the claimed estimate.

3. Fix $F \subseteq \mathbf{X}$ closed, and $\delta > 0$. Let $F^\delta = \{z \in \mathbf{X}\colon d(z,F) \le \delta\}$. We show that

$$\limsup_{\varepsilon\to 0} \varepsilon \ln \tilde{\mu}_\varepsilon(F) \le -\inf_{y\in F^\delta} I(y).$$

To see this, note that for $\varepsilon > 0$ we have

$$\tilde{\mu}_\varepsilon(F) \le \mu_\varepsilon(F^\delta) + P_\varepsilon(\Gamma_\delta).$$

Now apply the upper bound of the LDP for μ_ε to get

$$\begin{aligned}
\limsup_{\varepsilon\to 0} \varepsilon \ln \tilde{\mu}_\varepsilon(F) &\le \limsup_{\varepsilon\to 0} \varepsilon \ln \Big(\mu_\varepsilon(F^\delta) + P_\varepsilon(\Gamma_\delta)\Big) \\
&\le \Big(\limsup_{\varepsilon\to 0} \varepsilon \ln \mu_\varepsilon(F^\delta)\Big) \vee \Big(\limsup_{\varepsilon\to 0} \varepsilon \ln P_\varepsilon(\Gamma_\delta)\Big) \\
&\le \Big(-\inf_{y\in F^\delta} I(y), \limsup_{\varepsilon\to 0} \varepsilon \ln P_\varepsilon(\Gamma_\delta)\Big).
\end{aligned}$$

Since $(\mu_\varepsilon)_{\varepsilon>0}$ is exponentially equivalent to $(\tilde{\mu}_\varepsilon)_{\varepsilon>0}$, we obtain

$$\limsup_{\varepsilon\to 0} \varepsilon \ln \tilde{\mu}_\varepsilon(F) \le -\inf_{y\in F^\delta} I(y).$$

4. We prove that for $F \subseteq \mathbf{X}$ closed

$$\inf_{y\in F} I(y) = \lim_{\delta\to 0} \inf_{y\in F^\delta} I(y).$$

Note that, together with 3. this implies the upper bound for $F \subset \mathbf{X}$ closed.

Let $\eta > 0$. We need to show that

$$\gamma = \lim_{\delta\to 0} \inf_{y\in F^\delta} I(y) \geq \inf_{y\in F} I(y) - \eta.$$

Assume, avoiding a trivial case, that $\gamma < \infty$. Let $\alpha = \eta + \gamma$. Then for any $\delta > 0$ we have $\inf_{y\in F^\delta} I(y) \leq \alpha$, hence by definition

$$F^\delta \cap \Psi_I(\alpha) \neq \emptyset.$$

These sets being compact since I is a good rate function, we obtain

$$F \cap \Psi_I(\alpha) = \bigcap_{\delta>0} \left(F^\delta \cap \Psi_I(\alpha)\right) \neq \emptyset.$$

This translates into the relation

$$\inf_{y\in F} I(y) \leq \alpha,$$

and hence into the claimed inequality. □

2.3. Large deviations for Brownian motion

In this section, we shall establish a large deviation principle for d-dimensional Brownian motion. This will serve in the subsequent section to derive a large deviation principle for diffusions driven by additive noise, via the contraction principle. The large deviation principle for Brownian motion is usually referred to as Schilder's theorem. Our method to prove this Theorem takes its motivation from the Fourier series representation of Brownian motion discussed in Section 2.1. In fact, Ciesielski's isomorphism underlying this representation will enable us to reduce the argument for Schilder's theorem to the large deviation principle for one-dimensional Gaussian random variables.

2.3.1. Large deviations for one-dimensional Gaussian random variables. The large deviation rate for a standard one-dimensional Gaussian random variable can be directly calculated. Consider a random variable Z with standard normal law, and let μ_ε be the law of $\sqrt{\varepsilon}Z$. Then the following statement holds.

THEOREM 2.22. *Let*

$$I(x) = \frac{x^2}{2}, \quad x \in \mathbb{R}.$$

Then for any open set $G \subseteq \mathbb{R}$ and any closed set $F \subseteq \mathbb{R}$ we have

$$-\inf_{x\in G} I(x) \leq \liminf_{\varepsilon\to 0} \varepsilon \ln \mu_\varepsilon(G),$$

$$-\inf_{x\in F} I(x) \geq \limsup_{\varepsilon\to 0} \varepsilon \ln \mu_\varepsilon(F).$$

PROOF. We argue for a closed set $F \subseteq \mathbb{R}$. Let $a = \inf\{|x|: x \in F\}$. Note that the case $a = 0$ is trivial. We may therefore assume that $a > 0$. By symmetry we may further assume that there exists $b \geq a$ such that $F \subset (-\infty, -b] \cup [a, \infty)$. Hence for $\varepsilon > 0$

$$\mu_\varepsilon(F) \leq \mu_\varepsilon([a,\infty)) + \mu_\varepsilon((-\infty,-b]) \leq \frac{2}{\sqrt{2\pi}} \int_{\frac{a}{\sqrt{\varepsilon}}}^{\infty} e^{-x^2/2}\, dx.$$

For $u > 1$ we have

$$\int_u^\infty e^{-x^2/2}\,dx \le \int_u^\infty x e^{-x^2/2}\,dx = e^{-u^2/2}.$$

Hence for $\varepsilon < a^2$

$$\varepsilon \ln \mu_\varepsilon(F) \le \varepsilon\Big(\ln\Big(\frac{2}{\sqrt{2\pi}}\Big) - \frac{a^2}{2\varepsilon}\Big),$$

and therefore

$$\limsup_{\varepsilon\to 0} \varepsilon \ln \mu_\varepsilon(F) \le -\frac{a^2}{2} = -\inf_{x\in F} I(x).$$

For open sets we need a different inequality. In fact, integration by parts gives for $u > 1$

$$\int_u^\infty e^{-x^2/2}\,dx = \frac{1}{u}e^{-u^2/2} - \int_u^\infty \frac{1}{x^2}e^{-x^2/2}\,dx,$$

hence

$$\frac{1}{u}e^{-u^2/2} \le \Big(1+\frac{1}{u^2}\Big)\int_u^\infty e^{-x^2/2}\,dx$$

and

$$\frac{u}{1+u^2}e^{-u^2/2} \le \int_u^\infty e^{-x^2/2}\,dx.$$

Now let $G \subseteq \mathbb{R}$ be open, $y \in G$. By symmetry, we may assume that $y > 0$. Let, moreover, $a, b > 0$ such that $y \in (a,b) \subseteq G$. Then, for ε small enough we have

$$\begin{aligned}\mu_\varepsilon(G) &\ge \mu_\varepsilon((a,\infty)) - \mu_\varepsilon([b,\infty)) \ge \frac{1}{\sqrt{2\pi}}\Big(\frac{\frac{a}{\sqrt{\varepsilon}}}{1+\frac{a^2}{\varepsilon}}e^{-\frac{a^2}{2\varepsilon}} - e^{-\frac{b^2}{2\varepsilon}}\Big)\\ &\ge \frac{1}{\sqrt{2\pi}}\frac{\frac{a}{2\sqrt{\varepsilon}}}{1+\frac{a^2}{\varepsilon}}e^{-\frac{a^2}{2\varepsilon}}.\end{aligned}$$

Therefore

$$\liminf_{\varepsilon\to 0} \varepsilon \ln \mu_\varepsilon(G) \ge -\frac{a^2}{2} \ge -\frac{y^2}{2} = -I(y).$$

This implies the lower bound. □

2.3.2. Large deviations for one-dimensional Brownian motion in Hölder space. In this section we use the Fourier series decomposition of one-dimensional Brownian motion in Hölder space $\mathcal{C}_0^\alpha([0,1],\mathbb{R})$ of order $\alpha < \frac{1}{2}$, and Ciesielski's isomorphism mapping this space to a sequence space, to derive a large deviations principle for Brownian motion with respect to the topology of Hölder space. This remarkable approach was presented in Baldi and Roynette [**4**]. Let W be a one-dimensional Brownian motion indexed by $[0,1]$, described by

$$W = \sum_{n=0}^\infty Z_n \phi_n,$$

with a sequence $(Z_n)_{n\ge 0}$ of i.i.d. standard normal variables, and the Schauder functions $(\phi_n)_{n\ge 0}$, as described in Section 2.1. Recall the Haar functions $(\chi_n)_{n\ge 0}$ and the sequences $(c_n(\alpha))_{n\ge 0}$ appearing in Ciesielski's isomorphism in Theorem 2.2 for $0 < \alpha < 1$, given by

$$c_n(\alpha) = \begin{cases} 1, & n = 0,\\ 2^{k(\alpha-\frac{1}{2})+\alpha-1}, & n = 2^k + l,\ k \ge 0,\ 0 \le l \le 2^k - 1.\end{cases} \tag{2.20}$$

We investigate the asymptotic behavior of the family of probability measures $(\mu_\varepsilon)_{\varepsilon>0}$, where μ_ε is the law of $\sqrt{\varepsilon}W$, $\varepsilon > 0$. We remark that according to Theorem 2.6 for any $\varepsilon > 0$, $0 < \alpha < \frac{1}{2}$ we have

$$\mu_\varepsilon(\mathcal{C}_0^\alpha([0,1],\mathbb{R})) = 1. \tag{2.21}$$

Note also that the separability of $c_0(\mathbb{N}_0,\mathbb{R})$ is translated into separability of $\mathcal{C}_0^\alpha([0,1],\mathbb{R})$ by Ciesielski's isomorphism of Theorem 2.3, while Theorem 2.2 yields that $\mathcal{C}^\alpha([0,1],\mathbb{R})$ is not separable.

The large deviation rates for Brownian motion will crucially depend on the following function space, the *Cameron–Martin space* of absolutely continuous functions:

$$\begin{aligned}\mathcal{H}_1([0,1],\mathbb{R}) &= \mathcal{H}_1\\ &= \Big\{f\colon [0,1]\to\mathbb{R},\ f(0)=0,\ f \text{ is absolutely continuous with } \dot f \in L^2([0,1],\mathbb{R})\Big\}\\ &= \Big\{\int_0^t \dot f(s)\,ds,\ \dot f\in L^2([0,1],\mathbb{R})\Big\}.\end{aligned}$$

By means of (2.3.2) we can define the rate function for Brownian motion.

DEFINITION 2.23. Let

$$I(f) = \begin{cases} \dfrac{1}{2}\displaystyle\int_0^1 |\dot f(u)|^2\,du, & \text{if } f\in\mathcal{H}_1,\\ +\infty, & \text{otherwise.}\end{cases} \tag{2.22}$$

The following Theorem can be considered a version of Schilder's theorem with respect to a finer topology, proved for sets of a basis of the Hölder space topology first. For $\delta > 0$, $\psi \in \mathcal{C}_0^\alpha([0,1],\mathbb{R})$ denote the open ball

$$B_\delta^\alpha(\psi) = \Big\{f\in\mathcal{C}_0^\alpha([0,1],\mathbb{R})\colon \|f-\psi\|_\alpha < \delta\Big\}.$$

THEOREM 2.24. *Let $0<\alpha<1/2$, $\delta>0$ and $\psi\in(\mathcal{C}_0^\alpha([0,1],\mathbb{R}),\|\cdot\|_\alpha)$. Then with the rate function I defined by* (2.22)

$$\lim_{\varepsilon\to0}\varepsilon\ln\mu_\varepsilon(B_\delta^\alpha(\psi)) = -\inf_{f\in B_\delta^\alpha(\psi)} I(f), \tag{2.23}$$

$$\lim_{\varepsilon\to0}\varepsilon\ln\mu_\varepsilon(\overline{B_\delta^\alpha(\psi)}) = -\inf_{f\in\overline{B_\delta^\alpha(\psi)}} I(f). \tag{2.24}$$

PROOF. We give the arguments for (2.23). The proof of (2.24) is almost identical.

1. We use the Schauder representation of Brownian motion W and the function ψ given by

$$W = \sum_{n\ge0} Z_n\phi_n \quad\text{and}\quad \psi = \sum_{n\ge0}\frac{\xi_n}{c_n(\alpha)}\phi_n. \tag{2.25}$$

We recall the inverse of Ciesielski's isomorphism

$$T_{\alpha,0}^{-1}\colon c_0(\mathbb{N}_0,\mathbb{R})\to\mathcal{C}_0^\alpha([0,1],\mathbb{R}),\quad (\eta_n)_{n\ge0}\mapsto\sum_{n=0}^\infty\frac{\eta_n}{c_n(\alpha)}\phi_n,$$

and remark that the sequence $(\xi_n)_{n\geq 0}$ in the representation of ψ just satisfies $T_{\alpha,0}(\psi) = (\xi_n)_{n\geq 0}$, while $T_{\alpha,0}(\sqrt{\varepsilon}W) = (\sqrt{\varepsilon}c_n(\alpha)Z_n)_{n\geq 0}$. We therefore have

$$\sqrt{\varepsilon}W \in B_\delta^\alpha(\psi) \quad \text{if and only if} \quad \sup_{n\geq 0}\left|\sqrt{\varepsilon}c_n(\alpha)Z_n - \xi_n\right| < \delta.$$

Hence

$$(\sqrt{\varepsilon}W)^{-1}[B_\delta^\alpha(\psi)] = \bigcap_{n\geq 0}\Big\{\sqrt{\varepsilon}c_n(\alpha)Z_n \in (\xi_n - \delta, \xi_n + \delta)\Big\}.$$

Since $(Z_n)_{n\geq 0}$ is a family of independent random variables, we obtain for $\varepsilon > 0$

$$\mu_\varepsilon(B_\delta^\alpha(\psi)) = \prod_{n\geq 0}\mathbb{P}\Big(\sqrt{\varepsilon}c_n(\alpha)Z_n \in (\xi_n - \delta, \xi_n + \delta)\Big) =: \prod_{n\geq 0}P_n(\varepsilon). \tag{2.26}$$

We split the sequence of probabilities $(P_n(\varepsilon))_{n\geq 0}$ into four different parts to be treated separately:

$$\begin{aligned}
\Lambda_1 &= \Big\{n \geq 0\colon 0 \notin [\xi_n - \delta, \xi_n + \delta]\Big\},\\
\Lambda_2 &= \Big\{n \geq 0\colon \xi_n = \pm\delta\Big\},\\
\Lambda_3 &= \Big\{n \geq 0\colon [\xi_n - \delta, \xi_n + \delta] \supseteq \Big[-\frac{\delta}{2}, \frac{\delta}{2}\Big]\Big\},\\
\Lambda_4 &= \Lambda_3^c\backslash(\Lambda_1 \cup \Lambda_2).
\end{aligned}$$

Let us recall that $(\xi_n)_{n\geq 0} \in c_0(\mathbb{N}_0, \mathbb{R})$, so Λ_3 contains almost all $n \geq 0$, and hence the set $\Lambda_1 \cup \Lambda_2 \cup \Lambda_4 = (\Lambda_3)^c$ is finite.

2. Let us first discuss the contribution of Λ_3. Since $(Z_n)_{n\geq 0}$ are standard normal variables, we have

$$\begin{aligned}
\prod_{n\in\Lambda_3} P_n(\varepsilon) &\geq \prod_{n\in\Lambda_3}\mathbb{P}\Big(Z_n \in \Big[-\frac{\delta}{2c_n(\alpha)\sqrt{\varepsilon}}, \frac{\delta}{2c_n(\alpha)\sqrt{\varepsilon}}\Big]\Big)\\
&= \prod_{n\in\Lambda_3}\Big(1 - \sqrt{\frac{2}{\pi}}\int_{\delta/(2c_n(\alpha)\sqrt{\varepsilon})}^{\infty} e^{-u^2/2}\,du\Big).
\end{aligned}$$

Now according to (2.20) and our choice of α, $c_n(\alpha) \leq 1$, $\lim_{n\to\infty} c_n(\alpha) = 0$. Therefore, for $\varepsilon > 0$ such that $\varepsilon < \delta^2$ and all $n \geq 0$ we may estimate (see proof of Theorem 2.22)

$$\int_{\delta/(2c_n(\alpha)\sqrt{\varepsilon})}^{\infty} e^{-x^2/2}\,dx \leq \exp\Big(-\frac{\delta^2}{8c_n(\alpha)^2\varepsilon}\Big).$$

In order to prove that $\prod_{n\in\Lambda_3} P_n(\varepsilon)$ converges to 1 as $\varepsilon \to 0$, by the elementary inequality $-x \leq \ln(1-x)$ for $x \in (0, 1)$ it suffices to prove that $\sum_{n\geq 0}\exp(-\frac{\delta^2}{8c_n(\alpha)^2\varepsilon})$ converges to 0 as $\varepsilon \to 0$. This is indeed the case due to (2.20) and we deduce that

$$\lim_{\varepsilon\to 0}\prod_{n\in\Lambda_3} P_n(\varepsilon) = 1. \tag{2.27}$$

3. Next, we estimate the contribution of Λ_4. Indeed, $|\Lambda_4| < \infty$ and by definition $[\xi_n - \delta, \xi_n + \delta]$ contains a small neighborhood of the origin for any $n \in \Lambda_4$. Thus

$$\lim_{\varepsilon\to 0}\prod_{n\in\Lambda_4} P_n(\varepsilon) = 1. \tag{2.28}$$

4. Since $|\Lambda_2| < \infty$, its definition immediately gives

$$\lim_{\varepsilon \to 0} \prod_{n \in \Lambda_2} P_n(\varepsilon) = 2^{-|\Lambda_2|}. \tag{2.29}$$

5. Let us finally estimate the contribution of Λ_1. We define

$$\overline{\xi}_n = \begin{cases} \xi_n - \delta, & \text{if } \xi_n > \delta, \\ -(\xi_n + \delta), & \text{if } \xi_n < -\delta. \end{cases}$$

Since Z_n has a standard normal law for all $n \in \Lambda_1$, Theorem 2.22 implies

$$\lim_{\varepsilon \to 0} \varepsilon \ln P_n(\varepsilon) = -\frac{\overline{\xi}_n^2}{2c_n(\alpha)^2}.$$

Since $|\Lambda_1| < \infty$, we therefore have

$$\lim_{\varepsilon \to 0} \varepsilon \ln \prod_{n \in \Lambda_1} P_n(\varepsilon) = -\sum_{n \in \Lambda_1} \frac{\overline{\xi}_n^2}{2c_n(\alpha)^2}. \tag{2.30}$$

6. Using (2.27), (2.28), (2.29) and (2.30), we can deduce (2.23) if we are able to compare

$$\sum_{n \in \Lambda_1} \frac{\overline{\xi}_n^2}{2c_n(\alpha)^2} \quad \text{with} \quad \inf_{f \in B_\delta^\alpha(\psi)} I(f).$$

By Theorem 2.3 any function $f \in C_0^\alpha([0,1], \mathbb{R}) \cap \mathcal{H}_1$ has the Schauder representation

$$f = \sum_{n \geq 0} \frac{\eta_n}{c_n(\alpha)} \phi_n, \quad \text{with } (\eta_n)_{n \geq 0} \in c_0(\mathbb{N}_0, \mathbb{R}).$$

The derivative $\dot{f}$ satisfies $\dot{f} = \sum_{n \geq 0} \frac{\eta_n}{c_n(\alpha)} \chi_n$, and since $(\chi_n)_{n \geq 0}$ is an orthonormal system in $L^2([0,1], \mathbb{R})$, we obtain

$$\frac{1}{2} \int_0^1 |\dot{f}(s)|^2 \, ds = \sum_{n \geq 0} \frac{\eta_n^2}{2c_n(\alpha)^2}.$$

Hence the statement of the Theorem is an immediate consequence of the equality

$$\begin{aligned} \inf_{f \in B_\delta^\alpha(\psi) \cap \mathcal{H}_1} \frac{1}{2} \int_0^1 |\dot{f}(s)|^2 \, ds &= \inf \Big\{ \sum_{n \geq 0} \frac{\eta_n^2}{2c_n(\alpha)^2} \text{ with } \eta_n \in (\xi_n - \delta, \xi_n + \delta) \Big\} \\ &= \sum_{n \in \Lambda_1} \frac{\overline{\xi}_n^2}{2c_n(\alpha)^2}. \end{aligned}$$

□

Theorem 2.24 will allow us to derive an LDP for Brownian motion once we have established exponential tightness of the family $(\mu_\varepsilon)_{\varepsilon > 0}$.

THEOREM 2.25. *Let $0 < \alpha < \frac{1}{2}$. Then $(\mu_\varepsilon)_{\varepsilon > 0}$ is exponentially tight on the topological space $(\mathcal{C}_0^\alpha([0,1], \mathbb{R}), \|\cdot\|_\alpha)$. More precisely, for $\delta > 0$ and $0 < \beta < \alpha < \frac{1}{2}$, $\overline{B_\delta^\alpha(0)}$ is compact in $(\mathcal{C}_0^\beta([0,1], \mathbb{R}), \|\cdot\|_\beta)$, and we have*

$$\limsup_{\varepsilon \to 0} \varepsilon \ln \mu_\varepsilon \Big(\overline{B_\delta^\alpha(0)}^c \Big) \leq -\frac{\delta^2}{2}. \tag{2.31}$$

PROOF. Recall from the proof of Theorem 2.2 that

$$T_\alpha(\overline{B_\delta^\alpha(0)}) = \prod_{n=0}^{\infty}[-\delta,\delta] \cap c_0(\mathbb{N}_0,\mathbb{R}).$$

Moreover, for $0<\beta<\alpha<\frac{1}{2}$ we have

$$T_\beta(\overline{B_\delta^\alpha(0)}) = \prod_{n=0}^{\infty}\Big[-\frac{c_n(\beta)}{c_n(\alpha)}\delta, \frac{c_n(\beta)}{c_n(\alpha)}\delta\Big].$$

The space $\prod_{n=0}^{\infty}[-\frac{c_n(\beta)}{c_n(\alpha)}\delta, \frac{c_n(\beta)}{c_n(\alpha)}\delta]$ can easily be shown to be compact, for instance by establishing completeness and total boundedness, both of which are obviously satisfied. Since T_β is an isomorphism, we obtain that $\overline{B_\delta^\alpha(0)}$ is compact in $\mathcal{C}_0^\beta([0,1],\mathbb{R})$.

Using the decomposition (2.25), we define for δ, $\varepsilon>0$, $n\geq 0$

$$A_n = \Big\{Z_n \notin \Big[-\frac{\delta}{\sqrt{\varepsilon}c_n(\alpha)}, \frac{\delta}{\sqrt{\varepsilon}c_n(\alpha)}\Big]\Big\}.$$

Ciesielski's isomorphism of Theorem 2.3 allows to write

$$\mu_\varepsilon\Big(\overline{B_\delta^\alpha(0)}^c\Big) = \mathbb{P}\Big(\exists\, n\geq 0\colon \sqrt{\varepsilon}c_n(\alpha)Z_n \notin [-\delta,\delta]\Big) = \mathbb{P}\Big(\bigcup_{n\geq 0} A_n\Big) \leq \sum_{n\geq 0}\mathbb{P}(A_n).$$

Now for $n\geq 0$ by Theorem 2.22

$$\limsup_{\varepsilon\to 0}\varepsilon\ln\mathbb{P}(A_n) = -\frac{\delta^2}{2c_n(\alpha)^2}.$$

Observe that $c_n(\alpha)$ strictly decreases to 0, starting at $c_0(\alpha)=1$. Hence by monotone convergence using the ideas of the proof of Lemma 2.14

$$\limsup_{\varepsilon\to 0}\varepsilon\ln\Big(\sum_{n=0}^{\infty}\mathbb{P}(A_n)\Big) = \limsup_{\varepsilon\to 0}\varepsilon\sup_{n\geq 0}\ln\mathbb{P}(A_n) = -\frac{\delta^2}{2}.$$

This implies (2.31) and the proof is complete. □

We are ready to state the main result of this section, which is a version of Schilder's theorem with respect to a finer topology.

THEOREM 2.26 (Baldi–Roynette). *Let $0<\alpha<\frac{1}{2}$. For $\varepsilon>0$ let μ_ε be the law of $\sqrt{\varepsilon}W$ on the topological space $(\mathcal{C}_0^\alpha([0,1],\mathbb{R}),\|\cdot\|_\alpha)$. Then $(\mu_\varepsilon)_{\varepsilon>0}$ satisfies a large deviations principle with the following good rate function*

$$I(f) = \begin{cases} \frac{1}{2}\int_0^1 |\dot f(u)|^2\,du, & \text{if } f\in\mathcal{H}_1,\\ +\infty & \text{otherwise.}\end{cases} \tag{2.32}$$

PROOF. Combine Theorem 2.24 and Theorem 2.25 in Corollary 2.16. □

2.3.3. Large deviations for Brownian motion with respect to uniform norm. To obtain the classical result of Schilder's theorem from the large deviation principle with respect to the finer topologies in Theorem 2.26, we finally have to apply the contraction principle in the form of Theorem 2.17. It is then straightforward to extend the results to multi-dimensional Brownian motions indexed by $\mathbb{R}_+$.

THEOREM 2.27 (Schilder). *For $\varepsilon > 0$ let μ_ε be the law of $\sqrt{\varepsilon}W$ on the topological space $(C([0,1],\mathbb{R}), \|\cdot\|)$. Then $(\mu_\varepsilon)_{\varepsilon>0}$ satisfies a large deviations principle with the following good rate function*

$$I(f) = \begin{cases} \dfrac{1}{2}\displaystyle\int_0^1 |\dot f(u)|^2\, du, & \textit{if } f \in \mathcal{H}_1, \\ +\infty & \textit{otherwise.} \end{cases} \tag{2.33}$$

PROOF. According to Theorem 2.26, $(\mu_\varepsilon)_{\varepsilon>0}$ satisfies an LDP on the space $(\mathcal{C}_0^\alpha([0,1],\mathbb{R}), \|\cdot\|_\alpha)$ with rate function I. Since the Hölder topology is finer than the uniform topology, Theorem 2.17 applied to the identity map implies that $(\mu_\varepsilon)_{\varepsilon>0}$ satisfies an LDP on the space $(\mathcal{C}_0^\alpha([0,1],\mathbb{R}), \|\cdot\|)$ with the same rate function I. Finally we observe that the LDP is preserved under the identity map from $(\mathcal{C}_0^\alpha([0,1],\mathbb{R}), \|\cdot\|)$ to $(C([0,1],\mathbb{R}), \|\cdot\|)$, again as a consequence of the contraction principle of Theorem 2.17. □

It remains to extend Schilder's theorem to multi-dimensional Brownian motions.

Let therefore $W = (W^1, \dots, W^d)$ be a d-dimensional Brownian motion indexed by $[0,1]$. The Cameron–Martin space to be used in this context has to be d-dimensional as well, but still be denoted by the same symbol.

$$\begin{aligned} &\mathcal{H}_1([0,1],\mathbb{R}^d) = \mathcal{H}_1 \\ &= \Big\{ f\colon [0,1] \to \mathbb{R}^d,\ f(0) = 0,\ f \text{ is absolutely continuous with } |\dot f| \in L^2([0,1],\mathbb{R}) \Big\} \\ &= \Big\{ \int_0^t \dot f(s)\, ds,\ |\dot f| \in L^2([0,1],\mathbb{R}) \Big\}. \end{aligned}$$

Here, as usual, $|\cdot|$ denotes the Euclidean norm in $\mathbb{R}^d$.

The d-dimensional version of Schilder's theorem follows easily from the independence of the components of the Brownian motion W.

THEOREM 2.28 (Schilder). *Let W be a d-dimensional standard Brownian motion. For $\varepsilon > 0$ let μ_ε be the law of $\sqrt{\varepsilon}W$ on the topological space $(C([0,1],\mathbb{R}^d), \|\cdot\|)$. Then $(\mu_\varepsilon)_{\varepsilon>0}$ satisfies a large deviations principle with the following good rate function*

$$I(f) = \begin{cases} \dfrac{1}{2}\displaystyle\int_0^1 |\dot f(u)|^2\, du, & \textit{if } f \in \mathcal{H}_1([0,1],\mathbb{R}^d), \\ +\infty & \textit{otherwise.} \end{cases} \tag{2.34}$$

PROOF. For products of open sets in $C([0,1],\mathbb{R}^d)$ the result follows directly from Theorem 2.27 and the definition of I. This system of open sets forming a basis of the product topology, we may then argue using Corollary 2.16 and Theorem 2.25, which can be slightly extended to products of compact sets. □

2.4. The Freidlin–Wentzell theory

In this section, we shall extend the LDP for Brownian motion to diffusion processes that are obtained as the strong solutions of stochastic differential equations driven by additive Brownian motion. The basic idea for achieving this is simple. We map trajectories of the Brownian motion to trajectories of the solution of a given stochastic differential equation. Due to the additivity of the noise, this mapping is continuous. Therefore, a strict contraction principle in the sense of Section 2.2 is applicable to transfer the LDP.

2.4.1. The original theory. Denote $C_0([0,1],\mathbb{R}^d)$ the set of all functions $f \in C([0,1],\mathbb{R}^d)$ which satisfy $f(0) = 0$ (endowed with the uniform metric). Let $b\colon \mathbb{R}^d \to \mathbb{R}^d$ be a globally Lipschitz continuous function, i.e. there exists a constant B such that

$$|b(x) - b(y)| \leq B|x - y|, \quad x, y \in \mathbb{R}^d.$$

For $\varepsilon > 0$ and $x \in \mathbb{R}^d$ let $X^{\varepsilon,x}$ denote the unique strong solution process of the SDE

$$X_t^{\varepsilon,x} = x + \int_0^t b(X_s^{\varepsilon,x})\, ds + \sqrt{\varepsilon} W_t, \quad 0 \leq t \leq 1. \tag{2.35}$$

We argue that the trajectories of $X^{\varepsilon,x}$ are continuous images of the trajectories of $W^\varepsilon_\cdot = \sqrt{\varepsilon} W_\cdot$.

REMARK 2.29. Unique global solutions of (2.35) always exist. They can even be constructed pathwise by defining

$$Y^{\varepsilon,x} = X^{\varepsilon,x} - W^\varepsilon,$$
$$\tilde{b}\colon [0,1] \times \mathbb{R}^d \times \Omega \to \mathbb{R}^d, \quad \tilde{b}(t,y,\omega) := b(y + W_t^\varepsilon(\omega)).$$

Then $\tilde{b}$ inherits the global Lipschitz property from b, and therefore the random ordinary differential equation

$$Y_t^{\varepsilon,x}(\omega) = x + \int_0^t \tilde{b}(s, Y_s^{\varepsilon,x}(\omega), \omega)\, ds, \quad t \in [0,1],$$

is well known to possess unique global solutions pathwise. The solution processes $X^{\varepsilon,x}$ of (2.35) are related to their solution processes $Y^{\varepsilon,x}$ by the simple identity

$$X^{\varepsilon,x} = Y^{\varepsilon,x} + W^\varepsilon.$$

LEMMA 2.30. *Let $g \in C_0([0,1],\mathbb{R}^d)$, $x \in \mathbb{R}^d$. Then there exists a unique $f \in C([0,1],\mathbb{R})$ which satisfies*

$$f(t) = x + \int_0^t b(f(s))\, ds + g(t), \quad 0 \leq t \leq 1. \tag{2.36}$$

The mapping

$$F\colon C_0([0,1],\mathbb{R}^d) \to C([0,1],\mathbb{R}^d), \quad g \mapsto f$$

is continuous and one-to-one.

PROOF. 1. In order to prove the first claim about the unique solution, it suffices to show that $x \in \mathbb{R}$, $g \in C_0([0,1],\mathbb{R}^d)$, and $T \in [0,1]$ such that $BT < 1$ the equation (2.36) possesses a unique solution on the interval $[0,T]$. Once this is proved, we can repeat the procedure a finite number of times on finitely many adjacent intervals of length bounded by BT, with recursively chosen initial conditions.

To establish the existence of the unique solution on the interval $[0,T]$, define the mapping

$$\Gamma\colon C([0,T],\mathbb{R}^d) \to C([0,T],\mathbb{R}^d), \quad f(\cdot) \mapsto x + \int_0^\cdot b(f(s))\, ds + g(\cdot).$$

Using the global Lipschitz condition for b, we see that Γ is a contraction. Indeed, for $g \in C([0,1],\mathbb{R}^d)$ fixed, $f_1,\ f_2 \in C([0,1],\mathbb{R}^d)$ we have

$$\|\Gamma(f_1) - \Gamma(f_2)\| \le BT\|f_1 - f_2\|.$$

Since $(C([0,T],\mathbb{R}^d), \|\cdot\|)$ is a Banach space, the mapping Γ has a unique fixed point which we identify with f.

2. We show that F is continuous. For this purpose, fix $f_i = F(g_i)$, $i = 1,2$, and set $\delta = \|g_1 - g_2\|$, $e(t) = |f_1(t) - f_2(t)|$, $0 \le t \le 1$. Then for any $0 \le t \le 1$ by the Lipschitz continuity of b

$$e(t) \le B \int_0^t e(s)\, ds + \delta.$$

Hence by Gronwall's lemma for any $0 \le t \le 1$

$$e(t) \le \delta e^{Bt}, \text{ and globally } \|f_1 - f_2\| \le e^B \|g_1 - g_2\|.$$

This means that F is even Lipschitz continuous. □

Due to Lemma 2.30, the contraction principle is directly applicable and yields a LDP.

THEOREM 2.31. *For $\varepsilon > 0$, $x \in \mathbb{R}^d$ let $X^{\varepsilon,x}$ be a solution of (2.35), and $\mu_\varepsilon = \mathbb{P} \circ (X^{\varepsilon,x})^{-1}$. Then $(\mu_\varepsilon)_{\varepsilon>0}$ satisfies an LDP on $(C([0,1],\mathbb{R}^d), \|\cdot\|)$ with good rate function*

$$I_x(f) = \begin{cases} \dfrac{1}{2}\displaystyle\int_0^1 |\dot f(u) - b(f(u))|^2\, dt, & f - x \in \mathcal{H}_1, \\ +\infty, & f - x \notin \mathcal{H}_1. \end{cases}$$

PROOF. Let $x \in \mathbb{R}^d$ be fixed. According to Lemma 2.30, $F(W_\varepsilon) = X^{\varepsilon,x}$ pathwise. Hence Schilder's theorem 2.28 combines with the contraction principle 2.17 to yield an LDP for $(\mu_\varepsilon)_{\varepsilon>0}$ with good rate function

$$\overline{I}_x(f) = \inf_{\{g \in \mathcal{H}_1 :\, F(g) = f\}} \frac{1}{2}\int_0^1 |\dot g(u)|^2\, du.$$

To prove that $\overline{I}_x(f) = I_x(f)$ for $f - x \in \mathcal{H}_1$, note that F from Lemma 2.30 is a one-to-one mapping. Moreover, for $g \in \mathcal{H}_1$ we have that f is a.e. differentiable and

$$\dot f = b(f) + \dot g.$$

Hence we have for $t \in [0,1]$ by virtue of the global Lipschitz continuity of b

$$|\dot f(t)| \le |b(f(t))| + |\dot g| \le |b(0)| + B|f(t)| + |\dot g(t)|,$$

and thus by boundedness of continuous functions $f - x \in \mathcal{H}_1$ as well and we can write

$$\overline{I}_x(f) = \frac{1}{2}\int_0^1 |\dot f(u) - b(f(u))|^2\, du = I_x(f).$$

□

2.4.2. An extension of the Freidlin–Wentzell theory. The LDP due to Freidlin and Wentzell presented in the previous section requires global Lipschitz conditions which are typically imposed in standard existence and uniqueness theorems for stochastic differential equations. In the setting of diffusions with a drift of the type of a potential gradient studied in the framework of stochastic resonance besides depending on time the coefficients will not be globally Lipschitz. We therefore need some extensions of the classical LDP result. This extension can be carried out in a general context (see Azencott [**2**] and two subsequent papers by Priouret [**88**] and Baldi et al. [**3**]). We prefer to present a simpler proof which permits to get the desired result in our particular framework.

For $x \in \mathbb{R}^d$ and $\varepsilon > 0$ let us consider the family $X^{\varepsilon,x}$ of solutions of the SDE

$$X_t^{\varepsilon,x} = x + \int_0^t b(X_s^{\varepsilon,x})\, ds + \sqrt{\varepsilon} W_t, \qquad t \geq 0. \tag{2.37}$$

Here b is locally Lipschitz continuous and satisfies the following growth condition: there are constants η, $R_0 > 0$ such that

$$\langle x, b(x) \rangle < -\eta |x| \quad \text{for all } |x| \geq R_0. \tag{2.38}$$

As a first consequence of this condition, the existence of a unique strong solution for (2.37) follows (see, for instance [**102**], Theorem 10.2.2).

Secondly, the growth condition (also called *dissipativity condition*) implies that the diffusion essentially stays inside a big ball $B_R(0)$ of radius R with very high probability: the probability for the diffusion to leave $B_R(0)$ is exponentially small. Two essential conclusions can be drawn from this observation: the laws of the diffusion $X^{\varepsilon,x} = (X_t^{\varepsilon,x})_{t\in[0,1]}$ are exponentially tight in the space $C([0,1], \mathbb{R}^d)$, and satisfy a LDP with a good rate function.

Let us make precise the exponential tightness of the diffusion paths first. We are interested in the small noise behavior of the exit time from the ball $B_R(0)$, defined by

$$\sigma_R^{\varepsilon,x} := \inf\{t \geq 0 \colon |X_t^{\varepsilon,x}| \geq R\}.$$

The following Theorem provides an asymptotic bound for $\sigma_R^{\varepsilon,x}$. The arguments of its proof are borrowed from a treatment of self-attracting diffusions by means of large deviations techniques, see Raimond [**90**] and Herrmann and Roynette [**52**].

THEOREM 2.32. *Let $x \in \mathbb{R}^d$, $\delta > 0$, and let $r \colon (0, \delta) \to (0, \infty)$ be a function satisfying*

$$\lim_{\varepsilon \to 0} \frac{\varepsilon}{r(\varepsilon)} = 0.$$

Then there exist R_1, $\varepsilon_1 > 0$ and $C > 0$ such that for $R \geq R_1$, $\varepsilon < \varepsilon_1$

$$\mathbb{P}\Big(\sigma_R^{\varepsilon,x} \leq r(\varepsilon)\Big) \leq C\eta^2 \, \frac{r(\varepsilon)}{\varepsilon} \, e^{-\frac{\eta R}{\varepsilon}} \quad \textit{for} \quad |x| \leq \frac{R}{2}. \tag{2.39}$$

REMARK 2.33. The constants R_1, $\varepsilon_1 > 0$ and $C > 0$ are universal in the sense that they do not depend on the particular choice of the drift b, but only on the parameter η and R_0 introduced in the growth condition (2.38), and of course on the function r. Hence the bound is uniform in the class of all diffusions that satisfy (2.38).

PROOF. For convenience of notation, we suppress the superscripts in $X^{\varepsilon,x}$, $\sigma_R^{\varepsilon,x}$ etc.

Step 1: First we determine a diffusion process $(Z_t)_{t\geq 0}$ which takes positive values, is easier to handle than X, and dominates it, i.e. such that $|X_t| \leq Z_t$ almost surely, for all $t \geq 0$.

Choose a function $h \in C^2(\mathbb{R}^d, \mathbb{R})$ such that

$$\begin{cases} h(x) = |x|, & \text{if } x \geq R_0, \\ h(x) \leq R_0, & \text{if } |x| \leq R_0, \end{cases}$$

where R_0 is the constant given in the growth condition (2.38). By Itô's formula we have for $t \geq 0$

$$h(X_t) = h(x) + \sqrt{\varepsilon} \int_0^t \nabla h(X_s)\, dW_s + \int_0^t \langle \nabla h, b\rangle(X_s)\, ds + \frac{\varepsilon}{2} \int_0^t \Delta h(X_s)\, ds.$$

For $t \geq 0$, let

$$\xi_t := \int_0^t |\nabla h(X_s)|^2\, ds,$$

i.e. ξ is the quadratic variation of the continuous local martingale

$$M_t := \int_0^t \nabla h(X_s)\, dW_s.$$

Since $\nabla h(x) = \frac{x}{|x|}$ for $|x| \geq R_0$, we have $d\xi_t = dt$ on $\{|X_t| \geq R_0\}$.

We now introduce an auxiliary process Z that serves to control $|X|$. Let $0 < \tilde{\eta} < \eta$. According to Skorokhod's lemma (see e.g. Revuz and Yor [**93**, Chapter VI.2, Lemma (2.1)]) there is a unique pair of continuous adapted processes (Z, L) such that L is an increasing process (of finite variation) which increases only at times t for which $Z_t = R_0$, satisfies $Z \geq R_0$, and such that the equation

$$Z := R_0 \vee |x| + \sqrt{\varepsilon} M - \tilde{\eta}\xi + L$$

is valid. We show that there exists $\varepsilon_0 > 0$ such that for any $\varepsilon < \varepsilon_0$ and $t \geq 0$ we have $|X_t| \leq Z_t$ a.s. For that purpose, choose $f \in C^2(\mathbb{R}, \mathbb{R})$ such that

$$\begin{cases} f(x) > 0 \text{ and } f'(x) > 0 \text{ for all } x > 0, \\ f(x) = 0 \text{ for all } x \leq 0. \end{cases}$$

According to Itô's formula, for $t \geq 0$

$$\begin{aligned} f(h(X_t) - Z_t) = f(h(x) - R_0 \vee |x|) &+ \int_0^t f'(h(X_s) - Z_s)\, d(h(X) - Z)_s \\ &+ \frac{1}{2} \int_0^t f''(h(X_s) - Z_s)\, d\langle h(X) - Z\rangle_s. \end{aligned}$$

By definition of h and Z we have

$$\{h(X_t) > Z_t\} = \{h(X_t) > Z_t, |X_t| \leq R_0\} \cup \{h(X_t) > Z_t, |X_t| > R_0\} = \{|X_t| > Z_t\}.$$

Moreover by definition, $h(X) - Z$ is a process of finite variation, so that the bracket term in the preceding Itô formula vanishes. Hence the expression

$$\int_0^t f'\Big(|X_s| - Z_s\Big)\Big[\frac{1}{|X_s|}\langle X_s, b(X_s)\rangle + \frac{\varepsilon}{2}\Delta h(X_s) + \tilde{\eta}\Big]\, ds - \int_0^t f'(|X_s| - Z_s)\, dL_s$$

is an upper bound of $f(h(X_t)-Z_t)$. Furthermore, $\Delta h(x) = \frac{d-1}{|x|}$ for $|x| \geq R_0$, which by (2.38) implies

$$\frac{1}{|X_s|}\langle X_s, b(X_s)\rangle + \frac{\varepsilon}{2}\Delta h(X_s) + \tilde{\eta} < \frac{\varepsilon(d-1)}{2|X_s|} + \tilde{\eta} - \eta \quad \text{on} \quad \{|X_s| > Z_s\}.$$

The latter expression is negative if ε is small enough. Summarizing, we can find $\varepsilon_0 > 0$ such that $f(|X_t| - Z_t) \leq 0$ for $\varepsilon < \varepsilon_0$. This implies $|X_t| \leq Z_t$ a.s. by the definition of f, and Step 1 is established.

Step 2: We therefore can bound exit probabilities for X, expressed in terms of σ_R, by those for Z, expressed by an analogously defined stopping time τ_R. We have

$$\mathbb{P}(\sigma_R \leq r(\varepsilon)) \leq \mathbb{P}(\tau_R \leq r(\varepsilon))$$

for all $\varepsilon < \varepsilon_0$. We compute $\mathbb{P}(\tau_R \leq r(\varepsilon))$ for such ε. We have for any $\alpha > 0$

$$\mathbb{P}(\tau_R \leq r(\varepsilon)) \leq e^{\alpha r(\varepsilon)}\mathbb{E}e^{-\alpha\sigma_R}. \tag{2.40}$$

In order to find a bound for the right hand side of (2.40), let

$$K := \sup_{|x|\leq R_0} |\nabla h(x)|^2.$$

Then we have $\xi_t \leq Kt$ for all $t \geq 0$. Note that without loss of generality h can be chosen so that $K \leq 2R_0$. Now observe that, by Itô's formula, for any $\varphi \in C^2(\mathbb{R}, \mathbb{R})$

$$\begin{aligned} d\Big(\varphi(Z_t)\, e^{-\frac{\alpha}{K}\xi_t}\Big) &= \sqrt{\varepsilon}\, \varphi'(Z_t)\, e^{-\frac{\alpha}{K}\xi_t}\, dM_t + \varphi'(Z_t)\, e^{-\frac{\alpha}{K}\xi_t}\, dL_t \\ &\quad + e^{-\frac{\alpha}{K}\xi_t}\Big(\frac{\varepsilon}{2}\varphi''(Z_t) - \tilde{\eta}\varphi'(Z_t) - \frac{\alpha}{K}\varphi(Z_t)\Big)\, d\xi_t. \end{aligned}$$

Now let $R \geq R_0$. If we choose φ such that

$$\begin{cases} \frac{\varepsilon}{2}\varphi''(y) - \tilde{\eta}\varphi'(y) - \frac{\alpha}{K}\varphi(y) = 0 \text{ for } y \in [R_0, R], \\ \varphi'(R_0) = 0, \quad \varphi(R) = 1, \end{cases} \tag{2.41}$$

then $\varphi(Z_t)e^{-\frac{\alpha}{K}\xi_t}$ is a local martingale which is bounded up to time τ_R. Hence we are allowed to apply Doob's optional sampling theorem to obtain

$$\varphi(R_0 \vee |x|) = \mathbb{E}\Big[\varphi(Z_{\tau_R})e^{-\frac{\alpha}{K}\xi_{\tau_R}}\Big] = \mathbb{E}\, e^{-\frac{\alpha}{K}\xi_{\tau_R}}. \tag{2.42}$$

But since $\xi_{\tau_R} \leq K\tau_R$, which implies $\mathbb{E}\, e^{-\frac{\alpha}{K}\xi_{\tau_R}} \geq \mathbb{E}\, e^{-\alpha\tau_R}$, and we deduce from (2.40) that

$$\mathbb{P}(\sigma_R \leq r(\varepsilon)) \leq e^{\alpha r(\varepsilon)}\mathbb{E}\, e^{-\frac{\alpha}{K}\xi_{\tau_R}} \leq e^{\alpha r(\varepsilon)}\varphi(R_0 \vee |x|). \tag{2.43}$$

Step 3: We estimate φ satisfying (2.41). Solving the differential equation for φ yields

$$\varphi(x) = \frac{-\lambda^- e^{\lambda^+(x-R_0)} + \lambda^+ e^{\lambda^-(x-R_0)}}{-\lambda^- e^{\lambda^+(R-R_0)} + \lambda^+ e^{\lambda^-(R-R_0)}}, \quad x \in [R_0, R],$$

with

$$\lambda^{\pm} = \frac{1}{\varepsilon}\Big(\tilde{\eta} \pm \sqrt{\tilde{\eta}^2 + 2\frac{\alpha}{K}\varepsilon}\Big).$$

Hence

$$\varphi(x) \leq \frac{(\lambda^+ - \lambda^-)\, e^{\lambda^+(x-R_0)}}{(-\lambda^-)\, e^{\lambda^+(R-R_0)}}, \quad x \in [R_0, R].$$

Taking $\alpha = r(\varepsilon)^{-1}$ in (2.43) we obtain

$$\mathbb{P}(\sigma_R \le r(\varepsilon)) \le e^1\, \varphi(R_0 \vee |x|) \le \frac{\lambda^+ - \lambda^-}{-\lambda^-} \exp\Big(1 + \lambda^+(R_0 \vee |x| - R)\Big).$$

It is obvious that $\exp\Big(\lambda^+(R_0 \vee |x| - R)\Big) \le \exp\Big(-\frac{\tilde\eta R}{\varepsilon}\Big)$ for $R \ge 2(|x| \vee R_0)$, so it remains to comment on the prefactor. We have

$$\frac{\lambda^+ - \lambda^-}{-\lambda^-} = \frac{2\sqrt{\tilde\eta^2 + 2\frac{\alpha}{K}\varepsilon}}{\sqrt{\tilde\eta^2 + 2\frac{\alpha}{K}\varepsilon} - \tilde\eta} \le \frac{4\Big(\tilde\eta^2 + \frac{2\varepsilon}{Kr(\varepsilon)}\Big)}{\frac{2\varepsilon}{Kr(\varepsilon)}}.$$

Since $\lim_{\varepsilon\to 0} \frac{\varepsilon}{r(\varepsilon)} = 0$, the latter bound behaves like $2\tilde\eta^2 K \frac{r(\varepsilon)}{\varepsilon}$ as $\varepsilon \to 0$. Putting all estimates together yields the claimed asymptotic bound with $\tilde\eta$ instead of η. Finally, letting $\tilde\eta \to \eta$ establishes (2.39). □

REMARK 2.34. Theorem 2.32 can be easily extended to the context of time inhomogeneous diffusions. Let $X^{\varepsilon,x}$ be the solution process of

$$X_t^{\varepsilon,x} = x + \int_0^t b(s, X_s^{\varepsilon,x})\, ds + \sqrt{\varepsilon} W_t, \quad t \ge 0,\ x \in \mathbb{R}^d,$$

where b is locally Lipschitz continuous with respect to both variables and satisfies the following growth condition: there are constants η, $R_0 > 0$ such that

$$\langle x, b(t,x)\rangle < -\eta |x| \quad \text{for all } t \ge 0,\ |x| \ge R_0. \tag{2.44}$$

Then in the notation of Theorem 2.32 the exit time $\sigma_R^{\varepsilon,x}$ of $X^{\varepsilon,x}$ satisfies

$$\mathbb{P}\Big(\sigma_R^{\varepsilon,x} \le r(\varepsilon)\Big) \le C\eta^2 \frac{r(\varepsilon)}{\varepsilon}\, e^{-\frac{\eta R}{\varepsilon}} \quad \text{for} \quad |x| \le \frac{R}{2}.$$

The result of Theorem 2.32 is much sharper than what we really need to obtain large deviations estimates. It shall play a crucial rule in Chapter 4, where Remark 2.34 will be used. The importance of Theorem 2.32 for this section is linked to an immediate consequence obtained for $r(\varepsilon) = T$, $\varepsilon > 0$.

COROLLARY 2.35. For all $R \ge R_0$, $|x| \le \frac{R}{2}$ and $T > 0$ we have

$$\limsup_{\varepsilon\to 0} \varepsilon \ln \mathbb{P}(\sigma_R^{\varepsilon,x} \le T) \le -\eta R.$$

For $x \in \mathbb{R}^d$ and $\varepsilon > 0$ let μ_ε be the law of $X^{\varepsilon,x}$ on the function space $C([0,T],\mathbb{R}^d)$ endowed with the uniform norm. Then $(\mu_\varepsilon)_{\varepsilon>0}$ is exponentially tight.

PROOF. For convenience let $T = 1$, choose $x \in \mathbb{R}^d$, and denote $\nu_\varepsilon = \mathbb{P} \circ (\sqrt{\varepsilon} W)^{-1}$, $\varepsilon > 0$. For $\alpha > 0$ use Theorem 2.25 to choose a compact set $K \subset C([0,1],\mathbb{R}^d)$ such that

$$\limsup_{\varepsilon\to 0} \varepsilon \ln \nu_\varepsilon(K^c) \le -\alpha.$$

Next use Theorem 2.32 to choose $R > 0$ large enough to ensure $x \in B_{\frac{R}{2}}(0)$ and

$$\limsup_{\varepsilon\to 0} \varepsilon \ln \mathbb{P}(\sigma_R^{\varepsilon,x} \le 1) \le -\eta R < -\alpha.$$

Now choose a bounded vector field $b_R \colon \mathbb{R}^d \to \mathbb{R}^d$ which coincides with b on $B_R(0)$ and which is globally Lipschitz continuous. Let F_R be the continuous map on $C([0,1],\mathbb{R}^d)$ corresponding to b_R according to Lemma 2.30, and denote

$$Y_t^{\varepsilon,x} = x + \int_0^t b_R(Y_s^{\varepsilon,x})\, ds + \sqrt{\varepsilon} W_t, \quad t \ge 0.$$

Then by definition $X^{\varepsilon,x} = Y^{\varepsilon,x}$ on $\{\sigma_R^{\varepsilon,x} > 1\}$. Hence for ε small enough with the compact set $F_R(K) \subset C([0,1],\mathbb{R}^d)$

$$\begin{aligned}\mu_\varepsilon(F_R(K)^c) &\le \mathbb{P}\Big(X^{\varepsilon,x} \notin F_R(K), \sigma_R^{\varepsilon,x} > 1\Big) + \mathbb{P}(\sigma_R^{\varepsilon,x} \le 1)\\ &= \mathbb{P}\Big(Y^{\varepsilon,x} \notin F_R(K), \sigma_R^{\varepsilon,x} > 1\Big) + \mathbb{P}(\sigma_R^{\varepsilon,x} \le 1)\\ &\le \mathbb{P}\Big(Y^{\varepsilon,x} \notin F_R(K)\Big) + \mathbb{P}(\sigma_R^{\varepsilon,x} \le 1)\\ &= \nu_\varepsilon(K^c) + \mathbb{P}(\sigma_R^{\varepsilon,x} \le 1).\end{aligned}$$

This implies

$$\limsup_{\varepsilon\to 0} \varepsilon \ln \mu_\varepsilon(F_R(K)^c) \le \limsup_{\varepsilon\to 0} \varepsilon \ln \nu_\varepsilon(K^c) \vee \limsup_{\varepsilon\to 0} \varepsilon \ln \mathbb{P}(\sigma_R^{\varepsilon,x} \le 1) \le -\alpha,$$

which establishes the desired exponential tightness. □

With this at hand, we are in a position to state the large deviations result for the diffusion processes $X^{\varepsilon,x}$, $\varepsilon > 0$, $x \in \mathbb{R}^d$. We recall that $\mathcal{H}_1([0,1],\mathbb{R}^d)$ is the d-dimensional Cameron–Martin function space and $\|\cdot\|$ is the uniform norm.

THEOREM 2.36. *For $\varepsilon > 0$, $x \in \mathbb{R}^d$ let $X^{\varepsilon,x}$ be a solution of* (2.37) *on the time interval* $[0,1]$ *with* locally Lipschitz *drift term b that satisfies the growth condition* (2.44), *and $\mu_\varepsilon = \mathbb{P}\circ(X^{\varepsilon,x})^{-1}$. Then $(\mu_\varepsilon)_{\varepsilon>0}$ satisfies an LDP on $(C([0,1],\mathbb{R}^d),\|\cdot\|)$ with good rate function*

$$I_x(f) = \begin{cases} \dfrac{1}{2}\displaystyle\int_0^1 |\dot f(u) - b(f(u))|^2\,du, & f - x \in \mathcal{H}_1([0,1],\mathbb{R}^d),\\ +\infty, & f - x \notin \mathcal{H}_1([0,1],\mathbb{R}^d).\end{cases}$$

PROOF. By the exponential tightness of Corollary 2.35, it suffices to prove a weak LDP. More precisely, Lemma 2.14 explains that it is enough to establish the lower bound for compact sets and the upper bound for open sets. The proof is based on a localization technique.

Step 1. Upper bound. Let K be compact in $C([0,1],\mathbb{R}^d)$ with respect to the uniform metric. Then there exists $R > 0$ such that $\sup_{f\in K}\|f\| \le R$. We define a new drift b_R which is globally Lipschitz and equals b on $B_R(0)$, that is $b_R(x) = b(x)$ for $|x| \le R$. We again denote by Y the diffusion related to b_R. By definition we have

$$\mathbb{P}(X^{\varepsilon,x} \in K) = \mathbb{P}(Y^{\varepsilon,x} \in K).$$

Since b_R is globally Lipschitz continuous, Freidlin–Wentzell's theory yields a large deviation principle with rate function I_x^R which agrees with I_x on K, whence:

$$\limsup_{\varepsilon\to 0} \varepsilon \ln \mu_\varepsilon(K) = \limsup_{\varepsilon\to 0} \varepsilon \ln \mathbb{P}(Y^{\varepsilon,x} \in K) \le -\inf_{\varphi\in K} I_x^R(\varphi) = -\inf_{\varphi\in K} I_x(\varphi).$$

Step 2. Lower bound. The arguments are similar to the ones above. Let $G \subset C([0,1],\mathbb{R}^d)$ be open, and fix $f \in G$. There exists $0 < \delta < 1$ such that $G_f = \{g \in C([0,1],\mathbb{R}^d)\colon \|g - f\| < \delta\} \subset G$. By choosing R larger than $\sup_{g\in G_f}\|g\|$ and defining b_R and Y as in Step 1, we obtain

$$\mathbb{P}(X^{\varepsilon,x} \in G) \ge \mathbb{P}(X^{\varepsilon,x} \in G_f) = \mathbb{P}(Y^{\varepsilon,x} \in G_f).$$

The large deviations principle for the diffusion Y gives

$$\liminf_{\varepsilon\to 0} \varepsilon \ln \mu_\varepsilon(G_f) = \limsup_{\varepsilon\to 0} \varepsilon \ln \mathbb{P}(Y^{\varepsilon,x} \in G_f) \ge -\inf_{\varphi\in G_f} I_x^R(\varphi) = -\inf_{\varphi\in G_f} I_x(\varphi).$$

Due to the arbitrary choice of $f \in G$, we conclude that

$$\liminf_{\varepsilon \to 0} \varepsilon \ln \mu_\varepsilon(G) \geq - \inf_{\varphi \in G} I_x(\varphi).$$

The upper bound is satisfied and I_x is a good rate function since $(\mu_\varepsilon)_{\varepsilon>0}$ is exponentially tight. □

The large deviation principle presented in Theorem 2.36 depends on the initial position x of the diffusion process. This dependence can be analyzed and controlled. Indeed, the large deviation principle for the diffusion (2.37) is uniform with respect to its starting point. This fact which will also be very useful in the treatment of the asymptotic properties of the exit time will be established next.

PROPOSITION 2.37 (Uniform LDP). For a closed set $F \subset C([0,1], \mathbb{R}^d)$, an open set $G \subset C([0,1], \mathbb{R}^d)$, and $x \in \mathbb{R}^d$ we have

$$\text{a)} \quad \limsup_{\varepsilon \to 0, y \to x} \varepsilon \ln \mathbb{P}(X^{\varepsilon,y} \in F) \leq - \inf_{f \in F} I_x(f),$$
$$\text{b)} \quad \liminf_{\varepsilon \to 0, y \to x} \varepsilon \ln \mathbb{P}(X^{\varepsilon,y} \in G) \geq - \inf_{f \in G} I_x(f).$$

PROOF. According to Theorem 2.21 we have to show that for any family $(x_\varepsilon)_{\varepsilon>0}$ such that $\lim_{\varepsilon \to 0} x_\varepsilon = x$ the corresponding families of probability measures $\mu_\varepsilon = \mathbb{P} \circ (X^{\varepsilon,x})^{-1}$, $\varepsilon > 0$, and $\tilde{\mu}_\varepsilon = \mathbb{P} \circ (X^{\varepsilon,x_\varepsilon})^{-1}$, $\varepsilon > 0$, are exponentially equivalent in $(C([0,1], \mathbb{R}^d), \|\cdot\|)$.

To see this, fix $\varepsilon > 0$ and $x \in \mathbb{R}^d$. Then

$$\|X^{\varepsilon,x_\varepsilon} - X^{\varepsilon,x}\| \leq |x_\varepsilon - x| + \int_0^1 |b(X_u^{\varepsilon,x_\varepsilon}) - b(X_u^{\varepsilon,x})| \, du.$$

Let us now use a localization argument. We fix $R > 0$ and define the first exit times of the diffusions $X^{\varepsilon,x}$ and $X^{\varepsilon,x_\varepsilon}$ of $B_R(0)$ by

$$\sigma_R^{\varepsilon,x} = \inf\{t \geq 0 \colon X_t^{\varepsilon,x} \notin B_R(0)\},$$
$$\tau_R^{\varepsilon,x} = \inf\{t \geq 0 \colon X_t^{\varepsilon,x_\varepsilon} \notin B_R(0)\},$$

and $\tilde{\sigma}_R^{\varepsilon,x} = \sigma_R^{\varepsilon,x} \wedge \tau_R^{\varepsilon,x}$. Then on the event $\{\tilde{\sigma}_R^{\varepsilon,x} > 1\}$, the Lipschitz continuity of b on $B_R(0)$ with Lipschitz constant L_R implies

$$|b(X_u^{\varepsilon,x_\varepsilon}) - b(X_u^{\varepsilon,x})| \leq L_R |X_u^{\varepsilon,x_\varepsilon} - X_u^{\varepsilon,x}|, \quad 0 \leq u \leq 1.$$

Using Gronwall's lemma (see A.1), we obtain on the event $\{\tilde{\sigma}_R^{\varepsilon,x} > 1\}$ that

$$\|X^{\varepsilon,x_\varepsilon} - X^{\varepsilon,x}\| \leq |x_\varepsilon - x| e^{L_R}.$$

From this we deduce that the two sample paths are close together before one of them exits from $B_R(0)$. Let $\delta > 0$. Then, if we choose ε_0 small enough, for all $\varepsilon < \varepsilon_0$

$$\|X^{\varepsilon,x_\varepsilon} - X^{\varepsilon,x}\| \leq \delta \quad \text{on } \{\tilde{\sigma}_R^{\varepsilon,x} > 1\}.$$

Theorem 2.32 implies for R large enough so that for $\varepsilon < \varepsilon_0$ we have x, $x_\varepsilon \in B_{\frac{R}{2}}(0)$

$$\begin{aligned}\limsup_{\varepsilon \to 0} \varepsilon \ln \mathbb{P}(\|X^{\varepsilon,x_\varepsilon} - X^{\varepsilon,x}\| > \delta) &\leq \limsup_{\varepsilon \to 0} \varepsilon \ln \mathbb{P}(\tilde{\sigma}_R^{\varepsilon,x} \leq 1) \\ &\leq \limsup_{\varepsilon \to 0} \varepsilon \ln \mathbb{P}(\sigma_R^{\varepsilon,x} \leq 1) \vee \limsup_{\varepsilon \to 0} \varepsilon \ln \mathbb{P}(\tau_R^{\varepsilon,x} \leq 1) \leq -\eta R.\end{aligned}$$

Letting $R \to \infty$ allows to obtain the exponential equivalence between $(\mu_\varepsilon)_{\varepsilon>0}$ and $(\tilde{\mu}_\varepsilon)_{\varepsilon>0}$. □

We can draw a conclusion from Proposition 2.37 which will turn out to be of practical use later: an LDP which is uniform on compact sets of initial states for the diffusion.

COROLLARY 2.38. Let $K \subset \mathbb{R}^d$ be compact, $F \subset C([0,1],\mathbb{R}^d)$ closed, $G \subset C([0,1],\mathbb{R}^d)$ open. Then we have

$$\text{a)}\quad \limsup_{\varepsilon\to 0} \sup_{y\in K} \varepsilon \ln \mathbb{P}(X^{\varepsilon,y} \in F) \le - \inf_{y\in K, f\in F} I_y(f),$$

$$\text{b)}\quad \liminf_{\varepsilon\to 0} \inf_{y\in K} \varepsilon \ln \mathbb{P}(X^{\varepsilon,y} \in G) \ge - \sup_{y\in K} \inf_{f\in G} I_y(f).$$

PROOF. For similarity of arguments, we only show the upper bound. Let $-I_K$ denote the right hand side of the claimed inequality. For $\delta > 0$ let

$$I_K^\delta = (I_K - \delta) \wedge \frac{1}{\delta}.$$

Now fix $x \in K$. Then by Proposition 2.37 and lower semi-continuity of I_x there exists $\varepsilon_x > 0$ such that for any $\varepsilon \le \varepsilon_x$

$$\varepsilon \ln \sup_{y\in B_{\varepsilon_x}(x)} \mathbb{P}(X^{\varepsilon,y} \in F) \le -I_K^\delta.$$

Use compactness of K to choose $x_1, \ldots, x_m \in K$ such that $K \subset \bigcup_{i=1}^m B_{\varepsilon_{x_i}}(x_i)$. Then we have

$$\limsup_{\varepsilon\to 0} \varepsilon \ln \sup_{y\in K} \mathbb{P}(X^{\varepsilon,y} \in F) \le \max_{1\le i\le m} \limsup_{\varepsilon\to 0} \varepsilon \ln \sup_{y\in B_{\varepsilon_{x_i}}(x_i)} \mathbb{P}(X^{\varepsilon,y} \in F) \le -I_K^\delta.$$

It remains to let $\delta \to 0$. □

2.5. Diffusion exit from a domain

We now return to the problem sketched in the introduction. We shall present the treatment of the exit of a diffusion process from a domain in the simpler case of additive Gaussian noise via the theory by Freidlin and Wentzell. We roughly follow Dembo and Zeitouni [**25**]. In contrast to the treatment there, according to Section 2.4.2, we however allow local Lipschitz continuity instead of global Lipschitz continuity and boundedness of the vector field generating the underlying dynamical system.

2.5.1. Properties of the pseudopotential and statement of the main result.

Consider the SDE

$$X_t^{\varepsilon,x} = x + \int_0^t b(X_s^{\varepsilon,x})\, ds + \sqrt{\varepsilon} W_t, \quad x \in \mathbb{R}^d,\ t \ge 0. \tag{2.45}$$

Let $G \subset \mathbb{R}^d$ be a bounded domain. Suppose as in Section 2.4.2 that b is locally Lipschitz continuous and satisfies the following growth condition: there are constants η, $R_0 > 0$ such that

$$\langle x, b(x)\rangle < -\eta|x| \text{ for all } |x| \ge R_0. \tag{2.46}$$

Without loss of generality we may moreover suppose that $B_{R_0}(0)$ contains $\overline{G}$, that the distance from $\overline{G}$ to $B_{R_0}(0)^c$ is bigger than 1, and that the local Lipschitz constant as well as the uniform bound for b on the compact set $\overline{B_{R_0}(0)}$ is given by $B > 0$, i.e.

$$|b(x) - b(y)| \leq B|x - y|, \quad x, y \in \overline{B_{R_0}(0)}, \tag{2.47}$$

$$|b(x)| \leq B, \quad x \in \overline{B_{R_0}(0)}. \tag{2.48}$$

We view $X^{\varepsilon,x}$ as a perturbation of the ordinary differential equation

$$\dot{\phi}(t) = b(\phi(t)), \quad t \geq 0,$$

or the dynamical system associated with it. Suppose that the dynamical system has the following properties:

Assumption (A1)
The origin $0 \in G$ is the unique stable equilibrium point of $\dot{\phi} = b(\phi)$, and

$$\phi(0) \in G \text{ implies that } \phi(t) \in G \text{ for all } t > 0 \text{ and } \lim_{t\to\infty} \phi(t) = 0.$$

We are interested in the time it takes typically for the trajectories of $X^{\varepsilon,x}$ starting at $x \in G$ to leave the domain. More formally, we suppose that the filtration $(\mathcal{F}_t)_{t\geq 0}$ on our probability space $(\Omega, \mathcal{F}, \mathbb{P})$ satisfies the usual conditions, so that the objects of our main interest, namely the *exit times from* G

$$\tau^{\varepsilon,x} = \inf\{t > 0\colon X_t^{\varepsilon,x} \in \partial G\}, \ \varepsilon > 0, \ x \in \mathbb{R}^d,$$

are well defined stopping times. Chapter 2.4 provides explicit formulas for the good rate functions of LDP for the solution processes of (2.45) indexed by $[0, 1]$. For diffusions with time duration $t \geq 0$ we continue denoting by $\|\cdot\|_t$ the uniform norm on $C([0, t], \mathbb{R}^d)$ for any $t \geq 0$, and write $\|\cdot\|$ instead if this is unambiguous. The scaling property of Brownian motion allows a direct extension of the results of Chapter 2.4 to $[0, t]$ instead of $[0, 1]$. In this context the good rate functions are given by

$$I_{x,t}(f) = \begin{cases} \inf\limits_{\substack{g\in\mathcal{H}_1([0,t],\mathbb{R}^d):\\ f=x+\int_0^\cdot b(f(s))\,ds+g}} \dfrac{1}{2}\displaystyle\int_0^t |\dot{g}(s)|^2\,ds, & g \in \mathcal{H}_1([0, t], \mathbb{R}^d), \\ +\infty, & g \notin \mathcal{H}_1([0, t], \mathbb{R}^d). \end{cases}$$

We define the *cost function*

$$V(x, z, t) = \inf_{f\in C([0,1],\mathbb{R}^d)\colon f(t)=z} I_{x,t}(f), \quad x, z \in \mathbb{R}^d, \ t > 0.$$

The cost function quantifies the cost for forcing the system to z at time t when starting at x. We further define

$$V(x, z) = \inf_{t>0} V(x, z, t), \ x, z \in \mathbb{R}^d,$$

and call $V(0, z)$ *quasi-potential* of the system. The quasi-potential describes the minimal cost for the system to go to z when starting at the stable equilibrium. To derive the exit time law in terms of the quasi-potential, we shall need the following assumptions.

Assumption (A2)
If $\phi(0) \in \partial G$, then $\lim_{t\to\infty} \phi(t) = 0$.

Assumption (A3)

$$\overline{V} = \inf_{z\in\partial G} V(0, z) < \infty.$$

REMARK 2.39. The Assumption (A2) prevents the boundary ∂G to be *characteristic.*

We now prove an important *controllability* property of our diffusion. We shall see that within small neighborhoods of the boundary the smooth system can be controlled from any point x_0 into any other point y_0. Denote $d(x,F) = \inf\{|x - y|\colon y \in F\}$ for $x \in \mathbb{R}^d$, and a closed subset F in $\mathbb{R}^d$.

LEMMA 2.40. *1. For all $x, y \in \overline{B_{R_0}(0)}$*

$$V(x,y,|x-y|) \leq \frac{1}{2}(1+B)^2|x-y|.$$

2. For all $\rho > 0$, $x_0, y_0 \in \overline{B_{R_0}(0)}$, $z \in \partial G \cup \{0\}$ satisfying $|x_0 - z| + |z - y_0| \leq \rho$ there exists $u \in C([0,\rho],\mathbb{R}^d)$ such that $\|u\| \leq 1 + B$ and $\phi(\rho) = y_0$ for the solution ϕ of

$$\phi(t) = x_0 + \int_0^t b(\phi(s))\,ds + \int_0^t u(s)\,ds, \quad 0 \leq t \leq \rho.$$

PROOF. 1. Let $x, y \in \overline{B_{R_0}(0)}$ be given. Define

$$\phi(t) = x + t\frac{y-x}{|y-x|}, \quad 0 \leq t \leq |x-y|,$$

and

$$u(t) = \dot{\phi}(t) - b(\phi(t)), \quad 0 \leq t \leq |x-y|.$$

Then on $[0,|x-y|]$ we have $\dot{\phi} = b(\phi) + u$, and by assumptions on b we have

$$\|u\| \leq 1 + B.$$

Hence

$$V(x,y,|x-y|) \leq \frac{1}{2}\int_0^{|x-y|} |u(t)|^2\,dt \leq \frac{1}{2}(1+B)^2|x-y|.$$

2. Let $0 < \rho$, $x_0, y_0 \in \overline{B_{R_0}(0)}$, $z \in \partial G \cup \{0\}$, such that $|x_0 - z| + |y_0 - z| \leq \rho$. Define

$$\phi(t) = x_0 + t\frac{y_0 - x_0}{\rho}, \quad t \in [0,\rho],$$

and

$$u(t) = \dot{\phi}(t) - b(\phi(t)), \quad t \in [0,\rho].$$

Then for all $t \in [0,\rho]$ we have $|\phi(t) - z| \leq (1 - \frac{t}{\rho})|x_0 - z| + \frac{t}{\rho}|y_0 - z| \leq \rho$ and according to the calculation in part 1.

$$\|u\| \leq 1 + B.$$

□

From Lemma 2.40 we may deduce the following continuity property for the cost function.

LEMMA 2.41. *For any $\delta > 0$ there exists $\rho > 0$ such that*

(i) $$\sup_{|x|,|y| \leq \rho} \inf_{t \in [0,1]} V(x,y,t) < \delta;$$

(ii) $$\sup_{x,y\colon \inf_{z \in \partial G}(|x-z|+|y-z|) \leq \rho} \inf_{t \in [0,1]} V(x,y,t) < \delta.$$

PROOF. By the second part of Lemma 2.40 we have for two points x, y as specified in *(i)* or *(ii)* and $0 < \rho$

$$V(x, y, \rho) \leq \frac{1}{2}(1+B)^2 \rho.$$

Now the result follows. □

We state the main result of this Chapter.

THEOREM 2.42. *Assume that Assumptions **A1**–**A3** are satisfied. Then for any* $\delta > 0$, $x \in G$ *we have*

$$\lim_{\varepsilon \to 0} \mathbb{P}\Big(e^{\frac{\overline{V}-\delta}{\varepsilon}} < \tau^{\varepsilon,x} < e^{\frac{\overline{V}+\delta}{\varepsilon}}\Big) = 1.$$

Moreover, for all $x \in G$ *we have*

$$\lim_{\varepsilon \to 0} \varepsilon \ln \mathbb{E}\, \tau^{\varepsilon,x} = \overline{V}.$$

2.5.2. Proof of main result: the upper bound. In this part, we shall establish an upper bound for the exit rate of Theorem 2.42. This will be done in 3 steps. To begin, it will be necessary to give an exponential lower bound for the probability that $\tau^{\varepsilon,x}$ is small, uniformly over x in a small neighborhood of 0.

1. In the first step we prove the following lemma.

LEMMA 2.43. *Let* $\eta > 0$. *Then there exists* $\rho_0 > 0$ *such that for* $0 < \rho < \rho_0$ *there exists* $T_0 > 0$ *such that*

$$\liminf_{\varepsilon \to 0} \varepsilon \ln \inf_{|x| \leq \rho} \mathbb{P}(\tau^{\varepsilon,x} \leq T_0) > -(\overline{V} + \eta).$$

PROOF. Let $\delta = \frac{\eta}{6}$, and let ρ_0 be so small that the inequalities of Lemma 2.41 hold. Use Lemma 2.41*(i)* to choose for $x \in G$ such that $|x| \leq \rho$ a path $\psi^x \in C([0, t_x], \mathbb{R}^d)$, with $0 \leq t_x \leq 1$, satisfying $\psi^x(0) = x$, $\psi^x(t_x) = 0$, and

$$I_{x,t_x}(\psi^x) \leq \frac{\eta}{3}.$$

Use **A3** and Lemma 2.41*(ii)* to choose $z \in \mathbb{R}^d \backslash \overline{G}$, $T_1 > 0$ and $\phi \in C([0, T_1], \mathbb{R}^d)$ such that $\phi(0) = 0$, $\phi(T_1) = z$, and such that

$$I_{0,T_1}(\phi) \leq \overline{V} + \frac{\eta}{3}.$$

Let $\hat{\phi}$ be the solution of the differential equation $\dot{\hat{\phi}} = b(\hat{\phi})$ with $\hat{\phi}(0) = z$. Next, let $T_0 = T_1 + 1$, and set for $x \in G$ such that $|x| \leq \rho$

$$\phi_t^x = \begin{cases} \psi^x(t), & 0 \leq t \leq t_x, \\ \phi(t - t_x), & t_x \leq t \leq T_1 + t_x, \\ \hat{\phi}(t - T_1 - t_x), & T_1 + t_x \leq t \leq T_0. \end{cases}$$

Then we have

$$I_{x,T_0}(\phi) \leq I_{x,t_x}(\psi^x) + I_{0,T_1}(\phi) \leq \overline{V} + \frac{2\eta}{3}.$$

Now let $\Delta = d(z, \overline{G})$, and consider the open set

$$\Psi = \bigcup_{|x| \leq \rho} \Big\{\psi \in C([0, T_0], \mathbb{R}^d) \colon \|\psi - \phi^x\| < \frac{\Delta}{2}\Big\}.$$

Since ϕ^x visits z, by definition of Ψ every path $\psi \in \Psi$ leaves G before time T_0. Hence by virtue of Corollary 2.38 we have

$$\begin{aligned}\liminf_{\varepsilon\to 0} \varepsilon \ln \inf_{|x|\le\rho} \mathbb{P}(\tau^{\varepsilon,x} \le T_0) &\ge \liminf_{\varepsilon\to 0} \varepsilon \ln \inf_{|x|\le\rho} \mathbb{P}(X^{\varepsilon,x} \in \Psi) \\ &\ge -\sup_{|x|\le\rho} \inf_{\psi\in\Psi} I_{x,T_0}(\psi) \ge -\sup_{|x|\le\rho} I_{x,T_0}(\phi^x) \\ &\ge -\Big(\overline{V} + \frac{2\eta}{3}\Big) > -(\overline{V} + \eta).\end{aligned}$$

□

2. We next need to show that the probability that the diffusion stays inside G without hitting a small neighborhood of 0 is exponentially vanishing. For this purpose, for $\rho > 0$, $x \in \mathbb{R}^d$ such that $B_\rho(0) \subset G$ let

$$\sigma_\rho^x = \inf\Big\{t \ge 0 \colon |X_t^{\varepsilon,x}| \le \rho \text{ or } X_t^{\varepsilon,x} \in \partial G\Big\}.$$

LEMMA 2.44. *We have*

$$\lim_{t\to\infty} \limsup_{\varepsilon\to 0} \varepsilon \ln \sup_{x\in G} \mathbb{P}(\sigma_\rho^x > t) = -\infty.$$

PROOF. (i) For $t \ge 0$ consider the closed set in $C([0,t],\mathbb{R}^d)$

$$\Psi_t = \Big\{\phi \in C([0,t],\mathbb{R}^d) \colon \phi(s) \in \overline{G\backslash B_\rho(0)},\ s \in [0,t]\Big\}.$$

We show that

$$\limsup_{\varepsilon\to 0} \varepsilon \ln \sup_{x\in G} \mathbb{P}(\sigma_\rho^x > t) \le -\inf_{\psi\in\Psi_t} I_{\psi(0),t}(\psi).$$

Indeed, by definition and Corollary 2.38

$$\begin{aligned}\limsup_{\varepsilon\to 0} \varepsilon \ln \sup_{x\in G} \mathbb{P}(\sigma_\rho^x > t) &\le \limsup_{\varepsilon\to 0} \varepsilon \ln \sup_{x\in\overline{G\backslash B_\rho(0)}} \mathbb{P}(\sigma_\rho^x > t) \\ &\le -\inf_{x\in\overline{G\backslash B_\rho(0)}} \inf_{\psi\in\Psi_t} I_{x,t}(\psi) \\ &= -\inf_{\psi\in\Psi_t} I_{\psi(0),t}(\psi).\end{aligned}$$

(ii) It remains to prove that

$$\lim_{t\to\infty} \inf_{\psi\in\Psi_t} I_{\psi(0),t}(\psi) = \infty. \tag{2.49}$$

(ii.1) Let $(\Phi_t)_{t\ge 0}$ be the flow of the differential equation $\dot\phi = b(\phi)$. According to **A1** and **A2** for any $x \in \overline{G\backslash B_\rho(0)}$ there is $t_x \ge 0$ such that $\Phi_{t_x}(x) \in B_{\rho/2}(0)$. For $x \in \overline{G\backslash B_\rho(0)}$ define now

$$W_x = \Phi_{t_x}^{-1}\Big(B_{\rho/2}(0)\Big).$$

By definition, W_x is an open neighborhood of x. Now choose $x_1, \dots, x_n \in \overline{G\backslash B_\rho(0)}$ such that $\overline{G\backslash B_\rho(0)} \supset \bigcup_{1\le i\le n} W_{x_i}$ and set $T = \max_{1\le i\le n} t_{x_i}$. Then any solution trajectory of $\dot\phi = b(\phi)$ starting in $\overline{G\backslash B_\rho(0)}$ hits $B_{\rho/2}(0)$ before time T.
(ii.2) Assume that (2.49) does not hold. Then there exists $M > 0$ such that for any $n \in \mathbb{N}$ there exists $\psi^n \in \Psi_{nT}$ such that $I_{\psi^n(0),nT}(\psi^n) \le M$. Now for $0 \le k \le n-1$ let

$$\psi^{n,k}(t) = \psi^n(t - kT), \quad 0 \le t \le T.$$

Then $\psi^{n,k} \in \Psi_T$ and

$$M \geq I_{\psi^n(0),nT}(\psi^n) = \sum_{k=0}^{n-1} I_{\psi^n(kT),T}(\psi^{n,k}) \geq n \min_{0\leq k\leq n-1} I_{\psi^{n,k}(0),T}(\psi^{n,k}).$$

Hence there exists a sequence $(\phi^n)_{n\in\mathbb{N}}$ in Ψ_T such that

$$\lim_{n\to\infty} I_{\phi^n(0),T}(\phi^n) = 0.$$

By compactness of the set

$$\left\{\phi \in C([0,T],\mathbb{R}^d)\colon \phi(0) \in \overline{G\backslash B_\rho(0)},\ I_{\phi(0),T}(\phi) \leq 1\right\},$$

which follows directly from the goodness of the rate function, the sequence possesses a condensation point $\phi^* \in \Psi_T$. By lower semicontinuity of $\psi \mapsto I_{\psi(0),T}$ we obtain $I_{\phi^*(0),T}(\phi^*) = 0$. But this means that ϕ^* is a solution of the differential equation $\dot{\phi} = b(\phi)$, with $\phi^*(0) \in \overline{G\backslash B_\rho(0)}$. Hence by what has been proved in (ii.1), ϕ^* reaches $B_{\rho/2}(0)$ before time T. This contradicts $\phi^* \in \Psi_T$. □

3. We are now in a position to establish the upper bound. We show: for $x \in G$, $\delta > 0$ we have

$$\liminf_{\varepsilon\to 0} \varepsilon \ln \mathbb{P}\left(\tau^{\varepsilon,x} < e^{\frac{\overline{V}+\delta}{\varepsilon}}\right) = 1.$$

(i) To do this, we first show that for any $\delta > 0$ there exists $T > 0$ and $\varepsilon_0 > 0$ such that

$$\inf_{x\in G} \mathbb{P}(\tau^{\varepsilon,x} \leq T) \geq e^{-\frac{\overline{V}+\frac{\delta}{2}}{\varepsilon}}.$$

Use Lemma 2.43 to choose $T_0 > 0$ and $\rho > 0$ such that

$$\liminf_{\varepsilon\to 0} \varepsilon \ln \inf_{|x|\leq\rho} \mathbb{P}(\tau^{\varepsilon,x} \leq T_0) > -\left(\overline{V} + \frac{\delta}{4}\right),$$

and hence $\varepsilon_0 > 0$ such that for $\varepsilon < \varepsilon_0$

$$\inf_{|x|\leq\rho} \mathbb{P}(\tau^{\varepsilon,x} \leq T_0) \geq e^{-\frac{\overline{V}+\frac{\delta}{4}}{\varepsilon}}.$$

Next, use Lemma 2.44 to choose, for $\rho > 0$ given, $T_1 > 0$ such that

$$\limsup_{\varepsilon\to 0} \varepsilon \ln \sup_{x\in G} \mathbb{P}(\sigma_\rho^x > T_1) < 0,$$

and hence $r > 0$ and $\varepsilon_0 > 0$ (if necessary smaller than the one above) such that for $\varepsilon < \varepsilon_0$

$$\varepsilon \ln \sup_{x\in G} \mathbb{P}(\sigma_\rho^x > T_1) < -r,$$

and such that $1 - e^{-\frac{r}{\varepsilon}} > e^{-\frac{\delta}{4\varepsilon}}$. Now note that for $x \in G$ on the set $\{\sigma_\rho^x < \tau^{\varepsilon,x}\}$ we have

$$\tau^{\varepsilon,x} = \sigma_\rho^x + \tau^{\varepsilon, X^{\varepsilon,x}_{\sigma_\rho^x}} \circ \theta_{\sigma_\rho^x},$$

where θ_s denotes the usual shift on path space by time s. Hence by the strong Markov property for $\varepsilon < \varepsilon_0$

$$\begin{aligned}
\mathbb{P}(\tau^{\varepsilon,x} \leq T_0 + T_1) &\geq \mathbb{P}(\sigma_\rho^x \leq T_1) \cdot \mathbb{P}\left(\tau^{\varepsilon, X^{\varepsilon,x}_{\sigma_\rho^x}} \leq T_0\right) \\
&\geq \inf_{x\in G} \mathbb{P}(\sigma_\rho^x \leq T_1) \cdot \inf_{|x|\leq\rho} \mathbb{P}(\tau^{\varepsilon,x} \leq T_0) \\
&\geq e^{-\frac{\overline{V}+\frac{\delta}{4}}{\varepsilon}} (1 - e^{-\frac{r}{\varepsilon}}) \geq e^{-\frac{\overline{V}+\frac{\delta}{4}}{\varepsilon}} e^{-\frac{\delta}{4\varepsilon}} = e^{-\frac{\overline{V}+\frac{\delta}{2}}{\varepsilon}}.
\end{aligned}$$

It remains to set $T = T_0 + T_1$.

(ii) Abbreviate $q = \inf_{x\in G} \mathbb{P}(\tau^{\varepsilon,x} \leq T)$. For $k \in \mathbb{N}$, $x \in G$, $\varepsilon > 0$ consider the events $\{\tau^{\varepsilon,x} > kT\}$. Then, since $\tau^{\varepsilon,x} = \tau^{\varepsilon,X^{\varepsilon,x}_{kT}} \circ \theta_{kT} + kT$ on the event $\{\tau^{\varepsilon,x} > kT\}$, by conditioning on $\mathcal{F}_{kT}$ and the strong Markov property

$$\mathbb{P}(\tau^{\varepsilon,x} > (k+1)T) = \Big(1 - \mathbb{P}(\tau^{\varepsilon,x} \leq (k+1)T | \tau^{\varepsilon,x} > kT)\Big) \cdot \mathbb{P}(\tau^{\varepsilon,x} > kT)$$
$$\leq (1-q)\mathbb{P}(\tau^{\varepsilon,x} > kT).$$

Hence by recursion for $k \in \mathbb{N}$, $\varepsilon < \varepsilon_0$

$$\sup_{x\in G} \mathbb{P}(\tau^{\varepsilon,x} > kT) \leq (1-q)^k.$$

Therefore we obtain the following bound for the expected exit time

$$\sup_{x\in G} \mathbb{E}\,\tau^{\varepsilon,x} \leq T \sum_{k=0}^{\infty} \sup_{x\in G} \mathbb{P}(\tau^{\varepsilon,x} > kT) \leq T \sum_{k=0}^{\infty} (1-q)^k = \frac{T}{q}.$$

In particular, since $q \geq e^{-\frac{\overline{V}+\frac{\delta}{2}}{\varepsilon}}$, we have

$$\sup_{x\in G} \mathbb{E}\,\tau^{\varepsilon,x} \leq T e^{\frac{\overline{V}+\frac{\delta}{2}}{\varepsilon}}.$$

Finally, using Chebyshev's inequality, we arrive at

$$\mathbb{P}\Big(\tau^{\varepsilon,x} \geq e^{\frac{\overline{V}+\delta}{\varepsilon}}\Big) \leq e^{-\frac{\overline{V}+\delta}{\varepsilon}} \cdot \mathbb{E}\,\tau^{\varepsilon,x} \leq T e^{-\frac{\delta}{2\varepsilon}},$$

valid for all $x \in G$ and $\varepsilon < \varepsilon_0$.

2.5.3. Proof of main result: the lower bound. We now establish the lower bound for the exit rate of Theorem 2.42. Again we proceed in three main steps. We first need an estimate which shows that starting in G, the diffusion has high probability of being attracted to a small neighborhood of 0 before getting to ∂G.

1. Recall for $x \in G$, $\rho > 0$ such that $\overline{B_\rho(0)} \subset G$

$$\sigma^x_\rho = \inf\{t \colon X^{\varepsilon,x}_t \in \partial G \cup \overline{B_\rho(0)}\}.$$

LEMMA 2.45. *For any $x \in G$, $\rho > 0$ such that $\overline{B_\rho(0)} \subset G$ we have*

$$\lim_{\varepsilon\to 0} \mathbb{P}\Big(X^{\varepsilon,x}_{\sigma^x_\rho} \in \overline{B_\rho(0)}\Big) = 1.$$

PROOF. Given $\rho > 0$, we may and do consider $x \in G \backslash \overline{B_\rho(0)}$, for otherwise the claim is trivial. Let ϕ be the solution of the differential equation $\dot{\phi} = b(\phi)$, $\phi(0) = x$, and

$$T = \inf\{t \geq 0 \colon \phi(t) \in \overline{B_{\frac{\rho}{2}}(0)}\}.$$

According to **A2**, $T < \infty$. Further, according to **A1**, we have

$$\Delta = \rho \wedge d(\phi([0,T]), \partial G) > 0,$$

where $d(K, L)$ denotes the well defined distance between two compact sets $K, L \subset \mathbb{R}^d$. Then for $0 \leq t \leq T \wedge \sigma^x_\rho$ by the local Lipschitz continuity of b

$$|X^{\varepsilon,x}_{t\wedge\sigma^x_\rho} - \phi(t \wedge \sigma^x_\rho)| \leq B \int_0^{t\wedge\sigma^x_\rho} |X^{\varepsilon,x}_s - \phi(s)|\, ds + \sqrt{\varepsilon}|W_t|.$$

By Gronwall's lemma

$$\sup_{0\le t\le T\wedge\sigma_\rho^x} |X_t^{\varepsilon,x} - \phi(t)| \le e^{BT}\sqrt{\varepsilon}\sup_{0\le t\le T}|W_t|.$$

Therefore

$$\begin{aligned}\mathbb{P}\Big(X_{\sigma_\rho^x}^{\varepsilon,x} \notin \overline{K}_\rho(0)\Big) &\le \mathbb{P}\Big(\sup_{0\le t\le T\wedge\sigma_\rho^x} |X_t^{\varepsilon,x} - \phi(t)| > \frac{\Delta}{2}\Big)\\ &\le \mathbb{P}\Big(\sup_{0\le t\le T}|W_t| > \frac{\Delta}{2\sqrt{\varepsilon}}e^{-BT}\Big) \to 0\end{aligned}$$

as $\varepsilon \to 0$, due to Chebyshev's and Doob's inequalities. This implies the desired result. □

2. We next have to establish an auxiliary result which says that during bounded time intervals, the diffusion cannot get away too far from its starting point.

LEMMA 2.46. *Let $\rho > 0$, $c > 0$. Then there exists a constant $T(c,\rho)$ such that*

$$\limsup_{\varepsilon\to 0}\varepsilon\ln\sup_{x\in G}\mathbb{P}\Big(\sup_{0\le t\le T(c,\rho)}|X_t^{\varepsilon,x} - x| \ge \rho\Big) \le -c.$$

PROOF. Use Corollary 2.35 to choose $R > 0$ such $\overline{G} \subset B_{\frac{R}{2}}(0)$ and such that for the exit time $\sigma^{\varepsilon,x}$ of the diffusion starting at x, for all $|x| \le \frac{R}{2}$ and $T > 0$ we have

$$\limsup_{\varepsilon\to 0}\varepsilon\ln\mathbb{P}(\sigma_R^{\varepsilon,x} \le T) \le -\eta R \le -c.$$

Denote the Lipschitz constant of b on $B_R(0)$ as well as its global bound there again by B. Fix $\rho > 0$ and $\varepsilon > 0$. Then for $0 \le t \le T \le \frac{\rho}{2B}$, $x \in G$, on the set $\{\sigma^{\varepsilon,x} > T\}$

$$|X_t^{\varepsilon,x} - x| \le \int_0^t |b(X_s^{\varepsilon,x})|\,ds + \sqrt{\varepsilon}|W_t| \le \frac{\rho}{2} + \sqrt{\varepsilon}|W_t|.$$

Hence for $x \in G$, by the reflection principle for the Brownian motion and Theorem 2.22

$$\begin{aligned}\mathbb{P}\Big(\sup_{0\le t\le T}|X_t^{\varepsilon,x} - x| \ge \rho,\ \sigma^{\varepsilon,x} > T\Big) &\le \mathbb{P}\Big(\sup_{0\le t\le T}|W_t| \ge \frac{\rho}{2\sqrt{\varepsilon}}\Big)\\ = \mathbb{P}\Big(\sup_{0\le s\le 1}|W_s| \ge \frac{\rho}{2\sqrt{\varepsilon T}}\Big) &\le d\cdot\mathbb{P}\Big(\sup_{0\le s\le 1}|W_s^1| \ge \frac{\rho}{2d\sqrt{\varepsilon T}}\Big)\\ \le 2d\cdot\mathbb{P}\Big(\sup_{0\le s\le 1}W_s^1 \ge \frac{\rho}{2d\sqrt{\varepsilon T}}\Big) &= 4d\cdot\mathbb{P}\Big(W_1^1 \ge \frac{\rho}{2d\sqrt{\varepsilon T}}\Big)\\ &\le 4d\cdot e^{-\frac{\rho^2}{8d^2\varepsilon T}}.\end{aligned}$$

This finally implies for $0 \le T \le \frac{\rho}{2B}$

$$\begin{aligned}\limsup_{\varepsilon\to 0}\varepsilon\ln\sup_{x\in G}\mathbb{P}\Big(\sup_{0\le t\le T}|X^{\varepsilon,x} - x| \ge \rho\Big) &\le \limsup_{\varepsilon\to 0}\varepsilon\ln\Big(4de^{-\frac{\rho^2}{8d^2T\varepsilon}} + \mathbb{P}(\sigma^{\varepsilon,x} \le T)\Big)\\ &= -\frac{\rho^2}{8d^2T}\wedge c.\end{aligned}$$

For $T = T(c,\rho)$ small enough, the last line of the preceding inequality is bounded above by $-c$. □

3. We need a final auxiliary result relating the quasi-potential with the probability that, starting from the boundary of a small sphere centered at 0, to hit the boundary of G before hitting an even smaller sphere centered at 0.

LEMMA 2.47. *Let $F \subset \partial G$ be closed. Then*

$$\lim_{\rho\to 0}\limsup_{\varepsilon\to 0}\varepsilon \ln \sup_{|x|=2\rho} \mathbb{P}\Big(X^{\varepsilon,x}_{\sigma^x_\rho} \in F\Big) \le -\inf_{z\in F} V(0,z).$$

PROOF. For $\delta > 0$ fixed, let

$$V_F = \frac{1}{\delta} \wedge \Big(\inf_{z\in F} V(0,z) - \delta\Big).$$

Note first that by definition for $x, y, z \in \mathbb{R}^d$

$$V(x,z) \le V(x,y) + V(y,z).$$

Therefore, using Lemma 2.41*(i)* to estimate the last term in the second expression for ρ_0 small enough and $0 < \rho < \rho_0$

$$\inf_{z\in F,|y|=2\rho} V(y,z) \ge \inf_{z\in F} V(0,z) - \sup_{|y|=2\rho} V(0,y) \ge V_F.$$

Use Lemma 2.44 to choose $T > 0$ such that for any $0 < \rho < \rho_0$

$$\limsup_{\varepsilon\to 0}\varepsilon \ln \sup_{|y|=2\rho} \mathbb{P}(\sigma^y_\rho > T) < -V_F.$$

Consider the closed set

$$\Psi = \{\phi \in C([0,T],\mathbb{R}^d) \colon \phi(t) \in F \text{ for some } t \in [0,T]\}.$$

Then by Corollary 2.38 for $0 < \rho < \rho_0$

$$\begin{aligned}\limsup_{\varepsilon\to 0}\varepsilon \ln \sup_{|y|=2\rho} \mathbb{P}(X^{\varepsilon,y} \in \Psi) &\le -\inf_{|y|=2\rho}\inf_{\phi\in\Psi} I_{y,T}(\phi)\\ &\le -\inf_{|y|=2\rho, z\in F} V(y,z)) \le -V_F.\end{aligned}$$

We may summarize by stating that for $0 < \rho < \rho_0$

$$\begin{aligned}&\limsup_{\varepsilon\to 0}\varepsilon \ln \sup_{|x|=2\rho} \mathbb{P}(X^{\varepsilon,x}_{\sigma^x_\rho} \in F)\\ &\qquad \le \limsup_{\varepsilon\to 0}\varepsilon \ln\Big(\sup_{|y|=2\rho} \mathbb{P}(\sigma^y_\rho > T) + \sup_{|y|=2\rho} \mathbb{P}(X^{\varepsilon,y} \in \Psi)\Big) \le -V_F.\end{aligned}$$

Hence the claimed inequality follows. □

4. Let now $\overline{V} > 0$, $\delta > 0$, $x \in G$. We show that

$$\lim_{\varepsilon\to 0} \mathbb{P}\Big(\tau^{\varepsilon,x} \le e^{\frac{\overline{V}-\delta}{\varepsilon}}\Big) = 0.$$

To do this, choose $\rho > 0$ such that $\overline{B_{2\rho}(0)} \subset G$. Define recursively for $x \in G$ and $m \ge 0$

$$\begin{aligned}\theta^x_0 &= 0,\\ \tau^x_m &= \inf\{t \ge \theta^x_m : X^{\varepsilon,x}_t \in \overline{B_\rho(0)} \cup \partial G\},\\ \theta^x_{m+1} &= \begin{cases} +\infty, & \text{if } X^{\varepsilon,x}_{\tau^x_m} \in \partial G,\\ \inf\{t \ge \tau^x_m \colon |X^{\varepsilon,x}_t| = 2\rho\}, & \text{if } |X^{\varepsilon,x}_{\tau^x_m}| = \rho.\end{cases}\end{aligned}$$

The sequence $(X^{\varepsilon,x}_{\tau^x_m})_{m\geq 0}$ is a Markov chain, and we use the convention $X^{\varepsilon,x}_{\tau^x_m} = X^{\varepsilon,x}_{\tau^{\varepsilon,x}}$ if $\tau^x_m = \infty$. Fix $\delta > 0$. Using Lemma 2.47 applied to $F = \partial G$, choose $\rho_0 > 0$ such that for $0 < \rho < \rho_0$

$$\limsup_{\varepsilon\to 0} \varepsilon \ln \sup_{|x|=2\rho} \mathbb{P}\Big(X^{\varepsilon,x}_{\sigma^x_\rho} \in \partial G\Big) \leq -\overline{V} + \frac{\delta}{2}.$$

Now also fix $\rho < \rho_0$. Moreover, choose $T_0 = T(\overline{V}, \rho)$ according to Lemma 2.46. Then there exists $\varepsilon_0 > 0$ such that for any $0 \leq \varepsilon < \varepsilon_0$ and $m \geq 1$ by the strong Markov property

$$\sup_{x\in G} \mathbb{P}(\tau^{\varepsilon,x} = \tau^x_m) \leq \sup_{|y|=2\rho} \mathbb{P}\Big(X^{\varepsilon,y}_{\sigma^y_\rho} \in \partial G\Big) \leq e^{-\frac{\overline{V}-\frac{\delta}{2}}{\varepsilon}},$$

and also

$$\sup_{x\in G} \mathbb{P}(\theta^x_m - \tau^x_{m-1} \leq T_0) \leq \sup_{x\in G} \mathbb{P}\Big(\sup_{0\leq t\leq T_0} |X^{\varepsilon,x} - x| \geq \rho\Big) \leq e^{-\frac{\overline{V}-\frac{\delta}{2}}{\varepsilon}}.$$

Now let $k \in \mathbb{N}$. Then for $x \in G$

$$\{\tau^{\varepsilon,x} \leq kT_0\} \subseteq \{\tau^{\varepsilon,x} = \tau^x_0\} \cup \bigcup_{m=1}^{k} \Big(\{\tau^{\varepsilon,x} = \tau^x_m\} \cup \{\theta^x_m - \tau^x_{m-1} \leq T_0\}\Big).$$

Hence for $k \in \mathbb{N}$ and $x \in G$

$$\begin{aligned}\mathbb{P}(\tau^{\varepsilon,x} \leq kT_0) &\leq \mathbb{P}(\tau^{\varepsilon,x} = \tau^x_0) + \sum_{m=1}^{k} \Big(\mathbb{P}(\tau^{\varepsilon,x} = \tau^x_m) + \mathbb{P}(\theta^x_m - \tau^x_{m-1} \leq T_0)\Big) \\ &\leq \mathbb{P}(\tau^{\varepsilon,x} = \tau^x_0) + 2ke^{-\frac{\overline{V}-\frac{\delta}{2}}{\varepsilon}}.\end{aligned}$$

Now take

$$k = \left[\frac{e^{\frac{\overline{V}-\delta}{\varepsilon}}}{T_0}\right] + 1.$$

Then our estimate further yields for $x \in G$ and with the help of Lemma 2.45

$$\mathbb{P}\Big(\tau^{\varepsilon,x} \leq e^{\frac{\overline{V}-\delta}{\varepsilon}}\Big) \leq \mathbb{P}(\tau^{\varepsilon,x} \leq kT_0) \leq \mathbb{P}\Big(X^{\varepsilon,x}_{\sigma^x_\rho} \notin \overline{K}_\rho(0)\Big) + \frac{4}{T_0}e^{-\frac{\delta}{2\varepsilon}} \to 0$$

as $\varepsilon \to 0$. The lower bound for $\mathbb{E}\,\tau^{\varepsilon,x}$ now follows from Chebyshev's inequality.

5. We finally have to treat the case $\overline{V} = 0$. Let $\delta > 0$, $x \in G$. Choose $\rho > 0$ such that $\overline{K}_{2\rho}(0) \subset G$ and let $c > 0$. Then Lemma 2.45 and Lemma 2.46 combined with the Markov property allow us to choose $\varepsilon_0 > 0$ such that for $0 < \varepsilon < \varepsilon_0$

$$\mathbb{P}\Big(\tau^{\varepsilon,x} > e^{-\frac{\delta}{\varepsilon}}\Big) \geq \mathbb{P}\Big(X^{\varepsilon,x}_{\sigma^x_\rho} \in \overline{K}_\rho(0)\Big) \cdot \inf_{|y|\leq\rho} \mathbb{P}\Big(\sup_{0\leq t\leq T(c,\rho)} |X^{\varepsilon,y}_t - y| \leq \rho\Big) \to 1$$

as $\varepsilon \to 0$. This completes the proof of our main result.

CHAPTER 3

Semiclassical theory of stochastic resonance in dimension 1

In this Chapter, we study the stochastic resonance problem in the simplest possible setting, and mostly by analytical means. Indeed, we will consider the motion of an overdamped particle in a time-periodic potential landscape in dimension 1. Time periodicity will also be given in the simplest possible way: the potential function will switch between two time independent states every half-period. To fix ideas, let $U\colon \mathbb{R} \to \mathbb{R}$ be an at least continuously differentiable potential function with three critical points, two local minima at ± 1 and a saddle point at 0, assume $\lim_{x\to\pm\infty} U(x) = \infty$. Further assume that $U(-1) = -\frac{V}{2}$, $U(1) = -\frac{v}{2}$, $U(0) = 0$, where $0 < v < V$. To define the time periodic potential function, set

$$(3.1)\qquad U(t,x) = \sum_{k=0}^{\infty} \Big(U(x)\mathbb{I}_{[k,k+\frac{1}{2})}(t) + U(-x)\mathbb{I}_{[k+\frac{1}{2},k+1)}(t) \Big), \quad x \in \mathbb{R},\ t \geq 0.$$

This means that every half period the potential switches between the states $U(\cdot)$ and $U(-\cdot)$, and that the period length is 1.

The problem of stochastic resonance for diffusions with inhomogeneous time periodic drift will be solved by describing an optimal tuning rate between the period T of the time periodic drift and the noise intensity ε. In this Chapter we therefore consider diffusion processes which solve the SDE

$$(3.2)\qquad X_t^{\varepsilon,x} = x + \int_0^t b\Big(\frac{t}{T}, X_s^{\varepsilon,x}\Big)\, ds + \sqrt{\varepsilon} W_t, \quad t \geq 0,$$

where $b(t,x) = -\frac{\partial}{\partial x} U(t,x)$, $t \geq 0$, $x \in \mathbb{R}$, and where W is a one-dimensional Brownian motion. In fact, the treatment of the exit problem in Chapter 2 reveals that time scales in which exits and thus transitions between different wells of U, i.e. different domains of attraction of the underlying dynamical system, happen, are given by $T = T_\varepsilon = e^{\frac{\mu}{\varepsilon}}$, with a parameter $\mu \geq 0$ describing the energy needed to exit or transit at this scale. Freidlin [**39**] first realized the importance of these scales for stochastic resonance. In the following Section we shall present his results on asymptotic *quasideterministic motion* of a diffusion with time periodic potential in our simple case.

3.1. Freidlin's quasi-deterministic motion

Random trajectories of diffusion processes satisfying (3.2) will not be periodic. Still, the periodic changing of the position of the energetically most favorable point in the potential landscape will give it some periodic features. This will be related to the possibility of leaving or not leaving the well in which the diffusing particle is trapped at the chosen time scale T, in other words to the exit behavior of the diffusion from this well. Let us therefore first consider the diffusion in the time invariant potential landscape given by U. The corresponding SDE is described by

$$X_t^{\varepsilon,x} = x + \int_0^t b(X_s^{\varepsilon,x})\, ds + \sqrt{\varepsilon} W_t, \quad t \geq 0, \tag{3.3}$$

where $b = -U'$. According to Chapter 2 the exit behavior is essentially linked with the properties of the pseudo-potential, let us first recall the calculation of this function which in the underlying case is reflected in the classical Euler–Lagrange calculus of variations. Let us assume, as in Section 2.4.2 that $-U'$ is dissipative in the sense that there are constants $\eta > 0$ and $R_0 > 0$ such that

$$\langle x, b(x)\rangle < -\eta |x| \text{ for all } |x| \geq R_0. \tag{3.4}$$

Under these assumptions, Theorem 2.36 yields the following good rate function

$$I_x(f) = \begin{cases} \frac{1}{2}\int_0^1 |\dot{f}(s) + U'(f(s))|^2\, ds, & f - x \in \mathcal{H}_1([0,1],\mathbb{R}),\\ +\infty, & f - x \notin \mathcal{H}_1([0,1],\mathbb{R}),\end{cases}$$

for the time interval $[0,1]$, and analogously, if $t \geq 0$, for the time interval $[0,t]$

$$I_{x,t}(f) = \begin{cases} \frac{1}{2}\int_0^t |\dot{f}(s) + U'(f(s))|^2 ds, & f - x \in \mathcal{H}_1([0,t],\mathbb{R}),\\ +\infty, & f - x \notin \mathcal{H}_1([0,t],\mathbb{R}),\end{cases}$$

In these terms, we can evidently write

$$V(x,z,t) = \inf_{f \in \mathcal{H}_1([0,t],\mathbb{R}):\ f(t)=z} I_{x,t}(f), \quad x, z \in \mathbb{R},\ t > 0,$$

and

$$V(x,z) = \inf_{t>0} V(x,z,t), \quad x, z \in \mathbb{R}.$$

Let us consider our functional minimization problem in slightly more general terms. Let $F\colon \mathbb{R}\times\mathbb{R} \to \mathbb{R}$ be a continuous function, and

$$J_t(f) = \begin{cases} \int_0^t F(f(s), \dot{f}(s))\, ds, & f \in \mathcal{H}_1([0,t],\mathbb{R}),\\ +\infty, & f \notin \mathcal{H}_1([0,t],\mathbb{R}),\end{cases}$$

We call $f \in C_0([0,t],\mathbb{R})$ an *extremal* of $J_t(\cdot)$ if

$$\delta J_t(f) := \lim_{h\to 0} \frac{J_t(f + h\varphi) - J_t(f)}{h} = 0, \quad \varphi \in C_c^\infty((0,t)). \tag{3.5}$$

Here for an open set $G \subset \mathbb{R}$, we denote by $C_c^\infty(G,\mathbb{R})$ the set of all real valued infinitely smooth functions of compact support contained in G. The following Theorem states the classical Euler–Lagrange equations for the Lagrange functional J_t.

THEOREM 3.1. *Let $t \geq 0$. Assume that $F = F(x,y)$ is continuously differentiable, and for a continuously differentiable $f\colon [0,t] \to \mathbb{R}$, $f(0) = 0$, assume that $\delta J_t(f)$ is well-defined. Then the following are equivalent:*

(1) *f is an extremal of J_t;*

(2) $s \mapsto \frac{\partial}{\partial y} F(f(s), \dot{f}(s))$ *is continuously differentiable on* $(0, t)$ *and the* Euler–Lagrange equation *holds:*

$$\frac{d}{ds} \frac{\partial}{\partial y} F(f(s), \dot{f}(s)) = \frac{\partial}{\partial x} F(f(s), \dot{f}(s))$$

for almost all $s \in [0, t]$.

If in addition $F = F(x, y)$ *is twice continuously differentiable in* y *and for all* $s \in [0, t]$ *we have*

$$\frac{\partial^2}{\partial y^2} F(f(s), \dot{f}(s)) \neq 0,$$

then (2) *implies that* f *is twice continuously differentiable on* $(0, t)$.

PROOF. Let us first prove that (1) $\Rightarrow$ (2). By our regularity assumptions on F and f we may differentiate under the integral sign to obtain for $\varphi \in C_c^\infty((0, t), \mathbb{R})$

$$0 = \int_0^t \Big[\frac{\partial}{\partial x} F(f(s), \dot{f}(s)) \varphi(s) + \frac{\partial}{\partial y} F(f(s), \dot{f}(s)) \dot{\varphi}(s) \Big] \, ds. \tag{3.6}$$

Integration by parts yields

$$0 = \int_0^t \Big[- \int_0^s \frac{\partial}{\partial x} F(f(u), \dot{f}(u)) \, du + \frac{\partial}{\partial y} F(f(s), \dot{f}(s)) \Big] \dot{\varphi}(s) \, ds.$$

Abbreviate

$$g(s) := - \int_0^s \frac{\partial}{\partial x} F(f(u) \dot{f}(u)) \, du + \frac{\partial}{\partial y} F(f(s), \dot{f}(s)), \quad s \in [0, t].$$

Choose a nonnegative smooth function $\hat{\varphi}$ with compact support on $(0, t)$ such that $\int_0^t \hat{\varphi}(u) \, du = 1$, and define

$$\psi(s) = \int_0^s \Big[\varphi(u) - \hat{\varphi}(u) \int_0^t \varphi(v) \, dv \Big] \, du, \quad s \in [0, t].$$

Then by definition $\psi \in C_c^\infty((0, t), \mathbb{R})$ and with $c := \int_0^t g(s) \hat{\varphi}(s) \, ds$ we have

$$\begin{aligned} 0 = \int_0^t g(s) \dot{\psi}(s) \, ds &= \int_0^t g(s) \Big[\varphi(s) - \hat{\varphi}(s) \int_0^t \varphi(u) du \Big] \, ds \\ &= \int_0^t g(s) \varphi(s) \, ds - c \int_0^t \varphi(u) \, du \\ &= \int_0^t \Big[g(s) - c \Big] \varphi(s) \, ds. \end{aligned}$$

This easily implies that $g = c$ a.e., and by definition the Euler–Lagrange equation follows.

To show that (2) $\Rightarrow$ (1), we just have to note that the Euler–Lagrange equation and integration by parts yields (3.6). This means that the Fréchet derivative of J_t at f vanishes, and f is an extremal.

Assume finally that for $f \colon [0, t] \to \mathbb{R}$ continuously differentiable, and all $s \in (0, t)$ we have

$$\frac{\partial^2}{\partial y^2} F(f(s), \dot{f}(s)) \neq 0.$$

Then (2) implies that the function

$$s \mapsto \frac{\partial}{\partial y} F(f(s), \dot{f}(s))$$

is continuously differentiable. Hence by hypothesis, the function is strictly monotone, and therefore invertible, with a continuously differentiable inverse. This implies that $\dot{f}$ is continuously differentiable. □

In our situation, the Euler–Lagrange equation leads to a simple differential equation which allows to identify the quasi-potential.

THEOREM 3.2. *For $x, z \in \mathbb{R}$ we have*

$$V(x,z) = \begin{cases} 2(U(z) - U(x)) \vee 0, \\ \quad \textit{if there is no critical point between } x \textit{ and } z; \\ \sum_{i=1}^{k} V(x_{i-1}, x_i), \\ \quad \textit{if } x = x_0 < x_1 < \cdots < x_k = z \textit{ or } x = x_0 > x_1 > \cdots > x_k = z \\ \quad \textit{with } x_1, \dots, x_{k-1} \textit{ critical points of } U. \end{cases} \tag{3.7}$$

If U is twice continuously differentiable, $x, z \in \mathbb{R}$, and there is no critical point between x and z, there exists $t \geq 0$ such that the solution of the differential equation

$$\begin{aligned} \dot{f}(s) &= (-1)^{-\operatorname{sign}(U(z)-U(x))} U'(f(s)), \quad s \in (0,t), \\ f(0) &= x, \\ f(t) &= z, \end{aligned} \tag{3.8}$$

satisfies

$$V(x,z) = \inf_{f - x \in \mathcal{H}_1([0,t],\mathbb{R}):\ f(t)=z} I_{x,t}(f).$$

PROOF. By a general smoothing argument, we may assume that U is twice continuously differentiable. Let $x \in \mathbb{R}$, and

$$F(x_1, x_2) = \frac{1}{2}\Big|x_2 + U'(x + x_1)\Big|^2, \quad x_1, x_2 \in \mathbb{R}.$$

Then F satisfies the conditions of Theorem 3.1. To derive an ODE for a function f to minimize $I_{x,t}$ for an appropriate $t \geq 0$, assume that f is continuously differentiable, and t appropriate. Then $\dot{f}$ is continuously differentiable with derivative $\ddot{f}$, if the Euler–Lagrange equation is fulfilled, and it reads

$$\begin{aligned} \Big[\dot{f}(s) + U'(x + f(s))\Big] U''(x + f(s)) &= \frac{d}{ds}\Big[\dot{f}(s) + U'(x + f(s))\Big] \\ &= \ddot{f}(s) + U''(x + f(s))\dot{f}(s), \quad s \in (0,t). \end{aligned}$$

We have to consider two cases. All the other cases follow by concatenating adjacent trajectories in the potential landscape, or interchanging the order of x and z. Let us first suppose $x < z$ and that U is strictly decreasing between x and z. Then we can choose f such that

$$\dot{f}(s) = -U'(x + f(s)), \quad f(0) = 0,$$

and t appropriately, i.e. such that $f(t) = z - x$. This obviously gives $V(x,z) = 0$.

In the second case we suppose $x < z$ and that U is strictly increasing between x and z. In this case the Euler–Lagrange equation will be fulfilled by choosing

$$\dot{f}(s) = U'(x + f(s)), \quad f(0) = 0,$$

and t appropriately. In this case we have

$$\begin{aligned} V(x, z) = I_{x,t}(f) &= \frac{1}{2} \int_0^t \left|\dot{f}(s) + U'(x + f(s))\right|^2 ds \\ &= 2 \int_0^t \dot{f}(s) U'(x + f(s))\, ds = 2(U(z) - U(x)). \end{aligned}$$

□

COROLLARY 3.3. For the potential $U(\cdot)$ appearing in (3.1) we have

$$V(-1, 0) = V(-1, 1) = V, \quad V(1, 0) = V(1, -1) = v. \tag{3.9}$$

Equipped with the precise description of the quasipotential landscape of the underlying diffusion process given in Theorem 3.2 we can now describe the asymptotic behavior of its trajectories more precisely. We shall particularly be interested in their sojourns in different domains of attraction of the potential landscape. This involves exits from these domains, which are not covered by Theorem 2.42. We therefore have to extend Theorem 2.42 to this setting. In the one-dimensional case we are concerned with, this can be achieved by comparison arguments. To be in the framework of Theorem 2.42, we eventually have to work under the stronger hypotheses that $b = -U'$ is bounded and globally Lipschitz continuous. Note that this can always be assumed, by modifying b outside of large compact intervals in $\mathbb{R}$.

THEOREM 3.4. *Let $0 < \eta < 1$, $\varepsilon > 0$. For $x < 0$ consider*

$$\tau_\eta^{\varepsilon,x} = \inf\{t \geq 0 \colon X_t^{\varepsilon,x} \geq \eta\},$$

and for $x > 0$ consider

$$\tau_{-\eta}^{\varepsilon,x} = \inf\{t \geq 0 \colon X_t^{\varepsilon,x} \leq -\eta\}.$$

Then for any $\delta > 0$ and $x < 0$ we have

$$\lim_{\varepsilon \to 0} \mathbb{P}\Big(e^{\frac{V+\delta}{\varepsilon}} > \tau_\eta^{\varepsilon,x} > e^{\frac{V-\delta}{\varepsilon}}\Big) = 1,$$

and for $x > 0$ we have

$$\lim_{\varepsilon \to 0} \mathbb{P}\Big(e^{\frac{v+\delta}{\varepsilon}} > \tau_{-\eta}^{\varepsilon,x} > e^{\frac{v-\delta}{\varepsilon}}\Big) = 1.$$

PROOF. Let us consider the case $x < 0$, the other one being symmetric. To argue for the lower bound, let $\delta > 0$, $R < -1$ be such that $U(R) > 1$, and for any $-1 < \rho < 0$, first consider $G = (R, \rho)$. Then apply Theorem 2.42 to see that for any $\delta > 0$ the exit time $\tau^{\varepsilon,x}$ from G satisfies

$$\lim_{\varepsilon \to 0} \mathbb{P}\Big(\tau^{\varepsilon,x} > e^{\frac{V_\rho - \delta}{\varepsilon}}\Big) = 1,$$

where $V_\rho = 2(U(\rho) - U(-1))$. Since obviously $\tau^{\varepsilon,x} \leq \tau_\eta^{\varepsilon,x}$, we have

$$\lim_{\varepsilon \to 0} \mathbb{P}\Big(\tau_\eta^{\varepsilon,x} > e^{\frac{V_\rho - \delta}{\varepsilon}}\Big) = 1.$$

As $\rho \to 0$ by continuity $V_\rho \to V$. Hence also

$$\lim_{\varepsilon\to 0} \mathbb{P}\Big(\tau_\eta^{\varepsilon,x} > e^{\frac{V-\delta}{\varepsilon}}\Big) = 1.$$

For the upper bound, let again $\delta > 0$. We modify U and use a comparison argument. We know that the drift term $-U'$ is strictly negative on $(-1,0)$ and strictly positive on $(0,1)$, while $-U'(0) = 0$. We choose a potential function $\tilde{U}$ such that it is bounded and globally Lipschitz on $[\eta,\infty)$,

$$\begin{aligned} \tilde{U}(y) &= U(y), \quad y \le -\eta, \\ -\tilde{U}'(y) &\le -U'(y) \quad y \le \eta, \end{aligned}$$

$-\tilde{U}'$ is increasing and strictly negative on $[-\eta,\eta]$ and $V + \frac{\delta}{2} = 2(\tilde{U}(\eta) - U(-1))$. Let $\tilde{X}^{\varepsilon,x}$ be the diffusion process with the drift $\tilde{b} = -\tilde{U}'$, starting at x, and let $\tilde{\tau}_\eta^{\varepsilon,x}$ be the transition time corresponding to $\tau_\eta^{\varepsilon,x}$. Then by choice of $\tilde{U}$, Theorem 2.42 is applicable to $\tilde{X}^{\varepsilon,x}$ for $G = (R,\eta)$, $\frac{\delta}{2}$ with $R < -1$ such that $U(R) > 1$. By comparison, we have $\tau_\eta^{\varepsilon,x} \le \tilde{\tau}_\eta^{\varepsilon,x}$, and hence by Theorem 2.42 and exponential equivalence of the exit times of G and $\tilde{\tau}_\eta^{\varepsilon,x}$ we obtain

$$\lim_{\varepsilon\to 0} \mathbb{P}\Big(\tau_\eta^{\varepsilon,x} < e^{\frac{V+\delta}{\varepsilon}}\Big) \ge \lim_{\varepsilon\to 0} \mathbb{P}\Big(\tilde{\tau}_\eta^{\varepsilon,x} < e^{\frac{V+\delta}{\varepsilon}}\Big) = 1.$$

This provides the desired upper bound and finishes the proof. □

Assume for instance that initially the process is inside the domain of attraction of the shallower local minimum 1 of U. Corollary 3.3 indicates that if we consider the diffusion at time scales $T_\varepsilon = e^{\frac{\mu}{\varepsilon}}$, then in the small noise limit $\varepsilon \to \infty$ trajectories will exit this well if $\mu > v$, and will be trapped inside if $\mu < v$. To capture *quasi-deterministic behavior* in the limit, Freidlin [**39**] considers the set of times at which the process is close to one of the local minima. Denote by meas the one-dimensional Lebesgue measures on the Borel sets of (subsets of) the real line.

THEOREM 3.5. *Let $\mu > 0$ and $\varepsilon \mapsto T_\varepsilon$ be such that*

$$\lim_{\varepsilon\to 0} \varepsilon \ln T_\varepsilon = \mu.$$

Let δ, $T > 0$. Then for $x \neq 0$

$$\operatorname{meas}\{0 \le t \le T \colon |X_{tT_\varepsilon}^{\varepsilon,x} - \operatorname{sign}(x)| > \delta\} \to 0 \quad \textit{as } \varepsilon \to 0$$

if $\mu < v$, and for any $x \in \mathbb{R}$

$$\operatorname{meas}\{0 \le t \le T \colon |X_{tT_\varepsilon}^{\varepsilon,x} + 1| > \delta\} \to 0 \quad \textit{as } \varepsilon \to 0$$

if $\mu > v$.

PROOF. We shall derive the claims by using the exit time law of Theorem 2.42. Note first that by the Markov property and regularity, we may always assume $x \neq 0$. For convenience, we argue for $T = 1$.

1. Assume first that $\mu < v$, and let $x > 0$, i.e. in the domain of attraction of the shallower well. Let G be an open interval in $(0,\infty)$ containing x and the unique stable equilibrium point 1, and such that

$$v > \overline{V} = \inf_{z\in\partial G} V(1,z) > \mu.$$

Let $\delta > 0$ be such that $\overline{B}_\delta(1) \subset G$. Denote by $\tau^{\varepsilon,x}$ the exit time from G. Define as in Chapter 2, Section 2.5.3 recursively for $m \geq 0$

$$\begin{aligned}
\theta_0^x &= 0, \\
\tau_m^x &= \inf\{t \geq \theta_m^x \colon X_t^{\varepsilon,x} \in \overline{B}_{\frac{\delta}{2}}(1) \cup \partial G\}, \\
\theta_{m+1}^x &= \begin{cases} +\infty, & X_{\tau_m^x}^{\varepsilon,x} \in \partial G, \\ \inf\{t \geq \tau_m^x \colon |X_t^{\varepsilon,x} - 1| = \delta\}, & |X_{\tau_m^x}^{\varepsilon,x} - 1| = \frac{\delta}{2}. \end{cases}
\end{aligned}$$

Then $(X_{\tau_m^x}^{\varepsilon,x})_{m\geq 0}$ is a Markov chain, with the convention $X_{\tau_m^x}^{\varepsilon,x} = X_{\tau^{\varepsilon,x}}^{\varepsilon,x}$, if $\tau_m^x = +\infty$. Now let $\eta > 0$ be such that $\overline{V} - \eta > \mu$. Choose $\delta_0 > 0$ such that for $0 < \delta < \delta_0$, and $T_0 = T(\overline{V}, \delta)$ according to Lemma 2.46 there exists $\varepsilon_0 > 0$ such that for any $0 \leq \varepsilon < \varepsilon_0$, $m \geq 1$

$$\sup_{x\in G} \mathbb{P}(\theta_m^x - \tau_{m-1}^x \leq T_0) \leq \sup_{x\in G} \mathbb{P}\Big(\sup_{0\leq t\leq T_0} |X^{\varepsilon,x} - x| \geq \frac{\delta}{2}\Big) \leq e^{-\frac{\overline{V}-\frac{\eta}{2}}{\varepsilon}}.$$

Assume ε_0 is small enough to ensure that for $\varepsilon \leq \varepsilon_0$ we have

$$T_\varepsilon \leq e^{\frac{\mu+\frac{\eta}{2}}{\varepsilon}}.$$

Now let again $k \in \mathbb{N}$, $k \leq \frac{T_\varepsilon}{T_0}$. Then, since $|X_t^{\varepsilon,x} - 1| \leq \delta$ for $t \in [\tau_{m-1}^x, \theta_m^x]$, we have

$$\Big\{ \operatorname{meas}\{t \in [0, T_\varepsilon] \colon |X_t^{\varepsilon,x} - 1| \leq \delta\} < kT_0 \Big\} \cap \{T_\varepsilon < \tau^{\varepsilon,x}\} \subseteq \bigcup_{m=1}^k \{\theta_m^x - \tau_{m-1}^x \leq T_0\}.$$

Hence for $k \in \mathbb{N}$, $k \leq \frac{T_\varepsilon}{T_0}$

$$\begin{aligned}
&\mathbb{P}\Big(\Big\{ \operatorname{meas}\{t \in [0, T_\varepsilon] \colon |X_t^{\varepsilon,x} - 1| \leq \delta\} < kT_0 \Big\} \cap \{T_\varepsilon < \tau^{\varepsilon,x}\}\Big) \\
&\qquad\qquad \leq \sum_{m=1}^k \mathbb{P}(\theta_m^x - \tau_{m-1}^x \leq T_0) \leq k e^{-\frac{\overline{V}-\frac{\eta}{2}}{\varepsilon}}.
\end{aligned}$$

Now taking $k = [\frac{T_\varepsilon}{T_0}]$, and using Lemma 2.45 we obtain

$$\begin{aligned}
&\mathbb{P}\Big(\operatorname{meas}\{t \in [0, T_\varepsilon] \colon |X_t^{\varepsilon,x} - 1| \leq \delta\} < T_\varepsilon - 1\Big) \\
&\quad \leq \mathbb{P}\Big(\Big\{ \operatorname{meas}\{t \in [0, T_\varepsilon] \colon |X_t^{\varepsilon,x} - 1| \leq \delta\} < T_\varepsilon - 1\Big\} \cap \{T_\varepsilon < \tau^{\varepsilon,x}\}\Big) \\
&\quad + \mathbb{P}(T_\varepsilon \geq \tau^{\varepsilon,x}) \\
&\quad \leq \frac{4}{T_0} e^{\frac{\mu - (\overline{V}-\eta)}{\varepsilon}} + \mathbb{P}(T_\varepsilon \geq \tau^{\varepsilon,x}) \to 0
\end{aligned}$$

as $\varepsilon \to 0$ by means of Theorem 2.42. This means that

$$\operatorname{meas}\{t \in [0,1] \colon |X_{tT_\varepsilon}^{\varepsilon,x} - 1| > \delta\} \leq T_\varepsilon^{-1} = e^{-\frac{\mu}{\varepsilon}}$$

with a probability which converges to 1 exponentially fast, as $\varepsilon \to 0$.

2. Let finally $v < \mu$. Choose $0 < \eta < 1$, and $x \in \mathbb{R}$. Our arguments will be similar to the ones used in part 1. Let $G = (a, b)$ be an open interval containing the interval $[-1, 1]$ and the point $\{x\}$ and such that $\overline{V} = \inf_{z\in\partial G} V(-1, z) > \mu$. This

time, we define recursively for $m \geq 0$

$$\theta_0^x = 0,$$
$$\tau_m^x = \inf\{t \geq \theta_m^x : X_t^{\varepsilon,x} \in \overline{B}_{\frac{\delta}{2}}(-1) \cup \partial G\},$$
$$\theta_{m+1}^x = \begin{cases} +\infty, & X_{\tau_m^x}^{\varepsilon,x} \in \partial G, \\ \inf\{t \geq \tau_m^x : |X_t^{\varepsilon,x} + 1| = \delta\}, & |X_{\tau_m^x}^{\varepsilon,x} + 1| = \frac{\delta}{2}. \end{cases}$$

Then, with the conventions as above, $(X_{\tau_m^x}^{\varepsilon,x})_{m\geq 0}$ is a Markov chain.

Choose $\delta_0 > 0$ such that for $0 < \delta < \delta_0$, and $T_0 = T(\overline{V}, \delta)$ according to Lemma 2.46 there exists $\varepsilon_0 > 0$ such that for any $0 \leq \varepsilon < \varepsilon_0$, $m \geq 1$

$$\sup_{x\in G} \mathbb{P}(\theta_m^x - \tau_{m-1}^x \leq T_0) \leq \sup_{x\in G} \mathbb{P}\Big(\sup_{0\leq t\leq T_0} |X^{\varepsilon,x} - x| \geq \frac{\delta}{2}\Big) \leq e^{-\frac{\overline{V}-\frac{\eta}{2}}{\varepsilon}}.$$

Without loss of generality, we may assume that for $\varepsilon \leq \varepsilon_0$ we have

$$T_\varepsilon \leq e^{\frac{\mu+\frac{\eta}{2}}{\varepsilon}}.$$

Now let again $k \in \mathbb{N}$, $k \leq \frac{T_\varepsilon}{T_0}$, and define τ^G to be the exit time from G. Then, since $|X_t^{\varepsilon,x} + 1| \leq \delta$ for $t \in [\tau_{m-1}^x, \theta_m^x]$, we have

$$\Big\{ \text{meas}\{t \in [0, T_\varepsilon] : |X_t^{\varepsilon,x} + 1| \leq \delta\} < kT_0 \Big\} \cap \{T_\varepsilon < \tau^G\} \subseteq \bigcup_{m=1}^{k} \{\theta_m^x - \tau_{m-1}^x \leq T_0\}.$$

Hence for $k \in \mathbb{N}$, $k \leq \frac{T_\varepsilon}{T_0}$

$$\mathbb{P}\Big(\Big\{ \text{meas}\{t \in [0, T_\varepsilon] : |X_t^{\varepsilon,x} + 1| \leq \delta\} < kT_0 \Big\} \cap \{T_\varepsilon < \tau^G\}\Big)$$
$$\leq \sum_{m=1}^{k} \mathbb{P}(\theta_m^x - \tau_{m-1}^x \leq T_0) \leq k e^{-\frac{\overline{V}-\frac{\eta}{2}}{\varepsilon}}.$$

Now taking $k = [\frac{T_\varepsilon}{T_0}]$ and using Lemma 2.45 we obtain

$$\mathbb{P}\Big(\text{meas}\{t \in [0, T_\varepsilon] : |X_t^{\varepsilon,x} + 1| \leq \delta\} < T_\varepsilon - 1\Big)$$
$$\leq \mathbb{P}\Big(\Big\{ \text{meas}\{t \in [0, T_\varepsilon] : |X_t^{\varepsilon,x} + 1| \leq \delta\} < T_\varepsilon - 1\Big\} \cap \{T_\varepsilon < \tau^G\}\Big) + \mathbb{P}(T_\varepsilon \geq \tau^{\varepsilon,x})$$
$$\leq \frac{4}{T_0} e^{\frac{\mu - (\overline{V}-\eta)}{\varepsilon}} + \mathbb{P}(T_\varepsilon \geq \tau^G) \to 0$$

as $\varepsilon \to 0$ by means of Theorem 3.4 and the strong Markov property. This means that

$$\text{meas}\{t \in [0, 1] : |X_{tT_\varepsilon}^{\varepsilon,x} + 1| > \delta\} \leq T_\varepsilon^{-1} = e^{-\frac{\mu}{\varepsilon}}$$

with a probability which converges to 1 exponentially fast as $\varepsilon \to 0$. □

Let us now return to the situation in which the potential function underlying our dynamical system is time periodic. In fact we shall assume the simplest possible periodicity behavior, in which the potential changes every half period between two spatial states. More formally, let us assume that

$$(3.10) \qquad U(t,x) = \sum_{k=0}^{\infty} \Big(U(x)\mathbb{I}_{[k,k+\frac{1}{2})}(t) + U(-x)\mathbb{I}_{[k+\frac{1}{2},k+1)}(t) \Big), \quad x \in \mathbb{R},\ t \geq 0,$$

with a potential function as described in the introduction. Let us recall that the diffusion equation with a weakly periodic drift term we shall consider is given by

$$X_t^{\varepsilon,x} = x - \int_0^t \frac{\partial}{\partial x} U\Big(\frac{t}{T}, X_s^{\varepsilon,x}\Big)\, ds + \sqrt{\varepsilon} W_t, \quad t \geq 0, \tag{3.11}$$

where T is a given period length. Let us also recall that $-U'$ satisfies the dissipativity condition (3.4). In accordance with the previous Theorem we have the following quasi-deterministic behavior for large enough periodicity scales. It basically states that if for small $\varepsilon > 0$ the period T_ε is large enough so that the system has the chance to exit from the shallow well, i.e. if $\mu = \lim_{\varepsilon\to 0} \varepsilon \ln T_\varepsilon > v$, then on the scale T_ε the total time spent in a finite time interval a positive distance away from the position of the deeper well is asymptotically vanishing. This implies that the solution trajectories of 3.11 show some periodicity behavior, but for very large μ still may look erratic.

THEOREM 3.6. *For $\varepsilon > 0$ let $\varepsilon \mapsto T_\varepsilon$ be such that*

$$\lim_{\varepsilon\to 0} \varepsilon \ln T_\varepsilon = \mu.$$

Let $\delta > 0$ and $T > 0$ and define

$$\phi(t) = \sum_{k=0}^{\infty} \Big(- \mathbb{I}_{[k,k+\frac{1}{2})}(t) + \mathbb{I}_{[k+\frac{1}{2},k+1)}(t) \Big), \quad t \geq 0. \tag{3.12}$$

If $\mu < v$ then for $x \neq 0$

$$\text{meas}\{t \in [0,T] \colon |X_{tT_\varepsilon}^{\varepsilon,x} - \text{sign}(x)| > \delta\} \to 0$$

in probability as $\varepsilon \to 0$.

If $\mu > v$ then for any $x \in \mathbb{R}$

$$\text{meas}\{t \in [0,T] \colon |X_{tT_\varepsilon}^{\varepsilon,x} - \phi(t)| > \delta\} \to 0$$

in probability as $\varepsilon \to 0$.

PROOF. Consider the case $\mu > v$. For $\varepsilon > 0$ and $x \in \mathbb{R}$ consider the time homogeneous diffusions

$$Y_t^{\varepsilon,x} = x - \int_0^t U'(Y_s^{\varepsilon,x})\, ds + \sqrt{\varepsilon} W_t, \quad t \geq 0,$$

and

$$Z_t^{\varepsilon,x} = x - \int_0^t U'(-Z_s^{\varepsilon,x})\, ds + \sqrt{\varepsilon} W_t, \quad t \geq 0.$$

Further for $0 \leq k \leq [T]$, let $I_k = [k, (k+\frac{1}{2}) \wedge T]$, $J_k = [(k+\frac{1}{2}) \wedge T, (k+1) \wedge T]$. Then by the Markov property and definition of U for any $0 \leq k \leq [T]$

$$\begin{aligned}\text{Law}\Big(&\text{meas}\{t \in I_k \colon |X_{tT_\varepsilon}^{\varepsilon,x} - \phi(t)| > \delta\}\Big) \\ &= \text{Law}\Big(\text{meas}\{t \in [0, \tfrac{1}{2}] \colon |Y_{tT_\varepsilon}^{\varepsilon,x} + 1| > \delta\}\Big| x = X_{kT_\varepsilon}^{\varepsilon,x}\Big),\end{aligned}$$

while

$$\begin{aligned}&\text{Law}\Big(\text{meas}\{t \in I_k \colon X_{tT_\varepsilon}^{\varepsilon,x} - \phi(t)| > \delta\}\Big) \\ &= \text{Law}\Big(\text{meas}\{t \in [0, \tfrac{1}{2} \wedge (T-k-\tfrac{1}{2})^+ \colon |Z_{tT_\varepsilon}^{\varepsilon,x} - 1| > \delta\}\Big| x = X_{((k+\frac{1}{2})\wedge T)T_\varepsilon}^{\varepsilon,x}\Big),\end{aligned}$$

Now apply Theorem 3.5 individually for $0 \le k \le [T]$ to obtain

$$\begin{aligned} &\operatorname{meas}\{t \in I_k\colon |X^{\varepsilon,x}_{tT_\varepsilon} - \phi(t)| > \delta\} \to 0, \\ &\operatorname{meas}\{t \in J_k\colon |X^{\varepsilon,x}_{tT_\varepsilon} - \phi(t)| > \delta\} \to 0 \end{aligned}$$

in probability as $\varepsilon \to 0$. Now take the finite sum over all $0 \le k \le [T]$ to obtain the desired convergence result. The derivation of the case $\mu < v$ from Theorem 3.5 is simpler, and its proof is omitted. □

3.2. The reduced dynamics: stochastic resonance in two-state Markov chains

In this section we study a simple model of stochastic resonance. It can be seen as a reduction of diffusion to a Markov chain. A model of this type was first studied in McNamara and Wiesenfeld [74] and Pavlyukevich [**86**]. We follow here Imkeller and Pavlyukevich [**59**].

3.2.1. The Markov chain and its invariant measure. Consider a family of Markov chains $Y^{\varepsilon,T} = (Y^{\varepsilon,T}_t)_{t\ge 0}$ on the state space $\{-1, 1\}$. The variable $t \in \mathbb{R}_+$ denotes time, and $0 < \varepsilon,\ T < \infty$ parametrize the family in the following way. Let us introduce the transition probabilities

$$p_{ij}(s,t;\varepsilon,T) = \mathbb{P}(Y^{\varepsilon,T}_t = j | Y^{\varepsilon,T}_s = i), \quad i,j = \pm 1, \quad 0 \le s \le t. \tag{3.13}$$

To mimic the behaviour of the diffusion with a flipping potential we define time-periodic infinitesimal probabilities of $Y^{\varepsilon,T}$ as

$$Q_{\varepsilon,T}(Tt) = \begin{cases} Q_1 = \begin{pmatrix} -\varphi & \varphi \\ \psi & -\psi \end{pmatrix}, & 0 \le t \bmod 1 < \frac{1}{2}, \\ Q_2 = \begin{pmatrix} -\psi & \psi \\ \varphi & -\varphi \end{pmatrix}, & \frac{1}{2} \le t \bmod 1 < 1, \end{cases} \tag{3.14}$$

where we assume that

$$\begin{aligned} &\varphi = pe^{-V/\varepsilon}, \qquad \psi = qe^{-v/\varepsilon}, \\ &\varepsilon > 0, \qquad p,q > 0, \qquad 0 < v < V. \end{aligned} \tag{3.15}$$

On the intervals $[kT, (k+\frac{1}{2})T)$ and $[(k+\frac{1}{2})T, (k+1)T)$, $k \ge 0$, the process $Y^{\varepsilon,T}$ is time-homogeneous and its transition probabilities (3.13) can be expressed in terms of φ and ψ:

$$\begin{aligned} &p_{-1,1}(t, t+h; \varepsilon, T) = \varphi h + o(h), \\ &p_{1,-1}(t, t+h; \varepsilon, T) = \psi h + o(h), \quad 0 \le t \bmod T < T/2, \end{aligned} \tag{3.16}$$

and

$$\begin{aligned} &p_{-1,1}(t, t+h; \varepsilon, T) = \psi h + o(h), \\ &p_{1,-1}(t, t+h; \varepsilon, T) = \varphi h + o(h), \quad T/2 \le t \bmod T < T, \end{aligned} \tag{3.17}$$

The goal of this section is to study the 'periodic' properties of $Y^{\varepsilon,T}$ induced by the periodicity of the infinitesimal probabilities.

To determine the invariant law of the process $Y^{\varepsilon,T}$ we consider a new two-dimensional Markov process

$$\mathbf{Y}^{\varepsilon,T}_t = \left(Y^{\varepsilon,T}_{Tt}, \frac{t}{T} \bmod 1\right), \quad t \ge 0, \tag{3.18}$$

on the state space $\{-1,1\} \times S^1$, which topologically is the product of two unit length circles.

The process $\mathbf{Y}^{\varepsilon,T}$ is time-homogeneous. Note that we have compressed time: it is convenient to have the time scale independent of the parameter T. The infinitesimal generator of $\mathbf{Y}^{\varepsilon,T}$ is

$$B_{\varepsilon,T} f(x,\theta) = \lim_{h\to 0} \frac{\mathbb{E}_{x,\theta} f\Big(Y^{\varepsilon,T}_{T(\theta+h)}, \frac{\theta+h}{T} \bmod 1\Big) - f(x,\theta)}{h}, \qquad (x,\theta) \in \{\pm 1\} \times [0,1].$$

As $x = \pm 1$ we may think that $B_{\varepsilon,T}$ is defined on the space of vectors $f(\theta) = (f^-(\theta), f^+(\theta))^*$ with smooth components and such that $f(0) = f(1)$. Using (3.16) and (3.17) gives

$$B_{\varepsilon,T} f = \frac{1}{T}\frac{d}{d\theta} f + Q_{\varepsilon,T}(\theta) f.$$

Let the vector $\nu_{\varepsilon,T} = (\nu^-_{\varepsilon,T}(\theta), \nu^+_{\varepsilon,T}(\theta))^*$, $\theta \in [0,1]$, denote the invariant density of $\mathbf{Y}^{\varepsilon,T}$ w.r.t. the product of counting measure on $\{-1,1\}$ and Lebesgue measure on a circle S^1 normalized so that $\nu^-_{\varepsilon,T}(\theta) + \nu^+_{\varepsilon,T}(\theta) = 1$. We shall call $\nu_{\varepsilon,T}$ the *invariant law* of the process $Y^{\varepsilon,T}$. Indeed, for $\theta \in [0,1]$

$$\mathbb{P}_\nu(Y^{\varepsilon,T}_{T\theta} = \pm 1) = \nu^\pm_{\varepsilon,T}(\theta).$$

The invariant measure satisfies the forward Kolmogorov equation

$$B^*_{\varepsilon,T} \nu_{\varepsilon,T} = 0,$$

and the continuity condition $\nu_{\varepsilon,T}(0) = \nu_{\varepsilon,T}(1)$, where the adjoint operator is given by

$$B^*_{\varepsilon,T} f = -\frac{1}{T}\frac{d}{d\theta} f + Q^*_{\varepsilon,T}(\theta) f.$$

From the symmetry between Q_1 and Q_2 in (3.14) we deduce the following

PROPOSITION 3.7. The invariant measure of the process $\mathbf{Y}^{\varepsilon,T}$ has the following symmetry property: $\nu^\pm_{\varepsilon,T}(\theta) = \nu^\mp_{\varepsilon,T}(\theta + \frac{1}{2})$, $0 \le \theta \le \frac{1}{2}$.

PROOF. The statement follows easily from the fact that if, for example, $\nu = (\nu^-, \nu^+)^*$ is a solution of $-\frac{1}{T}\dot\nu + Q_1^*\nu = 0$, then $\overline{\nu} = (\nu^+, \nu^-)^*$ satisfies $-\frac{1}{T}\dot{\overline{\nu}} + Q_2^*\overline{\nu} = 0$, together with the continuity condition and the uniqueness of the invariant measure. Here, $\dot\nu = \frac{d}{d\theta}\nu$. □

It follows from Proposition 3.7 that in order to find the invariant measure, it is enough to solve the boundary value problem

$$(3.19) \qquad \begin{cases} -\dfrac{1}{T}\dfrac{d}{d\theta}\nu_{\varepsilon,T} + Q_1^*\nu_{\varepsilon,T} = 0, \\ \nu^-_{\varepsilon,T}(0) = \nu^+_{\varepsilon,T}(\frac{1}{2}), \\ \nu^-_{\varepsilon,T}(\theta) + \nu^+_{\varepsilon,T}(\theta) = 1, \\ \nu^\pm_{\varepsilon,T}(\theta) > 0, \quad \theta \in [0, \frac{1}{2}], \end{cases}$$

which is done in the following

PROPOSITION 3.8. For $T > 0$ and φ and ψ defined in (3.15), the invariant measure of the process $Y^{\varepsilon,T}$ equals

$$\begin{cases} \nu^-_{\varepsilon,T}(\theta) = \dfrac{\psi}{\varphi+\psi} + \dfrac{\varphi-\psi}{\varphi+\psi}\dfrac{e^{-(\varphi+\psi)T\theta}}{1+e^{-\frac{1}{2}(\varphi+\psi)T}}, \\ \nu^+_{\varepsilon,T}(\theta) = \dfrac{\varphi}{\varphi+\psi} - \dfrac{\varphi-\psi}{\varphi+\psi}\dfrac{e^{-(\varphi+\psi)T\theta}}{1+e^{-\frac{1}{2}(\varphi+\psi)T}}; \end{cases} \tag{3.20}$$
$$\begin{cases} \nu^-_{\varepsilon,T}(\theta+\frac{1}{2}) = \nu^+_{\varepsilon,T}(\theta), \\ \nu^+_{\varepsilon,T}(\theta+\frac{1}{2}) = \nu^-_{\varepsilon,T}(\theta), \quad 0 \le \theta \le \frac{1}{2}. \end{cases}$$

PROOF. Solving the differential equation in (3.19) and using the normalizing condition gives the following general solution

$$\nu^-_{\varepsilon,T}(\theta) = \frac{\psi}{\varphi+\psi} + Ae^{-(\varphi+\psi)T\theta}, \quad \nu^+_{\varepsilon,T}(\theta) = \frac{\varphi}{\varphi+\psi} - Ae^{-(\varphi+\psi)T\theta}$$

where A is an arbitrary constant. Applying the boundary condition leads to

$$A = \frac{\varphi-\psi}{\varphi+\psi}\frac{1}{1+e^{-\frac{1}{2}(\varphi+\psi)T}},$$

which concludes the proof. □

Note that $\nu^\pm_{\varepsilon,T}(\theta)$ is a sum of two parts. For $\theta \in [0, \frac{1}{2}]$, the time-independent pair $(\frac{\psi}{\varphi+\psi}, \frac{\varphi}{\varphi+\psi})$ is the invariant measure of the time-homogeneous Markov chain with infinitesimal generator Q_1, and $(\nu^-_{\varepsilon,T}(\frac{1}{2}), \nu^+_{\varepsilon,T}(\frac{1}{2})) \to (\frac{\psi}{\varphi+\psi}, \frac{\varphi}{\varphi+\psi})$ exponentially fast with rate $\varphi+\psi$ as $T \to \infty$. This is an illustration of a classical result about convergence of the law of a stochastic process to its invariant law.

On the second half-period, the Markov chain is governed by the infinitesimal generator Q_2 and therefore $(\nu^-_{\varepsilon,T}(1), \nu^+_{\varepsilon,T}(1)) \to (\frac{\varphi}{\varphi+\psi}, \frac{\psi}{\varphi+\psi})$, $T \to \infty$, which is the invariant law of the Markov process with generator Q_2.

The invariant measure found in Proposition 3.8 is our main tool in studying the periodic properties of the process $Y^{\varepsilon,T}$. In the following sections we shall introduce, study and compare several measures of goodness of stochastic resonance.

3.2.2. Spectral power amplification coefficient.

We define the *spectral power amplification* coefficient as the ratio between the component of period T of the power carried by the averaged trajectory of $Y^{\varepsilon,T}$ and the corresponding component of the power carried by the deterministic signal

$$I_T(t) = \begin{cases} V, & 0 \le t \bmod T < T/2, \\ v, & T/2 \le t \bmod T < T. \end{cases}$$

The signal I_T 'switches' the powers in the infinitesimal probabilities φ and ψ.

Precisely, we define the SPA coefficient as

$$\eta^Y(\varepsilon, T) = \left| \frac{\int_0^1 \mathbb{E}_\nu Y^{\varepsilon,T}_{Ts} e^{2\pi i s}\, ds}{\int_0^1 I_T(Ts) e^{2\pi i s}\, ds} \right|^2 = \frac{\pi^2}{(V-v)^2} \left| \int_0^1 \mathbb{E}_\nu Y^{\varepsilon,T}_{Ts} e^{2\pi i s}\, ds \right|^2, \tag{3.21}$$

where $\mathbb{E}_\nu$ denotes the expectation w.r.t. the invariant measure $\nu_{\varepsilon,T}$.

The goal of this section is to study the behaviour of the SPA coefficient as a function of ε and T.

PROPOSITION 3.9. a) The SPA coefficient is given by

$$\eta^Y(\varepsilon,T) = \frac{4}{(V-v)^2}\frac{T^2(\varphi-\psi)^2}{(\varphi+\psi)^2T^2+4\pi^2}, \tag{3.22}$$

with φ and ψ defined in (3.15).

b) $T\to\infty$ the mapping $\varepsilon\mapsto\eta^Y(\varepsilon,T)$ has a unique local maximum and the optimal tuning $T_\eta = T_\eta(\varepsilon)$ satisfies

$$T_\eta(\varepsilon) = \frac{\pi}{\sqrt{2pq}}\sqrt{\frac{v}{V-v}}e^{\frac{V+v}{2\varepsilon}}\left\{1+\mathcal{O}(e^{-\frac{V-v}{\varepsilon}})\right\}$$

and

$$\eta^Y(\varepsilon,T_\eta(\varepsilon)) \to \frac{4}{\pi^2(V-v)^2} \quad \text{as} \quad \varepsilon\to 0. \tag{3.23}$$

PROOF. a) We use (3.20) to obtain

$$\int_0^1 \mathbb{E}_\nu Y^{\varepsilon,T}_{Ts}e^{2\pi is}\,ds = \int_0^1(\nu^+_{\varepsilon,T}(s)-\nu^-_{\varepsilon,T}(s))e^{2\pi is}\,ds$$
$$= 2\int_0^{\frac12}(\nu^+_{\varepsilon,T}(s)-\nu^-_{\varepsilon,T}(s))e^{2\pi is}\,ds$$
$$= 2\frac{\varphi-\psi}{\varphi+\psi}\left(\frac{i}{\pi}+\frac{2}{2\pi i-(\varphi+\psi)T}\right)$$

which directly leads to (3.22).

b) Let $0<\beta=\frac{v}{V}<1$ and $T>0$ be fixed. Let us introduce the variable $t=\exp(-\frac{V}{\varepsilon})\in[0,1]$. Then $\varphi=pt$ and $\psi=qt^\beta$, and

$$\eta^Y(\varepsilon,T)=\eta^Y_T(t) = \frac{4}{(V-v)^2}\frac{T^2(pt-qt^\beta)^2}{(pt+qt^\beta)^2T^2+4\pi^2}.$$

Taking the derivative of $\eta^Y_T(t)$ in t we find that the extrema of $\eta^Y_T(t)$ are the roots of the equation

$$t^\beta(pt-qt^\beta)\left\{2\pi^2(q\beta-pt^{1-\beta})-pqT^2(1-\beta)t(qt^\beta+pt)\right\}=0. \tag{3.24}$$

The first parenthesis of (3.24) gives the coordinate of a local minimum of η^Y $t=(q/p)^{\frac{1}{1-\beta}}$ which does not depend on T, is less than 1 if $p>q$, and corresponds to $\widehat{\varepsilon}=(V-v)/\ln(\frac{p}{q})$.

Denote the second parenthesis of (3.24) by $F(t)$. Let us show that $F()t$ has exactly one root for big values of T. Note that $F(0)=2\pi^2q\beta>0$,

$$F'(t) = -2\pi^2p(1-\beta)t^{-\beta}-pq^2T^2(1-\beta^2)t^\beta-2p^2qT^2(1-\beta)t\le 0,$$

and consequently, $F(t)$ is monotonically decreasing on $(0,1]$. Hence, F has a unique root on $t\in(0,1]$ if and only if $F(1)\le 0$, i.e. if

$$2\pi^2(q\beta-p)-pqT^2(1-\beta)(q+p)\le 0, \tag{3.25}$$

which is obviously true for big T.

To find an 'optimal tuning' $T=T(\varepsilon)$ we note that $F(t)=0$ is a linear equation in T^2 from which we get that

$$T_\eta = \pi\sqrt{\frac{2}{pq}\frac{\beta}{1-\beta}}t^{-\frac{1+\beta}{2}}\left(\frac{1-\frac{p}{q}\frac{1}{\beta}t^{1-\beta}}{1+\frac{p}{q}t^{1-\beta}}\right)^{\frac12}. \tag{3.26}$$

It is clear that as $T \to \infty$ the noise parameter $\varepsilon \to 0$ and (3.26) can be rewritten as

$$T_\eta = \pi \sqrt{\frac{2}{pq} \frac{v}{V-v}} e^{\frac{V+v}{2\varepsilon}} \left\{ 1 + \mathcal{O}(e^{-\frac{V-v}{\varepsilon}}) \right\}. \tag{3.27}$$

The limit (3.23) is obtained by inserting (3.27) into (3.22). □

3.2.3. SPA-to-noise ratio. In this section we study another measure of quality of tuning, the SPA-to-noise ratio (SPN), which can be defined as the ratio of the 'amplitude' of the output signal and noise 'amplitude'. This notion is related to another popular measure of stochastic resonance, the so-called signal-to-noise ratio, see [**38, 43, 74**].

Formally, this measure of goodness is given by

$$\text{SPN}(\varepsilon, T) = \frac{1}{\varepsilon^2} \left| \int_0^1 \mathbb{E}_\nu Y^{\varepsilon,T}_{2Ts} e^{2\pi i s}\, ds \right|^2 = \frac{(V-v)^2}{\pi^2} \frac{\eta^Y(\varepsilon, T)}{\varepsilon^2}. \tag{3.28}$$

The SPA-to-noise ratio compares the spectral component of the random output with the noise intensity and measures the proportion of the noise energy transferred to the component of the output corresponding to period $2T$.

PROPOSITION 3.10. a) The SPA-to-noise ratio is given by

$$\text{SPN}(\varepsilon, T) = \frac{4}{\pi^2 \varepsilon^2} \frac{T^2(\varphi - \psi)^2}{(\varphi + \psi)^2 T^2 + 4\pi^2}, \tag{3.29}$$

with φ and ψ defined in (3.15).

b) For T large enough $\varepsilon \mapsto \text{SPN}(\varepsilon, T)$ has a local maximum, and the optimal tuning occurs for

$$T_{\text{SPN}}(\varepsilon) = 2 \frac{\pi \sqrt{v}}{q \sqrt{\varepsilon}} e^{\frac{v}{\varepsilon}} (1 + \mathcal{O}(\varepsilon)). \tag{3.30}$$

Moreover,

$$\text{SPN}(\varepsilon, T_{\text{SPN}}(\varepsilon)) = \frac{4}{\pi^2 \varepsilon^2} (1 + \mathcal{O}(\varepsilon)), \quad \text{as} \quad \varepsilon \to 0. \tag{3.31}$$

PROOF. The formula (3.29) is an obvious consequence of (3.22) and (3.28).

To study the extrema of the $\text{SPN}(\varepsilon, T)$ with respect to ε let us intoduce the variable $t = e^{-V/\varepsilon}$ taking values in $(0, 1]$. Then, $\varphi = pt$ and $\psi = qt^\beta$, where $\beta = \frac{v}{V} < 1$. In terms of t the formula (3.29) takes the form

$$\text{SPN}(t, T) = \frac{4}{\pi^2 V^2} \frac{T^2 (pt - qt^\beta)^2 \ln^2 t}{(pt + qt^\beta)^2 T^2 + 4\pi^2}, \tag{3.32}$$

and taking the derivative in t we find that the extrema of (3.32) are zeros of the following equation:

(3.33)

$$(pt - qt^\beta) \Big[4\pi^2 (q\beta t^\beta - pt) |\ln t| - 4\pi^2 (qt^\beta - pt) \\ - 2pq(1-\beta) T^2 t^{1+\beta} (qt^\beta + pt) |\ln t| - T^2 (qt^\beta - pt)(qt^\beta + pt)^2 \Big] = 0.$$

The expression in the first parenthesis gives the coordinate of the local minimum $\widehat{\varepsilon} = (V - v)/\ln\left(\frac{p}{q}\right)$ which is non-negative for $p > q$ and is also a root of $\text{SPN}(\varepsilon, T)$.

The equation in brackets can be solved for T. We get

$$T=2\left(\frac{\pi^2(q\beta t^\beta-pt)|\ln t|-\pi^2(qt^\beta-pt)}{(qt^\beta-pt)(qt^\beta+pt)^2+2pq(1-\beta)t^{1+\beta}(qt^\beta+pt)|\ln t|}\right)^{\frac{1}{2}}$$

$$=\frac{2\pi}{q}t^{-\beta}\sqrt{\beta|\ln t|}\left(\frac{1-\frac{p}{q\beta}t^{1-\beta}-\frac{1}{\beta|\ln t|}+\frac{p}{q\beta|\ln t|}t^{1-\beta}}{(1-\frac{p}{q}t^{1-\beta})(1+\frac{p}{q}t^{1-\beta})^2+2\frac{p}{q}(1-\beta)t^{1-\beta}(1+\frac{p}{q}t^{1-\beta})|\ln t|}\right)^{\frac{1}{2}}.$$

As $t\to 0$, the latter expression can be expanded as

$$T=\frac{2\pi}{q}t^{-\beta}\sqrt{\beta|\ln t|}\left(1+\mathcal{O}(|\ln t|^{-1})\right),$$

which yields (3.30). The limiting value (3.31) is obtained by inserting (3.30) into (3.29). □

As we see, the SPA-to-noise ratio has a local maximum in ε. This means that on the optimal noise level the noise energy feeds the periodic component of the averaged random output.

3.2.4. Energy. The SPA coefficient η^Y discussed in Section 3.2.2 describes the averaged power carried by the longest harmonic in the Fourier decomposition of $Y^{\varepsilon,T}$. Let us consider the *energy* carried by all harmonics, which, due to Parseval's equality, is given by

$$\mathrm{En}(\varepsilon,T)=\int_0^1\left(\mathbb{E}_\nu Y_{Ts}^{\varepsilon,T}\right)^2 ds.$$

PROPOSITION 3.11. a) The energy carried by the averaged Markov chain $Y^{\varepsilon,T}$ equals

$$\mathrm{En}(\varepsilon,T)=\left(\frac{\varphi-\psi}{\varphi+\psi}\right)^2\left[1-\frac{4}{T(\varphi+\psi)}\tanh\frac{T(\varphi+\psi)}{4}\right]. \tag{3.34}$$

b) For T large enough, $\varepsilon\mapsto\mathrm{En}(\varepsilon,T)$ has a local maximum. An optimal tuning rate at which the maximum is attained is given by

$$T_{\mathrm{En}}(\varepsilon)=\frac{1}{p}\frac{v}{V-v}e^{\frac{V}{\varepsilon}}\{1+\mathcal{O}(e^{-\frac{V-v}{\varepsilon}})\} \tag{3.35}$$

and

$$\mathrm{En}(\varepsilon,T_{\mathrm{En}}(\varepsilon))\to 1 \quad\text{as}\quad \varepsilon\to 0. \tag{3.36}$$

PROOF. The proof is analogous to the proof of Propositions 3.9 and 3.10. a) To obtain (3.34) we note that

$$\mathrm{En}(\varepsilon,T)=\int_0^1\left(\mathbb{E}_\nu Y_{2Ts}^{\varepsilon,T}\right)^2 ds=\int_0^1(\nu_{\varepsilon,T}^+(s)-\nu_{\varepsilon,T}^-(s))^2\,ds$$

$$=2\int_0^{\frac{1}{2}}(\nu_{\varepsilon,T}^+(s)-\nu_{\varepsilon,T}^-(s))^2\,ds$$

$$=2\left(\frac{\varphi-\psi}{\varphi+\psi}\right)^2\int_0^{\frac{1}{2}}\left(1-\frac{2e^{-(\varphi+\psi)Ts}}{1+e^{-\frac{1}{2}(\varphi+\psi)T}}\right)^2 ds$$

$$=\left(\frac{\varphi-\psi}{\varphi+\psi}\right)^2\left[1-\frac{4}{T(\varphi+\psi)}\tanh\frac{T(\varphi+\psi)}{4}\right].$$

b) To find extrema of $\mathrm{En}(\varepsilon, T)$ we denote $\beta = \frac{v}{V}$ and $t = e^{-\frac{V}{\varepsilon}} \in (0,1]$ and take the derivative of (3.34). A straightforward calculation yields the equation $(qt^\beta - pt)F_T(t) = 0$ for the extrema, where

$$\begin{aligned} F_T(t) &= T(p^2t^2 - q^2t^{2\beta})(pt + q\beta t^\beta)\operatorname{sech}^2\left[\tfrac{T}{4}(pt + qt^\beta)\right] \\ &\quad - 4Tpq(1-\beta)t^{1+\beta}(qt^\beta + pt) \\ &\quad - 4(p^2t^2 - q^2\beta t^{2\beta} + 5pq(1-\beta)t^{1+\beta})\tanh\left[\tfrac{T}{4}(pt + qt^\beta)\right]. \end{aligned} \tag{3.37}$$

The first parenthesis of the equation of the extrema gives the root $\widehat{\varepsilon} = (V - v)/\ln\left(\frac{p}{q}\right)$. Next, we parametrize $T_t = \frac{a}{t}$, $a > 0$, and consider $F_{T_t}(t)$ as $t \to 0$. It is clear that $\operatorname{sech}\left[\frac{T_t}{4}(qt^\beta + pt)\right] = \operatorname{sech}\left[\frac{a}{4}(qt^{\beta-1} + p)\right] \to 0$ exponentially fast, whereas $\tanh\left[\frac{T_t}{4}(qt^\beta + pt)\right] \to 1$ exponentially fast as $t \to 0$. This yields

$$F_{T_t}(t) = 4q^2t^{2\beta}\left(\beta - ap(1-\beta) + \mathcal{O}(t^{1-\beta})\right).$$

For $a < \frac{\beta}{p(1-\beta)}$ we have $F_{T_t}(t) > 0$, and for $a > \frac{\beta}{p(1-\beta)}$ we have $F_{T_t}(t) < 0$ as $t \to 0$, and we can localize the optimal value of T in the interval $\left[\frac{\beta}{2p(1-\beta)t}, \frac{2\beta}{p(1-\beta)t}\right]$. Therefore, in this interval we can asymptotically solve the equation for T with the result:

$$T = \frac{1}{p}\frac{\beta}{1-\beta}t^{-1}(1 + \mathcal{O}(t^{1-\beta})).$$

This gives (3.35). The limit (3.36) is obtained with the help of (3.34) and (3.35). □

3.2.5. Out-of-phase measure. The approach by Freidlin [**39**] is valid for a quite general class of stochastic systems. Theorem 3.6 describes the periodic responce of the diffusion. The measure of quality in this theorem is the Lebesgue measure of the time which the trajectory spends outside some neighbourhood of the minimum of the deep well, i.e. the total amount of time spent in the 'wrong' place. The same results hold for the Markov chain $Y^{\varepsilon,T}$ for small values of ε.

PROPOSITION 3.12 ([**39**]). *For $\varepsilon > 0$, let the half-period $T = T(\varepsilon)$ be such that*

$$\lim_{\varepsilon \to 0} \varepsilon \ln T(\varepsilon) = \lambda > 0.$$

Let the function

$$\phi(t) = \begin{cases} -1, & t \bmod 1 \in [0, \frac{1}{2}), \\ 1, & t \bmod 1 \in [\frac{1}{2}, 1), \end{cases}$$

be a periodic and deterministic function of time, and meas denote the Lebesgue measure on $[0,1]$. Then, if $\lambda > v$,

$$\operatorname{meas}\left\{t \in [0,1] \colon Y^{\varepsilon,T}_{T(\varepsilon)t} \neq \phi(t)\right\} \to 0 \tag{3.38}$$

in probability as $\varepsilon \to 0$.

If $\lambda < v$ then

$$\operatorname{meas}\left\{t \in [0,1] \colon Y^{\varepsilon,T}_{T(\varepsilon)t} \neq Y^{\varepsilon,T}_0\right\} \to 0$$

in probability as $\varepsilon \to 0$. □

Consider the Lebesgue measure from (3.38) as a function of ε and T

$$\Lambda(\varepsilon, T) = \operatorname{meas}\left\{t \in [0,1] \colon Y^{\varepsilon,T}_{Tt} \neq \phi(t)\right\}.$$

Note that $\Lambda(\varepsilon, T)$ is a random variable. We introduce the *out-of-phase measure* by

$$d(\varepsilon, T) = \mathbb{E}_\nu \Lambda(\varepsilon, T) = \mathbb{E}_\nu \int_0^1 \mathbb{I}(Y_{Ts}^{\varepsilon,T} \neq \phi(s))\, ds.$$

The out-of-phase measure describes how much time on average is spent by the Markov chain in the 'wrong' state.

PROPOSITION 3.13. a) The out-of-phase measure is given by

$$d(\varepsilon, T) = \frac{1}{\varphi + \psi}\left[\varphi - \frac{2}{T}\frac{\varphi - \psi}{\varphi + \psi}\tanh\frac{T(\varphi+\psi)}{4}\right]. \tag{3.39}$$

b) For T large enough, $\varepsilon \mapsto d(\varepsilon, T)$ has a local minimum. The optimal tuning rate, at which this minimum is attained is given by

$$T_d(\varepsilon) = \frac{2}{p}\frac{v}{V - v}e^{\frac{V}{\varepsilon}}\left\{1 + \mathcal{O}(e^{-\frac{V-v}{\varepsilon}})\right\} \tag{3.40}$$

and we have

$$d(\varepsilon, T_d(\varepsilon)) \to 0 \quad \text{as} \quad \varepsilon \to 0. \tag{3.41}$$

PROOF. a) To obtain (3.39) we notice that

$$\begin{aligned} d(\varepsilon, T) &= \mathbb{E}_\nu \int_0^1 \mathbb{I}(Y_{Ts}^{\varepsilon,T} \neq \phi(s))\, ds \\ &= \int_0^{\frac{1}{2}} \mathbb{P}_\nu(Y_{Ts}^{\varepsilon,T} = 1)\, ds + \int_{\frac{1}{2}}^1 \mathbb{P}_\nu(Y_{Ts}^{\varepsilon,T} = -1)\, ds \\ &= 2\int_0^{\frac{1}{2}} \nu_{\varepsilon,T}^+(s)\, ds = \frac{1}{\varphi+\psi}\left[\varphi - \frac{2}{T}\frac{\varphi-\psi}{\varphi+\psi}\tanh\left(\frac{T(\varphi+\psi)}{4}\right)\right]. \end{aligned}$$

b) After substituting $t = e^{-\frac{V}{\varepsilon}}$ into (3.39) and taking the derivative in t we find that the coordinates of extrema satisfy the equation $F_T(t) = 0$, where

$$\begin{aligned} F_T(t) =& T(q\beta t^\beta + pt)(q^2t^{2\beta} - p^2t^2)\operatorname{sech}^2\left[\tfrac{T}{4}(qt^\beta + pt)\right] \\ &+ 2Tpq(1-\beta)t^{1+\beta}(qt^\beta + pt) \\ &+ 4(p^2t^2 - q^2\beta t^{2\beta} - 3pq(1-\beta)t^{1+\beta})\tanh\left[\tfrac{T}{4}(qt^\beta + pt)\right], \end{aligned}$$

where $0 < \beta = \frac{v}{V} < 1$. Let us again localize the values of T such that $F_T(t) = 0$ has a root as $t \to 0$. We parametrize $T_t = \frac{a}{t}$, $a > 0$, and consider $F_{T_t}(t)$ as $t \to 0$. It is clear that $\operatorname{sech}\left[\frac{T_t}{4}(qt^\beta + pt)\right] = \operatorname{sech}\left[\frac{a}{4}(qt^{\beta-1} + p)\right] \to 0$ exponentially fast, whereas $\tanh\left[\frac{T_t}{4}(qt^\beta + pt)\right] \to 1$ exponentially fast as $t \to 0$. Hence, $F_{T_t}(t)$ can be asymptotically expanded as

$$F_{T_t}(t) = 2q^2t^{2\beta}(ap(1-\beta) - 2\beta) + \mathcal{O}(t^{1+\beta}).$$

Observe that $F_{T_t}(t) < 0$ if $a < \frac{2\beta}{p(1-\beta)}$, and $F_{T_t}(t) > 0$ if $a > \frac{2\beta}{p(1-\beta)}$ as $t \to 0$. In other words, as $t \to 0$, the half-period T we look for belongs to the interval $[\frac{\beta}{p(1-\beta)t}, \frac{3\beta}{p(1-\beta)t}]$. On this interval we can asymptotically solve the equation $F_T(t) = 0$ for T which results in

$$T = \frac{2\beta}{p(1-\beta)}t^{-1}(1 + \mathcal{O}(t^{1-\beta})) \quad \text{as} \quad t \to 0.$$

This is equivalent to (3.40).

The limit (3.41) is obtained directly from (3.39) and (3.40). □

REMARK 3.14. The out-of-phase measure can be rewritten as

$$d(\varepsilon, T) = \frac{1}{4}\int_0^1 \mathbb{E}_\nu\left(Y_{Ts}^{\varepsilon,T} - \phi(s)\right)^2 ds,$$

which represents the mean square deviation of $Y^{\varepsilon,T}$ from the deterministic function ϕ appearing in Proposition 3.12.

3.2.6. Relative entropy. Let us consider the point mass $\delta_{\phi(t)}$, $t \geq 0$, in $\phi(t)$ with ϕ according to the preceding section. We have $\delta_{\phi(t)} = \phi^-(t)\delta_{-1} + \phi^+(t)\delta_1$, where $\phi^-(t) = 1$, $\phi^+(t) = 0$ if $t \bmod 1 \in [0, \frac{1}{2})$ and $\phi^-(t) = 0$, $\phi^+(t) = 1$ if $t \bmod 1 \in [\frac{1}{2}, 1)$. In the language of diffusion, minimum of the deep well always has mass 1.

We consider the *relative entropy* of the invariant measure $\nu_{\varepsilon,T}$ with respect to δ_ϕ defined by

$$H_{\phi|\nu}(\varepsilon, T) = \int_0^1 \left(\phi^-(s)\ln\frac{\phi^-(s)}{\nu^-_{\varepsilon,T}(s)} + \phi^+(s)\ln\frac{\phi^+(s)}{\nu^+_{\varepsilon,T}(s)}\right) ds.$$

PROPOSITION 3.15. a) The relative entropy is given by

$$\begin{aligned} H_{\phi|\nu}(\varepsilon, T) = \frac{2}{T(\varphi+\psi)}\Bigg(&\ln\left[\frac{\psi-\varphi}{(1+\xi)\,\psi}\right]\ln\left[\frac{\varphi+\xi\psi}{(1+\xi)(\varphi+\psi)}\right] \\ &- \ln\left[\frac{\psi-\varphi}{(1+\xi^{-1})\,\psi}\right]\ln\left[\frac{\xi\varphi+\psi}{(1+\xi)(\varphi+\psi)}\right] \\ &+ \mathrm{Li}_2\left[\frac{\varphi+\xi\psi}{(1+\xi)\,\psi}\right] - \mathrm{Li}_2\left[\frac{\xi\varphi+\psi}{(1+\xi)\,\psi}\right]\Bigg), \\ \xi = e^{\frac{1}{2}T(\varphi+\psi)}&, \end{aligned} \tag{3.42}$$

where $\mathrm{Li}_2(\cdot)$ is the dilogarithm function defined by $\mathrm{Li}_2(x) = \int_x^0 \frac{\ln(1-y)}{y}\,dy$, $x \leq 1$.

b) For T large enough, $\varepsilon \mapsto H_{\phi|\nu}(\varepsilon, T)$ has a local minimum, and an optimal tuning rate is given by

$$T_{\phi|\nu}(\varepsilon) = \frac{\pi^2}{3p}\frac{v}{V-v}e^{\frac{V}{\varepsilon}}\left\{1 + \mathcal{O}(\varepsilon^{-1}e^{-\frac{V-v}{\varepsilon}})\right\}. \tag{3.43}$$

Moreover,

$$H_{\phi|\nu}(\varepsilon, T_{\phi|\nu}(\varepsilon)) \to 0 \quad \text{as} \quad \varepsilon \to 0. \tag{3.44}$$

PROOF. a) To obtain (3.42) we use (3.20) and the definition of the relative entropy. This gives

$$
\begin{aligned}
H_{\phi|\nu}(\varepsilon,T) &= \int_0^1 \sum_{\alpha=+,-} \phi^\alpha(s) \ln \frac{\phi^\alpha(s)}{\nu^\alpha_{\varepsilon,T}(s)}\, ds \\
&= \int_0^{\frac{1}{2}} \ln \frac{1}{\nu^-_{\varepsilon,T}(s)}\, ds + \int_{\frac{1}{2}}^1 \ln \frac{1}{\nu^+_{\varepsilon,T}(s)}\, ds \\
&= -2 \int_0^{\frac{1}{2}} \ln \nu^-_{\varepsilon,T}(s)\, ds \\
&= -2 \int_0^{\frac{1}{2}} \ln \left(\frac{\psi}{\varphi+\psi} + \frac{\varphi-\psi}{\varphi+\psi} \frac{e^{-(\varphi+\psi)Ts}}{1+e^{-\frac{1}{2}(\varphi+\psi)T}} \right) ds.
\end{aligned}
$$

The latter integral is evaluated by means of the formula:

$$
\begin{aligned}
\int \ln\left(a+be^{-2cs}\right) ds = &-\frac{1}{2c} \ln\left(a+be^{-2cs}\right) \ln\left(-\frac{be^{-2cs}}{a}\right) \\
&- \frac{1}{2c} \mathrm{Li}_2\left(1+\frac{be^{-2cs}}{a}\right).
\end{aligned}
$$

b) As usual, we denote $t = e^{-\frac{V}{\varepsilon}}$ and $\beta = \frac{v}{V}$ to find the derivative of $H_{\phi|\nu}$ with respect to t. This results in an algebraic equation $F_T(t) = 0$ for the extrema of $H_{\phi|\nu}$. For the sake of brevity we omit this equation here. To localize the zero of $F_T(t) = 0$ for $t \to 0$ we us parametrize $T_t = \frac{a}{t}$ for some $a > 0$ and consider F_{T_t} as $t \to 0$. We exploit the fact that as $t \to 0$, $\exp\left((pt+qt^\beta)T_t\right) \to \infty$ exponentially fast, and note that $\mathrm{Li}_2(1) = -\int_0^1 \frac{\ln(1-y)}{y}\, dy = \sum_{k=1}^\infty \frac{x^k}{k^2} = \frac{\pi^2}{6}$.

The terms in the equation $F_{T_t}(t) = 0$ which have a factor $\exp\left\{\frac{1}{2}(pt+qt^\beta)T_t\right\}$ will be dominating as $t \to 0$, and therefore the following expansion can be obtained:

$$
F_{T_t}(t) = e^{\frac{a}{2}(p+qt^{\beta-1})} q^2 t^{2\beta} \left(ap(1-\beta) - \frac{\pi^2}{3}\beta + \mathcal{O}(t^{1-\beta}\ln t) \right).
$$

Therefore, $F_{T_t}(t) < 0$ if $a < \frac{\pi^2\beta}{3p(1-\beta)}$, and $F_{T_t}(t) > 0$ if $a > \frac{\pi^2\beta}{3p(1-\beta)}$, as $t \to 0$. In other words, we have localized the optimal period T on the interval $\left[\frac{\pi^2\beta}{4p(1-\beta)t}, \frac{\pi^2\beta}{2p(1-\beta)t}\right]$ as $t \to 0$. In this interval we asymptotically solve the equation $F_T(t) = 0$ for T, which gives

$$
T = \frac{\pi^2\beta}{3p(1-\beta)} t^{-1}\left(1+\mathcal{O}(t^{1-\beta}\ln t)\right), \quad t \to 0.
$$

After rewriting it in terms of ε, the latter formula gives (3.43). The limit (3.44) follows from (3.42) and (3.43). □

3.2.7. Entropy. Define the *entropy* of the invariant measure $\nu^\pm_{\varepsilon,T}$ at time $t \in [0,1]$ as

$$
H_t(\varepsilon,T) = -\nu^-_{\varepsilon,T}(t)\ln\nu^-_{\varepsilon,T}(t) - \nu^+_{\varepsilon,T}(t)\ln\nu^+_{\varepsilon,T}(t), \quad t \in [0,1],
$$

and consider its *entire entropy* defined by

$$
H(\varepsilon,T) = \int_0^1 H_s(\varepsilon,T)\, ds.
$$

PROPOSITION 3.16. a) The entire entropy of $\nu_{\varepsilon,T}$ is given by

(3.45)

$$H(\varepsilon,T) = \ln(\varphi+\psi) - \frac{\varphi\ln\varphi+\psi\ln\psi}{\varphi+\psi}$$
$$+\frac{2\varphi}{T(\varphi+\psi)^2}\left[\mathrm{Li}_2\left(\frac{\varphi-\psi}{\varphi}\frac{1}{1+e^{-\frac{1}{2}T(\varphi+\psi)}}\right)-\mathrm{Li}_2\left(\frac{\varphi-\psi}{\varphi}\frac{e^{-\frac{1}{2}T(\varphi+\psi)}}{1+e^{-\frac{1}{2}T(\varphi+\psi)}}\right)\right]$$
$$+\frac{2\psi}{T(\varphi+\psi)^2}\left[\mathrm{Li}_2\left(\frac{\psi-\varphi}{\psi}\frac{1}{1+e^{-\frac{1}{2}T(\varphi+\psi)}}\right)-\mathrm{Li}_2\left(\frac{\psi-\varphi}{\psi}\frac{e^{-\frac{1}{2}T(\varphi+\psi)}}{1+e^{-\frac{1}{2}T(\varphi+\psi)}}\right)\right],$$

where $\mathrm{Li}_2(\cdot)$ is the dilogarithm function, $\mathrm{Li}_2(x) = \int_x^0 \frac{\ln(1-y)}{y}\,dy$, $x \le 1$.

b) For T large enough $H(\varepsilon,T)$ always has a local minimum, and an optimal tuning rate is given by

$$T_H(\varepsilon) = \frac{\pi^2}{3p}\frac{v\varepsilon}{(V-v)^2}e^{\frac{V}{\varepsilon}}\Big\{1+\mathcal{O}(\varepsilon)\Big\}. \tag{3.46}$$

Moreover,

$$H(\varepsilon, T_H(\varepsilon)) \to 0 \quad \text{as} \quad \varepsilon \to 0. \tag{3.47}$$

PROOF. a) To obtain (3.45) we use (3.20) and the integration formula

$$\int (a+be^{-2cs})\ln(a+be^{-2cs})\,ds = \frac{b}{2c}e^{-2cs}\left(1-\ln(a+be^{-2cs})\right)$$
$$+ as\ln a - \frac{a}{2c}\ln(a+be^{-2cs}) + \frac{a}{2c}\mathrm{Li}_2\left(-\frac{be^{-2cs}}{a}\right),$$
$$a>0, b>-a, c>0.$$

b) We denote as usual $t = e^{-\frac{V}{\varepsilon}}$ and $\beta = \frac{v}{V}$ so that $\varphi = pt$ and $\psi = qt^\beta$ and take the derivative of H in t to determine the extrema. The straightforward and tedious calculation yields that they are given by the roots of the equation $F_T(t) = 0$, with

$$F_T(t) = e^{\frac{1}{2}(pt+qt^\beta)T}\xi_2(t,T) + e^{\frac{1}{4}(pt+qt^\beta)T}\xi_1(t,T) + \xi_0(t,T),$$

where $\xi_2(t,T)$, $\xi_1(t,T)$ and $\xi_0(t,T)$ are combinations of powers of t and logarithmic and dilogarithmic functions. The functions ξ_0, ξ_1 and ξ_2 vanish if $pt = qt^\beta$. This yields the local maximum of the entropy at $\widehat{\varepsilon} = (V-v)/\ln(\frac{p}{q})$. Note that at $\widehat{\varepsilon}$ the infinitesimal probabilities φ and ψ are equal and it follows from (3.45) that $H(\widehat{\varepsilon},T) = \ln 2$ for all $T > 0$.

We shall determine the root of $F_T(t) = 0$ for small t. Let us parametrize $T_t = \frac{a}{t|\ln t|}$ for $a > 0$ and use the fact that as $t \to 0$, $\exp((pt+qt^\beta)T_t) \to \infty$ exponentially fast. Collecting the leading terms in ξ_2 then results in

$$F_{T_t}(t) = e^{a(p+qt^{\beta-1})|\ln t|^{-1}}q^3t^{3\beta}\left(\frac{\pi^2}{3}\beta - ap(1-\beta)^2 + \mathcal{O}(|\ln t|^{-1})\right)$$

Therefore, $F_{T_t} < 0$ for $a > \frac{\pi^2\beta}{3p(1-\beta)^2}$ and $F_{T_t} > 0$ for $a < \frac{\pi^2\beta}{3p(1-\beta)^2}$ as $t \to 0$. Hence we have localized the local minimum of the entropy, for example on the interval $[\frac{\pi^2\beta}{4p(1-\beta)^2t|\ln t|}, \frac{\pi^2\beta}{2p(1-\beta)^2t|\ln t|}]$, as $t \to 0$. Solving $F_T(t) = 0$ for T on this interval gives the asymptotic formula

$$T = \frac{\pi^2 \beta}{3p(1-\beta)^2}(t|\ln t|)^{-1}\left(1 + \mathcal{O}(|\ln t|^{-1})\right).$$

In terms of ε this is equivalent to (3.46). The limit (3.47) is an application of (3.46) to (3.45). □

3.2.8. Overview of measures of goodness. For the convenience of the reader we represent the optimal tuning rates for the quality measures studied in this Section in the following table (see also Figure 3.1).

	Measure of quality of stochastic resonance	Optimal tuning $T(\varepsilon)$ as $\varepsilon \to 0$
1.	SPA coefficient, η^Y	$T_\eta(\varepsilon) \approx \pi\sqrt{\frac{2}{pq}}\sqrt{\frac{v}{V-v}}e^{\frac{V+v}{2\varepsilon}}$
2.	SPA-to-noise ratio, SPN	$T_{\mathrm{SPN}}(\varepsilon) \approx \frac{2\pi\sqrt{v}}{q\sqrt{\varepsilon}}e^{\frac{v}{\varepsilon}}$
3.	Energy, En	$T_{\mathrm{En}}(\varepsilon) \approx \frac{1}{p}\frac{v}{V-v}e^{\frac{V}{\varepsilon}}$
4.	Out-of-phase, d	$T_d(\varepsilon) \approx \frac{2}{p}\frac{v}{V-v}e^{\frac{V}{\varepsilon}}$
5.	Relative entropy, $H_{\phi\|\nu}$	$T_{H_{\phi\|\nu}}(\varepsilon) \approx \frac{\pi^2}{3p}\frac{v}{V-v}e^{\frac{V}{\varepsilon}}$
6.	Entropy, H	$T_H(\varepsilon) \approx \frac{\pi^2}{3p}\frac{\varepsilon v}{(V-v)^2}e^{\frac{V}{\varepsilon}}$

Theorem 3.6 provides a lower bound for the exponential rate $T(\varepsilon)$ above which in the small noise limit $\varepsilon \to 0$ the randomly perturbed system produces trajectories with periodicity properties. Our results for the Markov chain obtained in this chapter all agree with this bound and determine the exact rates of growth for $T(\varepsilon)$ as $\varepsilon \to 0$ together with pre-factors.

It is interesting to remark that if the exponential rate is v, i.e. corresponds to the depth of the shallow well, then the pre-factors contain only the parameter q, which stands for the geometry of the shallow well. Analogously, if $T(\varepsilon) \sim e^{V/\varepsilon}$, then the pre-factor of the optimal tuning rate only contains p, the geometrical factor of the deep well, and not q. The only exception is the SPA coefficient η^Y whose optimal tuning rate contains all parameters of the system.

The measures of quality considered can be subdivided into two groups. The SPA coefficient, the SPA-to-noise ratio and the energy belong to the first group. They are based on the interpretation of the Markov chain as a random amplifier and describe the spectral properties of the averaged output. The key result which may seem counterintuitive is as follows: an optimal transfer of a deterministic periodic signal through a random system is not guaranteed by the elimination of noise, but rather by tuning it in on some essentially non-zero intensity level.

The same result holds for the second group of measures which can be considered as measures of stabilization. Indeed, the out-of-phase measure and the relative entropy determine the deviation of the random output from a deterministic function

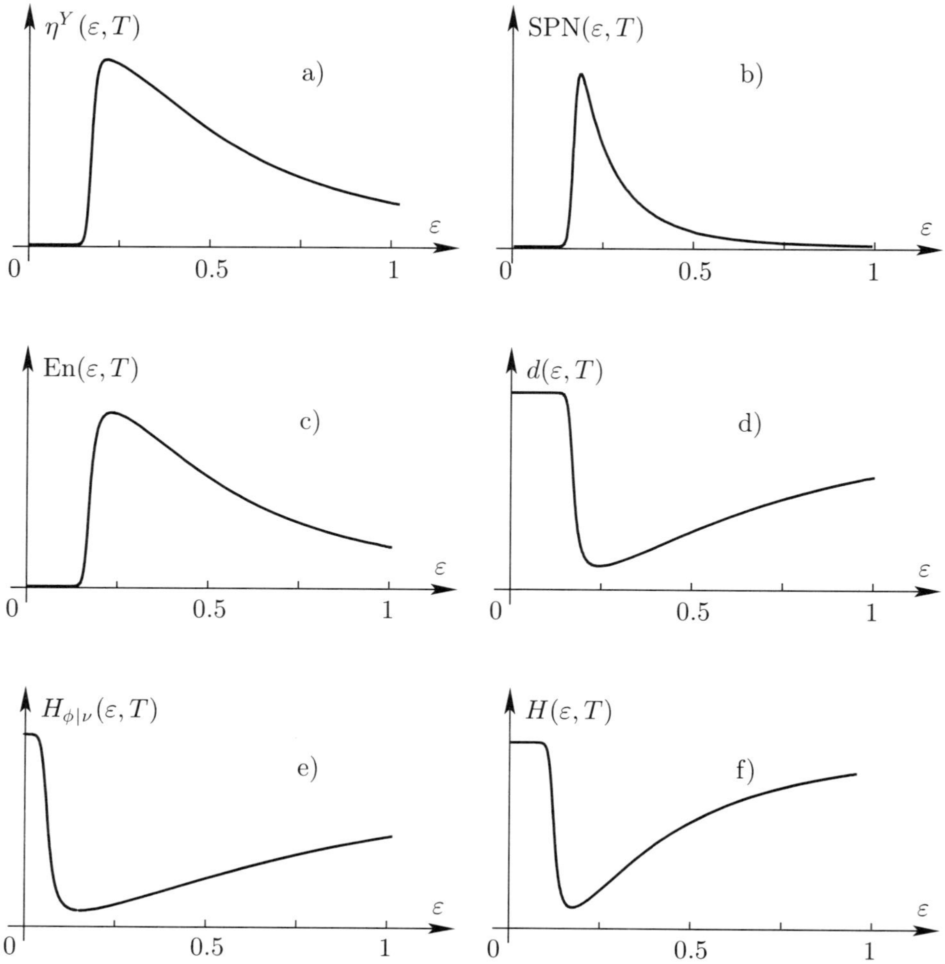

FIGURE 3.1. Measures of quality of stochastic resonance of the Markov chain Y for the parameters $p = q = 1$, $V = 3$, $v = 2$ and $T = 10^6$: a) the SPA coefficient η^Y; b) the SPA-to-noise ratio; c) the energy; d) the out-of-phase measure; e) the relative entropy; f) the entropy.

ϕ. Again, *increasing* the noise makes the random output *less random*. The final measure — the entropy of the invariant measure of the process — is, roughly speaking, the measure of randomness. The bigger the entropy, the more chaotic is the system. The fact that non-zero noise minimizes the entropy is a very good example of noise-induced order.

Let us next briefly discuss the particular noise level $\widehat{\varepsilon} = (V - v)/\ln\left(\frac{p}{q}\right)$. At this level, the infinitesimal probabilities φ and ψ are equal, the generators Q_1 and Q_2 are indistinguishable from each other for all $t, T > 0$, and the process $Y^{\widehat{\varepsilon}, T}$ is time-homogeneous with invariant measure $\left(\frac{1}{2}, \frac{1}{2}\right)$. In other words, at $\widehat{\varepsilon}$ the Markov chain is a symmetric telegraph process which lives in -1 and 1 with equal probability. No periodicity is observed.

The out-of-phase measure and the relative entropy, of course, do not vanish at $\widehat{\varepsilon}$, but are far from their maxima. Indeed, the process which spends only half of the time in the 'right' state does not constitute the worst case for these measures. The worst case is rather the deterministic process $1-\phi(t)$ which is in the 'wrong' place for all t. The entropy of the invariant measure takes its maximum at $\widehat{\varepsilon}$ because for any t the random variable $Y_t^{\widehat{\varepsilon},T}$ has a symmetric Bernoulli law which is the most 'chaotic' among binary laws.

3.3. Spectral analysis of the infinitesimal generator of small noise diffusion

In this chapter we study spectral properties of the infinitesimal generator of a one-dimensional diffusion with small noise level in a double-well potential. The first section is designed to motivate our approach. In particular, we give heuristic derivations of the first eigenvalues and eigenfunctions. Moreover, we employ the theory of Schrödinger operators for potentials increasing at infinity to prove the discreteness of the spectrum and the existence of a complete system of eigenfunctions. Variational principles are evoked for the study of the eigenvalues and eigenfunctions. Eigenvalue and eigenfunction of order zero are easy to derive exactly, while the first eigenvalue λ_1 is determined approximately. It is known to be exponentially small in ε. To find a good estimate of the first eigenfunction Φ_1, we use a series representation of Φ_1 in powers of λ_1. In fact, we show that there is a spectral gap between λ_1 and a second eigenvalue λ_2: the second eigenvalue λ_2 is bounded from below by some positive constant not depending on ε. In the last section we briefly discuss the multi-dimensional case. The exposition of this Section follows Pavlyukevich [**86**] and Imkeller and Pavlyukevich [**59**].

3.3.1. The heuristics. In this section we present the heuristic derivation of the first eigenvalue and eigenfunction of the differential operator

$$L_\varepsilon = \frac{\varepsilon}{2}\frac{d^2}{dx^2} - U'\frac{d}{dx} \tag{3.48}$$

which is the infinitesimal generator of a time-homogeneous diffusion driven by the SDE

$$dX_t^\varepsilon = -U'(X_t^\varepsilon)\,dt + \sqrt{\varepsilon}\,dW_t, \quad t \geq 0, \varepsilon > 0, \tag{3.49}$$

with a purely space dependent potential function U.

For convenience of notation, in this and the next chapter we fix and omit the subscript ε in the notation of the operators L_ε, L_ε^* etc. and their spectral data.

The results about the spectrum of the operator L we shall derive are needed in Chapter 3.4, where we have to solve the forward Kolmogorov (Fokker–Planck) equation for the invariant density of the diffusion (3.2) with time-periodic potential. The equation to solve is given by

$$\frac{1}{2T}\frac{\partial}{\partial t} f(x,t) = L^* f(x,t), \tag{3.50}$$

where

$$L^* f = \frac{\varepsilon}{2}\frac{d^2}{dx^2} f + U'\frac{d}{dx} f + U'' f$$

is the formal adjoint of L, i.e. a differential operator such that for all $f, g \in C_0^\infty(\mathbb{R}, \mathbb{R})$

$$\int_{\mathbb{R}} Lf \cdot g\, dx = \int_{\mathbb{R}} f \cdot L^* g\, dx$$

and U a temporally periodic potential function switching between two spatially symmetric states, recall Figure 1.10. Thus, equation (3.50) is time-homogeneous on half-intervals. To obtain its solution for all t we will continuously knit the solutions corresponding to time-independent states of the potential. In the time-independent setting of the present section, U is supposed to take one of these two states at all times.

Assume that we know that the spectrum of the operator L^* is discrete, and let us denote $\{-\lambda_k, \Psi_k\}_{k\geq 0}$ its eigenvalues and eigenfunctions. We assume that the spectrum is non-positive, i.e. $0 \leq \lambda_0 < \lambda_1 < \cdots$, and has no finite condensation point. If such a spectrum exists the method of separation of variables (or the Fourier method) can be applied and a solution of (3.50) can be given in the form of

$$f(x,t) = \sum_{k=0}^{\infty} a_k \Psi_k(x) e^{-2T\lambda_k t}, \quad x \in \mathbb{R}, t \geq 0,$$

where the real numbers a_k are determined from the initial and boundary conditions (see, e.g. [**34, 92**]).

Thus, we face the problem of describing the spectrum of the time-independent operator L^*, at least for small ε. It turns out that it can be solved by studying the spectral properties of L. Indeed, one easily notes that

$$L^* f = e^{-\frac{2U}{\varepsilon}} L(f e^{\frac{2U}{\varepsilon}}).$$

This equality means that L^* and L have the same eigenvalues, and the eigenfunctions of the operator L^* can be obtained as $\Psi_k = e^{-\frac{2U}{\varepsilon}} \Phi_k$, where Φ_k are the eigenfunctions of the diffusion generator L, $k \geq 0$.

Spectral analysis for generators of small noise diffusions was addressed in a big number of papers. Diffusions in bounded domains are particularly well understood [**19, 18, 22, 23, 24, 40, 56, 57, 110**]. In a nutshell, the essential result states that in the small noise limit the lowest eigenvalue corresponds to the inverse of the expected diffusion exit time from the bounded domain. The corresponding eigenfunction is almost constant on any compact subset of the domain.

Our state space in contrast is the non-compact real line. Here the situation is somewhat more complicated.

We start with the following basic assumptions about the potential U:

(**S**) $U \in C^3(\mathbb{R}, \mathbb{R})$;

(**G**) there exists $R > 0$ such that $U(x) = x^4/4$ for $|x| \geq R$;

(**M**) U has exactly two local minima at $x = \pm 1$ and one local maximum at $x = 0$; moreover,

$$U(-1) = -\frac{V}{2}, \quad U(0) = 0, \quad U(1) = -\frac{v}{2},$$
$$\frac{1+\varkappa}{2} < \frac{v}{V} < 1; \text{ for some } \varkappa > 0;$$

the extrema are non-degenerate, i.e.

$$U''(\pm 1) = \omega_\pm > 0, \quad U''(0) = -\omega_0 < 0.$$

The condition $\frac{1+\varkappa}{2} < \frac{v}{V} < 1$ is a rather weak condition. It allows the depths of the two potential wells to be arbitrarily close and just requires the depths not to be too different. This agrees with the interpretation of stochastic resonance as optimal amplification of weak periodic signals by noise. Here 'weakness' means small amplitude of periodic signal which in our case corresponds to $\frac{V-v}{2}$ (see Section 3.2).

The operator L can be extended to a self-adjoint operator on the domain $\mathcal{D}_L$ consisting of functions from $L^2(\mathbb{R},\mathbb{R};\rho^{-1}dx)$, $\rho = e^{\frac{2U}{\varepsilon}}$ with locally absolutely continuous first derivative and square-integrable second derivative. Self-adjointness and the fast increase of the potential U at infinity imply the existence of the discrete non-positive spectrum and an orthogonal system of eigenfunctions lying in $L^2(\mathbb{R},\mathbb{R};\rho^{-1}dx)$.

The zeroth eigenvalue can be found directly from the variational principle [**55**]:

$$-\lambda_0 = \inf_{0\neq\varphi\in\mathcal{D}_L} \frac{-\int_{\mathbb{R}} \varphi L\varphi e^{-\frac{2U}{\varepsilon}}\,dx}{\int_{\mathbb{R}} \varphi^2 e^{-\frac{2U}{\varepsilon}}\,dx} = \inf_{0\neq\varphi\in\mathcal{D}_L} \frac{\varepsilon}{2}\frac{\int_{\mathbb{R}}(\varphi')^2 e^{-\frac{2U}{\varepsilon}}\,dx}{\int_{\mathbb{R}} \varphi^2 e^{-\frac{2U}{\varepsilon}}\,dx} = 0.$$

To verify the latter formula it is enough to take $\varphi = \text{Const}$. Thus, the zeroth eigenfunction Φ_0 can be chosen to be identically equal to 1.

To determine the first eigenfunction Φ_1 and eigenvalue λ_1 for small values of ε we use the intuitive ideas displayed in [**96, 71**].

The only a priori knowledge about λ_1 we use follows from [**61**] or the Freidlin–Wentzell theory [**40**]. It states that the first eigenvalue is exponentially small in ε and $\lim_{\varepsilon\to 0}\varepsilon\ln\lambda_1 = -v$.

Let us next consider the eigenvalue problem for the first eigenvalue λ_1

$$\frac{\varepsilon}{2}\varphi''(x) - U'(x)\varphi'(x) = -\lambda_1\varphi(x), \quad x\in\mathbb{R} \tag{3.51}$$

for $\varepsilon\to 0$.

The smallness of λ_1 implies that the terms containing φ'' and φ in (3.51) vanish as $\varepsilon \to 0$, and the equation (3.51) reduces to

$$U'\varphi' = 0. \tag{3.52}$$

Equation (3.52) expresses the fact that φ is approximately constant on the intervals of constant sign of U'.

Let $\delta > 0$ be a small number tending to 0 together with ε, and let us study the consequences of the previous equation by determining the constants $c_1,\dots,c_4$ in the following ansatz for an approximation $\Phi_1^{(\delta)}$ of Φ_1:

$$\Phi_1^{(\delta)}(x) = \begin{cases} c_1, & x < -1-\delta,\\ c_2, & -1+\delta < x < \delta,\\ c_3, & \delta < x < 1-\delta,\\ c_4, & x > 1+\delta.\end{cases}$$

If, for example, $c_1 \neq c_2$ then the so-called *internal layer* may exist in the δ-neighbourhood of the point $x=-1$. In this neighbourhood we perform the stretching transformation by introducing the new coordinate

$$\xi = \frac{x+1}{\sqrt{\varepsilon}}, \quad |x+1|\le\delta, \quad \varepsilon > 0.$$

This change of variables gives $\frac{d}{dx}\varphi(x) = \frac{1}{\sqrt{\varepsilon}}\frac{d}{d\xi}\varphi(\xi)$ and $\frac{d^2}{dx^2}\varphi(x) = \frac{1}{\varepsilon}\frac{d^2}{d\xi^2}\varphi(\xi)$. The drift term can be expanded in a Taylor series with respect to x. Since -1 is a critical point, this leads to $U'(x) = \omega_-(x+1) + \mathcal{O}((x+1)^2)$, where $\omega_- > 0$ is the curvature of U in -1. In terms of ξ this gives $U'(\xi) = \omega_-\sqrt{\varepsilon}\xi + \mathcal{O}(\varepsilon)$.

Using this asymptotics in (3.51) together with the exponential smallness of λ_1, we therefore obtain

$$\frac{1}{2}\varphi''(\xi) - \omega_-\xi\varphi'(\xi) = 0 \tag{3.53}$$

and the continuity conditions

$$\begin{aligned} \varphi(-\delta/\sqrt{\varepsilon}) &\to c_1, \\ \varphi(\delta/\sqrt{\varepsilon}) &\to c_2, \quad \text{as} \quad \varepsilon \to 0. \end{aligned}$$

We solve equation (3.53) to obtain the general solution

$$\varphi(\xi) = d_1 \int_0^\xi e^{\omega_- y^2}\, dy + d_2$$

with constants d_1, d_2. The continuity conditions immediately imply that $d_1 = 0$, and $c_1 = d_2 = c_2$. Hence there is no internal layer near -1.

Analogously, one rules out the existence of an internal layer near 1. Denote now the common value of c_1, c_2 by C_1, the common value of c_3, c_4 by C_2.

We may further determine asymptotically C_1, C_2 in the ansatz

$$\Phi_1^{(\delta)}(x) = \begin{cases} C_1, & x < -\delta, \\ C_2, & x > \delta. \end{cases}$$

In a δ-neighbourhood of 0 it follows from (3.51) that the equation for the possible internal layer takes the form

$$\frac{1}{2}\varphi''(\xi) + \omega_0\xi\varphi'(\xi) = 0,$$

where $U''(0) = -\omega_0 < 0$. The continuity conditions become

$$\begin{aligned} \varphi(-\delta/\sqrt{\varepsilon}) &\to C_1, \\ \varphi(\delta/\sqrt{\varepsilon}) &\to C_2, \quad \text{as} \quad \varepsilon \to 0. \end{aligned}$$

Solving this equation gives the general solution

$$\varphi(\xi) = d_1 \int_0^\xi e^{-\omega_0 y^2}\, dy + d_2,$$

with some constants d_1, d_2 to be determined from the continuity conditions. We use the limiting value of the integral

$$\lim_{\xi\to+\infty} \int_0^\xi e^{-\omega_0 y^2}\, dy = \frac{1}{2}\int_{-\infty}^{\infty} e^{-\omega_0 y^2}\, dy = \frac{1}{2}\sqrt{\frac{\pi}{\omega_0}}.$$

Hence the continuity conditions give

$$\begin{aligned} C_1 &= -\frac{1}{2}\sqrt{\frac{\pi}{\omega_0}}d_1 + d_2, \\ C_2 &= \frac{1}{2}\sqrt{\frac{\pi}{\omega_0}}d_1 + d_2. \end{aligned}$$

This results in the following formula for the internal layer near 0:

$$\begin{aligned}\varphi_\varepsilon(x) &= \sqrt{\frac{\omega_0}{\pi}}(C_2 - C_1)\int_0^{x/\sqrt{\varepsilon}} e^{-\omega_0 y^2}\,dy + \frac{C_1 + C_2}{2}\\ &= \sqrt{\frac{\omega_0}{\pi}}(C_2 - C_1)\int_0^{x} e^{-\frac{\omega_0 y^2}{\varepsilon}}\,dy + \frac{C_1 + C_2}{2}.\end{aligned}$$

Thus, we have determined an approximation of a solution of (3.51) by

$$\Phi_1^{(\delta)}(x) = \begin{cases} C_1, & x < -\delta,\\ \varphi_\varepsilon(x), & x \in [-\delta, \delta],\\ C_2, & x > \delta.\end{cases} \tag{3.54}$$

Let us now specify asymptotic values for the constants C_1 and C_2. First, we normalise $\Phi_1^{(\delta)}$ by putting $C_2 = 1$. Second, we use the orthogonality condition

$$\langle \Phi_0, \Phi_1^{(\delta)}\rangle_{L^2(\mathbb{R},\mathbb{R};\rho^{-1}dx)} = 0,$$

which is equivalent to

$$\int_{\mathbb{R}} \Phi_1^{(\delta)}(y) e^{-\frac{2U(y)}{\varepsilon}}\,dy = 0.$$

Let us rewrite the latter equation as

$$C_1\int_{-\infty}^{-\delta} e^{-\frac{2U(y)}{\varepsilon}}\,dy + \int_{-\delta}^{\delta} \varphi_\varepsilon(y) e^{-\frac{2U(y)}{\varepsilon}} + \int_{\delta}^{\infty} e^{-\frac{2U(y)}{\varepsilon}}\,dy = 0. \tag{3.55}$$

Using Laplace's method (see Appendix B) we asymptotically evaluate the first and the third summands in the previous formula. The second summand can be estimated by

$$\left|\int_{-\delta}^{\delta} \varphi_\varepsilon(y) e^{-\frac{2U(y)}{\varepsilon}}\right| \le 2\delta(|C_1| + 1) e^{-\frac{2}{\varepsilon}\min_{[-\delta,\delta]} U} = \mathcal{O}(e^{\frac{\delta'}{\varepsilon}})$$

for some $\delta' > 0$, which can be made arbitrary small by reducing δ. Equation (3.55) takes the form

$$C_1\sqrt{\frac{\pi\varepsilon}{\omega_-}} e^{\frac{V}{\varepsilon}}(1 + \mathcal{O}(\varepsilon)) + \mathcal{O}(e^{\frac{\delta'}{\varepsilon}}) + \sqrt{\frac{\pi\varepsilon}{\omega_+}} e^{\frac{v}{\varepsilon}}(1 + \mathcal{O}(\varepsilon)) = 0.$$

This leads to the following value of C_1

$$C_1 \approx -\sqrt{\frac{\omega_-}{\omega_+}} e^{-\frac{V - v}{\varepsilon}}.$$

So finally we obtained some heuristic knowledge of asymptotic properties of the first eigenfunction Φ_1 in the small noise limit (see Figure 3.2). It is constant and exponentially small in ε for negative x, equals 1 for positive x, and there exists an internal layer which makes Φ_1 smooth on the whole axis. Note that our arguments to this point just used the information that λ_1 is exponentially small. We can continue along these lines to heuristically deduce its pre-exponential order, using our knowledge of Φ_1. We apply the variational principle to obtain

$$\lambda_1 = \frac{\varepsilon}{2}\frac{\int_{\mathbb{R}}(\Phi_1'(y))^2 e^{-\frac{2U(y)}{\varepsilon}}\,dy}{\int_{\mathbb{R}}(\Phi_1(y))^2 e^{-\frac{2U(y)}{\varepsilon}}\,dy}.$$

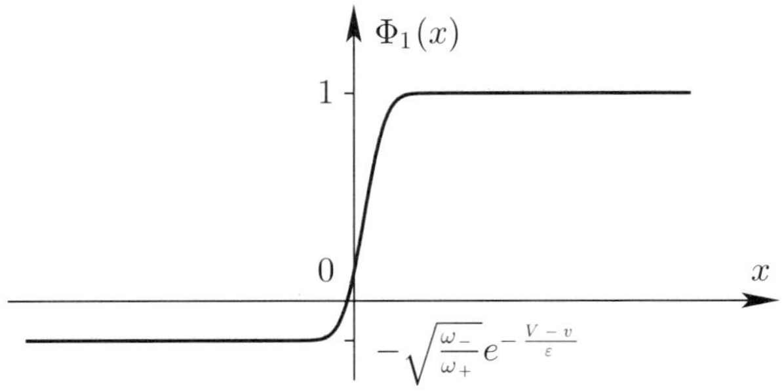

FIGURE 3.2. The first eigenfunction Φ_1 of the operator L_ε for small values of ε.

A straightforward application of Laplace's method gives

$$\int_{\mathbb{R}} (\Phi_1'(y))^2 e^{-\frac{2U(y)}{\varepsilon}}\, dy = \frac{\omega_0}{\pi}(1-C_1)^2 \int_{-\delta}^{\delta} e^{-\frac{\omega_0 y^2}{\varepsilon}}\, dy \approx \frac{\omega_0}{\pi}\sqrt{\frac{\pi\varepsilon}{\omega_0}} = \sqrt{\frac{\omega_0\varepsilon}{\pi}},$$

$$\int_{\mathbb{R}} (\Phi_1(y))^2 e^{-\frac{2U(y)}{\varepsilon}}\, dy \approx \sqrt{\frac{\pi\varepsilon}{\omega_+}} e^{\frac{v}{\varepsilon}},$$

and therefore finally

$$\lambda_1 \approx \frac{\sqrt{\omega_0\omega_+}}{2\pi} e^{-\frac{v}{\varepsilon}}.$$

This result is in agreement with the Freidlin–Wentzell theory and Theorem 3.24 by S. Jacquot as far as the exponential order of the related quantities is concerned. Furthermore, we have also found a constant pre-factor. This result reminds us of Kramers' rate [**65**] describing the probability with which a Brownian particle overcomes a potential barrier of height $\frac{v}{2}$.

The argument presented in this section is not rigorous. The existence of the discrete spectrum of L is assumed and not proved. The passage from (3.51) to (3.52) and (3.53) is mathematically not sound. The use of internal layers near critical points also has to be made precise. The mathematical justification of these qualitatively correct results will be the main task of the following sections.

3.3.2. A Schrödinger operator and the discreteness of its spectrum. On the domain $\mathcal{D}_0 = C_0^\infty(\mathbb{R},\mathbb{R})$ we consider the operator $(L,\mathcal{D}_0)$, where L is the differential operator of second order

$$L = \frac{\varepsilon}{2}\frac{d^2}{dx^2} - U'(x)\frac{d}{dx}. \tag{3.56}$$

The double-well potential U is supposed to fulfil the properties (**S**), (**G**) and (**M**) described in the previous section.

As was already mentioned above, the formal adjoint of the operator $(L,\mathcal{D}_0)$ with respect to $L^2(\mathbb{R},\mathbb{R})$ is given by the differential operator

$$L^* = \frac{\varepsilon}{2}\frac{d^2}{dx^2} + U'(x)\frac{d}{dx} + U''(x) \tag{3.57}$$

on the domain $\mathcal{D}_0$. Indeed, for any $f, g \in C_0^\infty(\mathbb{R}, \mathbb{R})$ integration by parts yields

$$\begin{aligned}\langle Lf, g\rangle &= \int_{\mathbb{R}} Lf \cdot g\,dx = \int_{\mathbb{R}} \Big(\frac{\varepsilon}{2} f'' - U'f'\Big) g\,dx \\ &= -\frac{\varepsilon}{2}\int_{\mathbb{R}} f'g'\,dx + \int_{\mathbb{R}} f(U'g' + U''g)\,dx \\ &= \int_{\mathbb{R}} f\Big(\frac{\varepsilon}{2} g'' + U'g' + U''g\Big)\,dx = \int_{\mathbb{R}} f \cdot L^* g\,dx = (f, L^* g).\end{aligned}$$

A simple straightforward calculation shows that these operators are related in the following way

$$Lf = e^{\frac{2U}{\varepsilon}} L^*(f e^{-\frac{2U}{\varepsilon}}), \quad f, g \in \mathcal{D}_0.$$

The operator $(L, \mathcal{D}_0)$ is symmetric in $L^2(\mathbb{R}, \mathbb{R}, e^{-\frac{2U}{\varepsilon}}\,dx)$, the operator $(L^*, \mathcal{D}_0)$ is symmetric in $L^2(\mathbb{R}, \mathbb{R}, e^{\frac{2U}{\varepsilon}}\,dx)$, so they can be closed in these spaces. We denote the closures as $(L, \mathcal{D}_L)$ and $(L^*, \mathcal{D}_{L^*})$. In order to determine the spectral properties of the operators we consider still another operator $(l, \mathcal{D}_0)$ with

$$l = \frac{d^2}{dx^2} - w(x),$$

where the function

$$w = \frac{(U')^2}{\varepsilon^2} - \frac{U''}{\varepsilon} \in C^1(\mathbb{R}, \mathbb{R}). \tag{3.58}$$

The operator $(l, \mathcal{D}_0)$ is a one-dimensional Schrödinger operator. The function w is therefore called Schrödinger potential.

The Schrödinger operator $(l, \mathcal{D}_0)$ is symmetric in $L^2(\mathbb{R}, \mathbb{R})$, i.e.

$$\langle lf, g\rangle = \int_{\mathbb{R}} lf \cdot g\,dx = \int_{\mathbb{R}} f \cdot lg\,dx = \langle f, lg\rangle, \quad f, g \in \mathcal{D}_0.$$

The set $C_0^\infty(\mathbb{R}, \mathbb{R})$ is dense in $L^2(\mathbb{R}, \mathbb{R})$, therefore $(l, \mathcal{D}_0)$ can be closed in $L^2(\mathbb{R}, \mathbb{R})$. We denote the closure as $(l, \mathcal{D}_l)$.

In what follows, for the sake of brevity, we refer to the closures $(l, \mathcal{D}_l)$, $(L, \mathcal{D}_L)$ and $(L^*, \mathcal{D}_{L^*})$ as l, L and L^* respectively.

The Schrödinger operator l is related to L and L^* by the formulae

$$lf = \frac{2}{\varepsilon} e^{-\frac{U}{\varepsilon}} L(f e^{\frac{U}{\varepsilon}}), \tag{3.59}$$

$$lf = \frac{2}{\varepsilon} e^{\frac{U}{\varepsilon}} L^*(f e^{-\frac{U}{\varepsilon}}), \quad f \in \mathcal{D}_l. \tag{3.60}$$

These equations imply that the eigenvalues of l have to be multiplied by $\frac{\varepsilon}{2}$ to obtain those of L or L^*, and the eigenfunctions of L and L^* can be obtained from those of l by multiplying by $e^{\frac{U}{\varepsilon}}$ or $e^{-\frac{U}{\varepsilon}}$ respectively.

The Schrödinger potential w is smooth on the whole line, and

$$w(x) \to \infty, \quad |x| \to \infty. \tag{3.61}$$

More precisely, $w(x) = \frac{x^6}{\varepsilon^2} - \frac{3x^2}{\varepsilon} \approx \frac{x^6}{\varepsilon^2}$, as $|x| \to \infty$.

The problem of determining the spectrum of l is called a *singular Sturm-Liouville problem for the operator with increasing potential.* The theory for these problems is well developed. We formulate the main facts concerning the spectral properties of l following [**68, 7, 55, 34, 33**].

The following theorem by Sears gives a sufficient condition for l to be self-adjoint.

THEOREM 3.17 ([**98, 7**]). *Let w satisfy*

$$w(x) \geq -Q(x), \quad x \in \mathbb{R},$$

where Q is a positive, even and continuous function on $\mathbb{R}$ that is non-decreasing on $[0, \infty)$ and satisfies

$$\int_{-\infty}^{\infty} \frac{dx}{\sqrt{Q(2x)}} = \infty.$$

Then $(l, \mathcal{D}_0)$ is essentially self-adjoint (that is, its closure l is self-adjoint).

The conditions for the Sears theorem obviously hold for the potential w given by (3.58) due to assumption (**G**). Indeed, w is continuous and increases at infinity. So Q can be chosen, for example, to be identical to the constant $Q = |\min_{x\in\mathbb{R}} w(x)|+1$.

Moreover, we can state that the domain D_l consists of all continuous square-integrable functions f with absolutely continuous first derivative on any finite interval, such that $f'' - w(x)f$ is also square-integrable on the whole line. From this also follows that f' is also square-integrable function on $\mathbb{R}$, see [**68**].

The general theory of self-adjoint operators imply that the eigenvalues of l are real and the eigenfunctions corresponding to different eigenvalues are orthogonal in $L^2(\mathbb{R}, \mathbb{R})$.

The *regular case* of the Sturm-Liouville problem is well understood. On the closed finite interval $[a, b]$, the operator l under the boundary conditions $y(a)\cos\alpha + y'(a)\sin\alpha = 0$, $y(b)\cos\beta + y'(b)\sin\beta = 0$ has a complete orthonormal system of eigenfunctions f_k, $k \geq 0$. The corresponding eigenvalues a_k are simple, $a_0 > a_1 > \cdots$, $a_k \to -\infty$ as $k \to \infty$, and f_k has exactly k simple zeros on the open interval (a, b).

More generally, in the *singular case*, the operator l may have a continuous spectrum, but the increase of the potential w at infinity guaranteed by condition (3.61) makes the eigenvalue problem very similar to the regular one.

THEOREM 3.18 ([**7**]). *Assume that the condition* (3.61) *is satisfied. Then the following holds.*

1. The operator l has a discrete spectrum. More precisely, there is a complete orthonormal system of eigenfunctions y_k of l belonging to $L^2(\mathbb{R}, \mathbb{R})$ whose eigenvalues μ_k, $k \geq 0$, tend to $-\infty$ as $k \to \infty$.

2. If μ is an eigenvalue of l, then its eigenspace is one-dimensional.

3. Every eigenfunction of l has a finite number of zeros. Moreover, if y_1, y_2 are eigenfunctions with eigenvalues μ_1, μ_2 and n_1, n_2 are the numbers of zeros, then $\mu_2 < \mu_1$ implies $n_2 > n_1$.

Let the eigenfunctions

$$y_0, y_1, y_2, \ldots, y_k, \ldots \tag{3.62}$$

be arranged in order of increasing number of their roots, that is, if n_k is the number of zeros of y_k, then

$$n_0 < n_1 < n_2 < \cdots < n_k < \cdots \tag{3.63}$$

The corresponding eigenvalues satisfy

$$\mu_0 > \mu_1 > \mu_2 > \cdots > \mu_k > \cdots \tag{3.64}$$

THEOREM 3.19 ([**7**]). *Let* (3.61) *be satisfied. Arrange the eigenfunctions* (3.62) *so that* (3.63) *and* (3.64) *hold. Then* y_k *has precisely* k *zeros, that is, in* (3.63) *we have* $n_k = k$.

The above formulated statements are easily transferred to the operators L and L^* with the help of (3.59) and (3.60). More precisely, the $L^2(\mathbb{R}, \mathbb{R}; e^{-\frac{2U}{\varepsilon}} dx)$-closure of L is self-adjoint on the domain $\mathcal{D}_L$ which consists of all continuous functions f from $L^2(\mathbb{R}, \mathbb{R}; e^{-\frac{2U}{\varepsilon}} dx)$, such that f' is locally absolutely continuous and $f'' - w(x)f$ is also in $L^2(\mathbb{R}, \mathbb{R}; e^{-\frac{2U}{\varepsilon}} dx)$.

L has a discrete spectrum $\{-\lambda_k\}_{k\geq 0}$ and its eigenfunctions $\{\Phi_k\}_{k\geq 0}$ are orthogonal in $L^2(\mathbb{R}, \mathbb{R}; e^{-\frac{2U}{\varepsilon}} dx)$, that is

(3.65)
$$\left\langle \frac{\Phi_i}{\|\Phi_i\|_{\rho^{-1}}}, \frac{\Phi_j}{\|\Phi_j\|_{\rho^{-1}}} \right\rangle_{\rho^{-1}} = \int_{\mathbb{R}} \frac{\Phi_i(y)}{\|\Phi_i\|_{\rho^{-1}}} \frac{\Phi_j(y)}{\|\Phi_j\|_{\rho^{-1}}} e^{-\frac{2U(y)}{\varepsilon}} dy = \delta_{ij}, \quad i,j = 0,1,2,\ldots$$

It will turn out to be convenient for us to consider non-normalised eigenfunctions.

The operator L^* has the same eigenvalues $\{-\lambda_k\}_{k\geq 0}$ and its eigenfunctions $\Psi_k = e^{-\frac{2U}{\varepsilon}} \Phi_k$ are orthogonal in the space $\mathcal{L}^2(\mathbb{R}, \mathbb{R}; e^{\frac{2U}{\varepsilon}} dx)$.

It is necessary to mention that the diffusion operator L is negatively definite, i.e. $\langle Lf, f\rangle_{\rho^{-1}} \leq 0$. Indeed, by using integration by parts on $\mathcal{D}_L$ one can obtain

$$\langle Lf, f\rangle_{\rho^{-1}} = \int_{\mathbb{R}} \left(\frac{\varepsilon}{2} f'' - U'f'\right) f e^{-\frac{2U}{\varepsilon}} dy = -\frac{\varepsilon}{2} \int_{\mathbb{R}} (f')^2 e^{-\frac{2U}{\varepsilon}} dy \leq 0. \tag{3.66}$$

The results of this section give us only general information about the self-adjointness and spectrum of L. In the next section we formulate two variational principles which allow us to determine the first two eigenvalues and eigenfunctions of L (and consequently of L^*) for small values of the parameter ε.

3.3.3. Variational principles for the eigenvalues.

Denote $\{\Phi_k, -\lambda_k\}_{k\geq 0}$ the eigensystem of the operator L, where the real numbers λ_k are arranged in increasing order. The eigenvalues and eigenfunctions of L can be obtained as solutions of the following variational problems.

THEOREM 3.20 (Recursive definition, [**55**]). *The zeroth eigenvalue and eigenfunction of* L *satisfy*

$$\lambda_0 = \frac{\langle -L\Phi_0, \Phi_0\rangle_{\rho^{-1}}}{\langle \Phi_0, \Phi_0\rangle_{\rho^{-1}}} = \inf_{\substack{f\in\mathcal{D}_L \\ f\neq 0}} \frac{\langle -Lf, f\rangle_{\rho^{-1}}}{\langle f, f\rangle_{\rho^{-1}}}. \tag{3.67}$$

For $k \geq 1$ *we have*

(3.68)
$$\lambda_1 = \frac{\langle -L\Phi_1, \Phi_1\rangle_{\rho^{-1}}}{\langle \Phi_1, \Phi_1\rangle_{\rho^{-1}}} = \inf_{\substack{f\in\mathcal{D}_L \\ \langle \Phi_0, f\rangle_{\rho^{-1}}=0 \\ f\neq 0}} \frac{\langle -Lf, f\rangle_{\rho^{-1}}}{\langle f, f\rangle_{\rho^{-1}}},$$
$$\ldots$$
$$\lambda_k = \frac{\langle -L\Phi_k, \Phi_k\rangle_{\rho^{-1}}}{\langle \Phi_k, \Phi_k\rangle_{\rho^{-1}}} = \inf_{\substack{f\in\mathcal{D}_L \\ \langle \Phi_j, f\rangle_{\rho^{-1}}=0 \\ 0\leq j\leq k-1 \\ f\neq 0}} \frac{\langle -Lf, f\rangle_{\rho^{-1}}}{\langle f, f\rangle_{\rho^{-1}}}.$$

THEOREM 3.21 (Courant's minimax principle, [**7, 33**]). *The eigenvalues of the operator L satisfy*

$$\lambda_0 = \inf_{\substack{f\in\mathcal{D}_L\\ f\neq 0}} \frac{\langle -Lf, f\rangle_{\rho^{-1}}}{\langle f, f\rangle_{\rho^{-1}}},$$

$$\lambda_k = \sup_{\substack{M_k\\ \dim M_k = k}} \inf_{\substack{f\in\mathcal{D}_L\cap M_k^\perp\\ 0\leq j\leq k\\ f\neq 0}} \frac{\langle -Lf, f\rangle_{\rho^{-1}}}{\langle f, f\rangle_{\rho^{-1}}}, \quad k\geq 1,$$

where M_k is a subspace in $L^2(\mathbb{R},\mathbb{R};\rho^{-1}dx)$, $M_k^\perp$ denotes the orthogonal complement of M_k in $L^2(\mathbb{R},\mathbb{R};\rho^{-1}dx)$.

REMARK 3.22. The variational principles can be also formulated in terms of the operator $(L,\mathcal{D}_0)$. In this case we have to consider the inf over $\mathcal{D}_0 = C_0^\infty(\mathbb{R},\mathbb{R})$ instead of $\mathcal{D}_L$, that is, over a domain on which $(L,\mathcal{D}_0)$ is essentially self-adjoint.

Using Theorem 3.20 we easily determine the zeroth eigenfunction and eigenvalue of L.

THEOREM 3.23. *The zeroth eigenvalue of the operator L equals 0, and the corresponding eigenfunction is given by $\Phi_0 = 1$.*

PROOF. It follows from (3.66) for any function in $\mathcal{D}_L$

$$\frac{\langle -Lf, f\rangle_{\rho^{-1}}}{\langle f, f\rangle_{\rho^{-1}}} = \frac{\varepsilon}{2}\frac{\langle f', f'\rangle_{\rho^{-1}}}{\langle f, f\rangle_{\rho^{-1}}} \geq 0$$

and consequently $\lambda_0 \geq 0$. Taking $f = \text{Const}$ yields $\lambda_0 = 0$. Hence, any constant function can be taken as the zeroth eigenfunction of L. For convenience we choose $\Phi_0 = 1$. □

Recalling that the eigenfunctions Ψ_k of the operator L^* are expressed in terms of Φ_k as $\Psi_k = e^{-\frac{2U}{\varepsilon}}\Phi_k$ we conclude that $\Psi_0 = e^{-\frac{2U}{\varepsilon}}$. It is helpful to remark that

$$\mu_0(x) = \frac{\Psi_0(x)}{\|\Psi_0\|_\rho^2} = \frac{e^{-\frac{2U(x)}{\varepsilon}}}{\int_{\mathbb{R}} e^{-\frac{2U(y)}{\varepsilon}}\,dy}, \quad x\in\mathbb{R},$$

is the invariant density of the diffusion (3.49).

3.3.4. The first eigenfunction and eigenvalue of L. In this section we shall investigate the first eigenvalue λ_1 and the first eigenfunction Φ_1 for small values of ε. Our study is again based on Theorem 3.20. The first eigenfunction of L is uniquely determined up to a normalizing factor by the equation

$$\frac{\varepsilon}{2}\varphi''(x) - U'(x)\varphi'(x) = -\lambda_1\varphi(x), \quad x\in\mathbb{R}, \tag{3.69}$$

and its orthogonality to Φ_0 in $L^2(\mathbb{R},\mathbb{R};\rho^{-1})$, i.e.

$$\int_{\mathbb{R}} \varphi(y) e^{-\frac{2U(y)}{\varepsilon}}\,dy = 0. \tag{3.70}$$

We also have some a priori information about the behaviour of λ_1.

THEOREM 3.24 ([**61**]). *Let the potential U satisfy the conditions* (**S**), (**G**) *and* (**M**). *Then*

$$\lim_{\varepsilon\to 0}\varepsilon\ln\lambda_1 = -v.$$

This means that for any $v' < v$ we have $\lambda_1 = o(e^{-v'/\varepsilon})$ as $\varepsilon \to 0$.

As we see, in the small noise limit λ_1 is the smallest parameter in (3.69). This suggests that it can be possible to look for the solution of (3.69) in the form of a power series in λ_1.

3.3.4.1. *Series solution for* Φ_1. We look for the solution of (3.69) in the form of a power series in the exponentially small parameter λ_1, formally

$$\varphi = \sum_{k=0}^{\infty} \varphi_k \lambda_1^k. \tag{3.71}$$

Substituting (3.71) into (3.69) induces ordinary differential equations for φ_k. More precisely, φ_0 satisfies the homogeneous equation

$$\frac{\varepsilon}{2}\varphi_0''(x) - U'(x)\varphi_0'(x) = 0, \quad x \in \mathbb{R} \tag{3.72}$$

and for $k \geq 1$

$$\frac{\varepsilon}{2}\varphi_k''(x) - U'(x)\varphi_k'(x) = -\varphi_{k-1}(x), \quad x \in \mathbb{R} \tag{3.73}$$

The heuristics of Section 3.3.1 suggests that the first eigenfunction takes constant values outside of some neighbourhood of zero, and there is an internal layer which glues these constant parts together (see Figure 3.2). Recall also that the first approximation of the eigenfunction (3.54) gave us good asymptotics of the first eigenvalue.

We first show that for arbitrary constants a and b there exists a solution of (3.69) such that $\varphi(-1) = a$, $\varphi(1) = b$, $\varphi \in L^2(\mathbb{R}, \mathbb{R}, \rho^{-1}dx)$. It subsequently turns out that the constants a and b can be chosen in such a way that the orthogonality condition (3.70) holds. This yields the first eigenfunction Φ_1.

In other words we look for the solution of the problem

$$\begin{cases} \frac{\varepsilon}{2}\varphi''(x) - U'(x)\varphi'(x) = -\lambda_1\varphi(x), \quad x \in \mathbb{R}, \\ \varphi(-1) = a, \ \varphi(1) = b, \\ \varphi \in L^2(\mathbb{R}, \mathbb{R}; \rho^{-1}dx). \end{cases} \tag{3.74}$$

For any $k \geq 0$ let us recursively determine functions h_k, f_k and g_k defined on the intervals $(-\infty, -1]$, $[-1, 1]$ and $[1, +\infty)$ respectively such that for $x \in \mathbb{R}$

$$\varphi_k(x) = h_k(x)\mathbb{I}_{(-\infty,-1]}(x) + f_k(x)\mathbb{I}_{[-1,1]}(x) + g_k(x)\mathbb{I}_{[1,+\infty)}(x). \tag{3.75}$$

Then, (3.72) can be rewritten in the form

$$\begin{cases} \frac{\varepsilon}{2}h_0'' - U'h_0' = 0, \\ \text{on } (-\infty, -1], \\ h_0(-1) = a, \\ h_0 \in L^2((-\infty, -1], \mathbb{R}; \rho^{-1}dx); \end{cases} \quad \begin{cases} \frac{\varepsilon}{2}f_0'' - U'f_0' = 0, \\ \text{on } [-1, 1], \\ f_0(-1) = a, \\ f_0(1) = b; \end{cases} \quad \begin{cases} \frac{\varepsilon}{2}g_0'' - U'g_0' = 0, \\ \text{on } [1, +\infty), \\ g_0(1) = b, \\ g_0 \in L^2([1, +\infty), \mathbb{R}; \rho^{-1}dx). \end{cases} \tag{3.76}$$

The solution of (3.76) allows us to assemble the first approximation φ_0 which is continuous and lies in $L^2(\mathbb{R}, \mathbb{R}; \rho^{-1}dx)$. The higher order terms of (3.71) are obtained by solving (3.73) on the three domains which results in the following boundary value problems for $k \geq 1$:

$$
(3.77)\qquad
\begin{cases}
\frac{\varepsilon}{2} h_k'' - U' h_k' = -h_{k-1}, \\
\text{on } (-\infty, -1], \\
h_k(-1) = 0, \\
h_k \in L^2((-\infty, -1], \mathbb{R}; \rho^{-1} dx);
\end{cases}
\qquad
\begin{cases}
\frac{\varepsilon}{2} f_k'' - U' f_k' = -f_{k-1}, \\
\text{on } [-1, 1], \\
f_k(-1) = 0, \\
f_k(1) = 0;
\end{cases}
$$
$$
\begin{cases}
\frac{\varepsilon}{2} g_k'' - U' g_k' = -g_{k-1}, \\
\text{on } [1, +\infty), \\
g_k(1) = 0, \\
g_k \in L^2([1, +\infty), \mathbb{R}; \rho^{-1} dx).
\end{cases}
$$

LEMMA 3.25. *For any $a, b \in \mathbb{R}$ the problem* (3.76) *has the following solution*

$$
\begin{aligned}
h_0(x) &= a, \quad x \in (-\infty, -1], \\
f_0(x) &= a + (b - a) \frac{\int_{-1}^{x} e^{\frac{2U(y)}{\varepsilon}} dy}{\int_{-1}^{1} e^{\frac{2U(y)}{\varepsilon}} dy}, \quad x \in [-1, 1], \\
g_0(x) &= b, \quad x \in [1, +\infty).
\end{aligned}
$$

We have $\max_{x \in [-1,1]} |f_0(x)| \leq |b - a|$.

PROOF. The statement of the lemma easily follows from the formula for the general solution of equation (3.72) which is

$$
\varphi_0(x) = A + B \int_0^x e^{\frac{2U(y)}{\varepsilon}} dy,
$$

with constants $A, B \in \mathbb{R}$. On the interval $[-1, 1]$ the constants A and B have to be appropriately chosen for the boundary conditions to hold. On the infinite intervals B must equal zero; otherwise the corresponding solution increases too fast at infinity and does not belong to the corresponding L^2 space. The estimate for $|f_0|$ follows from the monotonicity of f_0. $\square$

Let $-1 < x_- < 0 < x_+ < 1$ be such that $U(x_-) = -\frac{1+\varkappa}{4} V$ and $U(x_+) = -\frac{1+\varkappa}{4} v$. The existence of these points follows from the geometry of U.

LEMMA 3.26. *There exist constants $c > 0$ and $\varepsilon_0 > 0$ such that for all $0 < \varepsilon \leq \varepsilon_0$ the following inequalities hold*

$$
\max_{x \in [-1, x_-]} |f_0(x) - a| \leq c\sqrt{\varepsilon} |b - a| e^{-\frac{1+\varkappa}{2\varepsilon} V},
$$
$$
\max_{x \in [x_+, 1]} |f_0(x) - b| \leq c\sqrt{\varepsilon} |b - a| e^{-\frac{1+\varkappa}{2\varepsilon} v}
$$

PROOF. Laplace's method (see Appendix B) yields for small ε that

$$
\int_{-1}^{1} e^{\frac{2U(y)}{\varepsilon}} dy = \sqrt{\frac{\pi \varepsilon}{\omega_0}} (1 + \mathcal{O}(\varepsilon))
$$

and

$$
\int_{-1}^{x_-} e^{\frac{2U(y)}{\varepsilon}} dy = \frac{\varepsilon}{2U'(x_-)} e^{-\frac{1+\varkappa}{2\varepsilon} V} (1 + \mathcal{O}(\sqrt{\varepsilon})).
$$

By definition of f_0, this results in the following inequality

$$\max_{x\in[-1,x_-]}|f_0(x)-a| \le |b-a|e^{-\frac{1+\varkappa}{2\varepsilon}V}\frac{\varepsilon}{2U'(x_-)}\sqrt{\frac{\omega_0}{\pi\varepsilon}}(1+\mathcal{O}(\sqrt{\varepsilon})) \le c_1\sqrt{\varepsilon}|b-a|e^{-\frac{1+\varkappa}{2\varepsilon}V}$$

for some $c_1 > 0$. A similar inequality holds on the interval $[x_+, 1]$ for some positive constant c_2. Taking $c = \max\{c_1, c_2\}$ completes the proof. □

LEMMA 3.27. *Let w be a bounded continuous function on $\mathbb{R}$. Then there exists $C > 0$ and $\varepsilon_0 > 0$ such that for $0 < \varepsilon \le \varepsilon_0$*

$$\sup_{x\in(-\infty,-1]}\left|\int_x^{-1} e^{\frac{2U(y)}{\varepsilon}}\int_{-\infty}^{y} e^{-\frac{2U(z)}{\varepsilon}}w(z)\,dz\,dy\right| \le C\sqrt{\varepsilon}\sup_{x\in(-\infty,-1]}|w(x)|, \tag{3.78}$$

$$\sup_{x\in[-1,1]}\left|\int_0^{x} e^{\frac{2U(y)}{\varepsilon}}\int_{0}^{y} e^{-\frac{2U(z)}{\varepsilon}}w(z)\,dz\,dy\right| \le \sup_{x\in[-1,1]}|w(x)|, \tag{3.79}$$

$$\sup_{x\in[1,+\infty)}\left|\int_1^{x} e^{\frac{2U(y)}{\varepsilon}}\int_{y}^{\infty} e^{-\frac{2U(z)}{\varepsilon}}w(z)\,dz\,dy\right| \le C\sqrt{\varepsilon}\sup_{x\in[1,\infty)}|w(x)|. \tag{3.80}$$

PROOF. We start with the proof of (3.78). First, we notice that

$$\left|\int_x^{-1} e^{\frac{2U(y)}{\varepsilon}}\int_{-\infty}^{y} e^{-\frac{2U(z)}{\varepsilon}}w(z)\,dz\,dy\right| \le \sup_{x\in(-\infty,-1]}|w(x)|\int_x^{-1} e^{\frac{2U(y)}{\varepsilon}}\int_{-\infty}^{y} e^{-\frac{2U(z)}{\varepsilon}}\,dz\,dy.$$

The function $v(x) = \int_x^{-1} e^{\frac{2U(y)}{\varepsilon}}\int_{-\infty}^{y} e^{-\frac{2U(z)}{\varepsilon}}\,dz\,dy$ is non-negative and decreasing on $[-\infty,-1]$, hence $v(x) \le v(-\infty)$. Moreover, for $\delta > 0$ we have

$$v(-\infty) = \int_{-1-\delta}^{-1} v'(y)\,dy + \int_{-\infty}^{-1-\delta} v'(y)\,dy \le \max_{x\in[-1-\delta,-1]} v(x) + \int_{-\infty}^{-1-\delta} v'(y)\,dy,$$

where $v'(x) = e^{\frac{2U(x)}{\varepsilon}}\int_{-\infty}^{x} e^{-\frac{2U(y)}{\varepsilon}}\,dy$, $x \in (-\infty,-1]$. For $x \in [-1-\delta,-1]$ we have, using Laplace's method for $0 < \varepsilon \le \varepsilon_0$

$$v(x) = v'(-1)(x+1)+\mathcal{O}(x+1) = e^{-\frac{V}{\varepsilon}}e^{\frac{V}{\varepsilon}}\sqrt{\frac{\pi\varepsilon}{\omega_-}}(1+\mathcal{O}(\sqrt{\varepsilon}))(x+1)+\mathcal{O}(x+1) \le c_1\sqrt{\varepsilon}$$

for some $c_1 > 0$.

For x near $-\infty$ we have $U(x) = \frac{x^4}{4}$. Applying de l'Hôspital's rule yields

$$v'(x) = \frac{\int_{-\infty}^{x} e^{-\frac{2U(y)}{\varepsilon}}\,dy}{e^{-\frac{2U(x)}{\varepsilon}}} \approx \frac{e^{-\frac{2U(x)}{\varepsilon}}}{-\frac{2}{\varepsilon}U'(x)e^{-\frac{2U(x)}{\varepsilon}}} = \frac{\varepsilon}{2|x|^3}, \quad \text{as} \quad x \to -\infty,$$

hence $\int_{-\infty}^{-1-\delta} v'(y)\,dy \le c_2\varepsilon$ for some $c_2 > 0$. Therefore, (3.78) is proved with the constant $C_1 = \max\{c_1, c_2\}$.

To show (3.79) we note that $U(y) - U(z) \le 0$ for $-1 \le y \le z \le 0$ and for $0 \le z \le y \le 1$. Hence

$$\left|\int_0^{x} e^{\frac{2U(y)}{\varepsilon}}\int_{0}^{y} e^{-\frac{2U(z)}{\varepsilon}}w(z)\,dz\,dy\right| \le$$

$$\sup_{x\in[-1,1]}|w(x)|\left(\int_{-1}^{0}\int_{y}^{0} e^{\frac{2(U(y)-U(z))}{\varepsilon}}\,dz\,dy + \int_0^1\int_0^y e^{\frac{2(U(y)-U(z))}{\varepsilon}}\,dz\,dy\right) \le$$

$$\sup_{x\in[-1,1]}|w(x)|\left(\int_{-1}^{0}\int_{y}^{0} dz\,dy + \int_0^1\int_0^y dz\,dy\right) \le \sup_{x\in[-1,1]}|w(x)|.$$

Analogously to (3.78), the inequality (3.80) holds for some positive constant C_2. Taking $C = \max\{C_1, C_2\}$ completes the proof. □

LEMMA 3.28. *For $k \geq 1$ the problem* (3.77) *has the following recursively defined solution*

$$h_k(x) = \frac{2}{\varepsilon}\int_x^{-1} e^{\frac{2U(y)}{\varepsilon}} \int_{-\infty}^{y} e^{-\frac{2U(z)}{\varepsilon}} h_{k-1}(z)\,dz\,dy, \quad x \in (-\infty,-1], \tag{3.81}$$

$$\begin{aligned} f_k(x) = &-\frac{2}{\varepsilon}\int_{-1}^{1} e^{\frac{2U(y)}{\varepsilon}} \int_0^y e^{-\frac{2U(z)}{\varepsilon}} f_{k-1}(z)\,dz\,dy \cdot \frac{\int_x^1 e^{\frac{2U(y)}{\varepsilon}}\,dy}{\int_{-1}^1 e^{\frac{2U(y)}{\varepsilon}}\,dy} \\ &+\frac{2}{\varepsilon}\int_x^1 e^{\frac{2U(y)}{\varepsilon}} \int_0^y e^{-\frac{2U(z)}{\varepsilon}} f_{k-1}(z)\,dz\,dy, \quad x \in [-1,1], \end{aligned} \tag{3.82}$$

$$g_k(x) = \frac{2}{\varepsilon}\int_1^{x} e^{\frac{2U(y)}{\varepsilon}} \int_{y}^{\infty} e^{-\frac{2U(z)}{\varepsilon}} h_{k-1}(z)\,dz\,dy, \quad x \in [1,+\infty). \tag{3.83}$$

Moreover, for $0 < \varepsilon \leq \varepsilon_0$

$$|h_k(x)| \leq |a|\left(\frac{2C}{\sqrt{\varepsilon}}\right)^k, \quad x \in (-\infty,-1], \tag{3.84}$$

$$|f_k(x)| \leq |b-a|\left(\frac{2}{\varepsilon}\right)^k, \quad x \in [-1,1], \tag{3.85}$$

$$|g_k(x)| \leq |b|\left(\frac{2C}{\sqrt{\varepsilon}}\right)^k, \quad x \in [1,+\infty), \tag{3.86}$$

where ε_0 and C are as in Lemma 3.27.

PROOF. The general solution of equation (3.73) is given by

$$\varphi_k(x) = A + B\int_0^x e^{\frac{2U(y)}{\varepsilon}}\,dy - \frac{2}{\varepsilon}\int_0^x e^{\frac{2U(y)}{\varepsilon}} \int_0^y e^{-\frac{2U(z)}{\varepsilon}} \varphi_{k-1}(z)\,dz\,dy,$$

with real constants A, B. On the interval $[-1,1]$ we determine A and B from the boundary conditions $f_k(\pm 1) = 0$ which results in (3.82).

In case $x \in (-\infty,-1]$, the general solution can be written in the form

$$h_k(x) = A + B\int_x^{-1} e^{\frac{2U(y)}{\varepsilon}}\,dy + \frac{2}{\varepsilon}\int_x^{-1} e^{\frac{2U(y)}{\varepsilon}} \int_{-1}^y e^{-\frac{2U(z)}{\varepsilon}} h_{k-1}(z)\,dz\,dy.$$

It follows from the condition $h_k(-1) = 0$ that $A = 0$. The second constant B is determined by the condition that $h_k \in \mathcal{L}^2((-\infty,-1], \rho^{-1}dx)$. Let us write

$$h_k(x) = \frac{2}{\varepsilon}\int_x^{-1} e^{\frac{2U(y)}{\varepsilon}} \left(\frac{\varepsilon}{2}B + \int_{-1}^y e^{-\frac{2U(z)}{\varepsilon}} h_{k-1}(z)\,dz\right) dy$$

Setting $B = \frac{2}{\varepsilon}\int_{-\infty}^{-1} e^{-\frac{2U(z)}{\varepsilon}} h_{k-1}(z)\,dz$, we obtain the solution (3.81), which is bounded by Lemma 3.27.

The case of the function g_k on the interval $[1,\infty)$ is studied analogously.

The estimates (3.84), (3.85) and (3.86) are obtained by applying Lemma 3.27 k times and recalling the definition of h_0, f_0, g_0. □

Now we have enough information to study the closeness of the solution φ of (3.69) to the constants a and b and the first approximation φ_0.

LEMMA 3.29. *There exist* $\varepsilon_0 > 0$ *and* $A > 0$ *such that for* $0 < \varepsilon \le \varepsilon_0$ *the following inequalities hold:*

$$\max_{x\in(-\infty,-1]} |\varphi(x) - a| \le A|a|\frac{\lambda_1}{\sqrt{\varepsilon}},$$

$$\max_{x\in[-1,x_-]} |\varphi(x) - a| \le A|b-a|\sqrt{\varepsilon}e^{-\frac{2V}{3\varepsilon}},$$

$$\max_{x\in[x_-,x_+]} |\varphi(x) - f_0(x)| \le A|b-a|\frac{\lambda_1}{\varepsilon},$$

$$\max_{x\in[x_+,1]} |\varphi(x) - b| \le A|b-a|\sqrt{\varepsilon}e^{-\frac{2v}{3\varepsilon}},$$

$$\max_{x\in[1,\infty)} |\varphi(x) - b| \le A|b|\frac{\lambda_1}{\sqrt{\varepsilon}},$$

PROOF. Choose $\varepsilon_0 > 0$ small enough according to the previous lemmas and such that $\max\{\frac{2C}{\sqrt{\varepsilon}}\lambda_1, \frac{2}{\varepsilon}\lambda_1\} \le \frac{1}{2}$ for $0 < \varepsilon \le \varepsilon_0$. This is possible by Theorem 3.24.

For $x \le -1$, $0 < \varepsilon \le \varepsilon_0$, we obtain using (3.84)

$$|\varphi(x) - a| = |h(x) - h_0(x)| = \left|\sum_{k=1}^{\infty} h_k(x)\lambda_1^k\right| \le \sum_{k=1}^{\infty} |h_k(x)|\lambda_1^k$$

$$\le \sum_{k=1}^{\infty} |a| \left(\frac{2C}{\sqrt{\varepsilon}}\right)^k \lambda_1^k = |a|\frac{\frac{2C}{\sqrt{\varepsilon}}\lambda_1}{1 - \frac{2C}{\sqrt{\varepsilon}}\lambda_1} \le A_1|a|\frac{\lambda_1}{\sqrt{\varepsilon}}$$

for some $A_1 > 0$ independent of ε.

For $x \in [-1, x_-]$, $0 < \varepsilon \le \varepsilon_0$, we use Lemma 3.26, (3.85), and the assumption that $\frac{1+\varkappa}{2} < \frac{v}{V} < 1$ to obtain

$$|\varphi(x) - a| \le |f_0(x) - a| + |f(x) - f_0(x)| \le c|b-a|\sqrt{\varepsilon}e^{-\frac{1+\varkappa}{2\varepsilon}V} + \sum_{k=1}^{\infty} |f_k(x)|\lambda_1^k$$

$$\le c|b-a|\sqrt{\varepsilon}e^{-\frac{1+\varkappa}{2\varepsilon}V} + \sum_{k=1}^{\infty} |b-a| \left(\frac{2}{\varepsilon}\right)^k \lambda_1^k$$

$$= c|b-a|\sqrt{\varepsilon}e^{-\frac{1+\varkappa}{2\varepsilon}V} + |b-a|\frac{\frac{2}{\varepsilon}\lambda_1}{1 - \frac{2}{\varepsilon}\lambda_1} \le A_2|b-a|\sqrt{\varepsilon}e^{-\frac{1+\varkappa}{2\varepsilon}V}$$

for some $A_2 > 0$ independent of ε.

The estimates on the remaining intervals are obtained analogously. Choosing A to be the maximal constant appearing in the inequalities we complete the proof of the lemma. □

Let us now determine the constants a and b in the small noise limit so that $\varphi \perp 1$ in $L^2(\mathbb{R}, \mathbb{R}; \rho^{-1}dx)$. This will provide a good approximation of the first eigenfunction Φ_1. It is clear that φ can be determined up to a constant factor. For definiteness we put $b = 1$ and look for a.

LEMMA 3.30. *There exists a function* $a\colon \mathbb{R}_+ \to \mathbb{R}$ *such that*

$$a = -\sqrt{\frac{\omega_-}{\omega_+}}e^{-\frac{V-v}{\varepsilon}}(1 + \mathcal{O}(\varepsilon))$$

and $\varepsilon_0 > 0$ such that with $b = 1$ the function Φ_1 chosen according to the previous lemmas for $a(\varepsilon)$ and b is orthogonal to $\Phi_0 = 1$ in $L^2(\mathbb{R}, \mathbb{R}; \rho^{-1}dx)$ for $0 < \varepsilon \leq \varepsilon_0$. In sequel we suppress the dependence of $a(\varepsilon)$ on ε and continue to use the symbol a instead.

PROOF. Choose ε_0 according to Lemma 3.29 and let $0 < \varepsilon \leq \varepsilon_0$. In order to determine a we shall solve two inequalities $\int \Phi_1 \Phi_0 \, e^{-\frac{2U}{\varepsilon}} \, dy \geq 0$ and $\int \Phi_1 \Phi_0 \, e^{-\frac{2U}{\varepsilon}} \, dy \leq 0$.

We use the inequalities of Lemma 3.29 and Laplace's method to obtain for $\varepsilon \leq \varepsilon_0$

$$\begin{aligned}
0 \leq \int_{\mathbb{R}} \Phi_1(y) e^{-\frac{2U(y)}{\varepsilon}} \, dy &\leq a \int_{-\infty}^{x_-} e^{-\frac{2U(y)}{\varepsilon}} \, dy + \int_{x_-}^{x_+} f_0(y) e^{-\frac{2U(y)}{\varepsilon}} \, dy + \int_{x_+}^{\infty} e^{-\frac{2U(y)}{\varepsilon}} \, dy \\
&+ A|a| \frac{\lambda_1}{\sqrt{\varepsilon}} \int_{-\infty}^{-1} e^{-\frac{2U(y)}{\varepsilon}} \, dy + A|1-a| \sqrt{\varepsilon} e^{-\frac{1+\varkappa}{2\varepsilon} V} \int_{-1}^{x_-} e^{-\frac{2U(y)}{\varepsilon}} \, dy \\
&+ A|1-a| \frac{\lambda_1}{\varepsilon} \int_{x_-}^{x_+} e^{-\frac{2U(y)}{\varepsilon}} \, dy + A|1-a| \sqrt{\varepsilon} e^{-\frac{1+\varkappa}{2\varepsilon} v} \int_{x_+}^{1} e^{-\frac{2U(y)}{\varepsilon}} \, dy \\
&+ A \frac{\lambda_1}{\sqrt{\varepsilon}} \int_1^{\infty} e^{-\frac{2U(y)}{\varepsilon}} \, dy \\
&\leq a \sqrt{\frac{\pi\varepsilon}{\omega_-}} e^{\frac{V}{\varepsilon}} (1 + \mathcal{O}(\varepsilon)) + (1+|a|) \frac{\varepsilon}{2U'(x_-)} e^{\frac{1+\varkappa}{2\varepsilon} V} (1 + \mathcal{O}(\varepsilon)) + \sqrt{\frac{\pi\varepsilon}{\omega_+}} e^{\frac{v}{\varepsilon}} (1 + \mathcal{O}(\varepsilon)) \\
&+ A|a| \frac{\lambda_1}{2} \sqrt{\frac{\pi}{\omega_-}} e^{\frac{V}{\varepsilon}} (1 + \mathcal{O}(\sqrt{\varepsilon})) + A(1+|a|) \frac{\varepsilon}{2} e^{-\frac{1+\varkappa}{2\varepsilon} V} \sqrt{\frac{\pi}{\omega_-}} e^{\frac{V}{\varepsilon}} (1 + \mathcal{O}(\sqrt{\varepsilon})) \\
&+ A(1+|a|) \frac{\lambda_1}{2U'(x_-)} e^{\frac{1+\varkappa}{2\varepsilon} V} (1 + \mathcal{O}(\varepsilon)) + A(1+|a|) \frac{\varepsilon}{2} e^{-\frac{1+\varkappa}{2\varepsilon} v} \sqrt{\frac{\pi}{\omega_+}} e^{\frac{v}{\varepsilon}} (1 + \mathcal{O}(\sqrt{\varepsilon})) \\
&+ A \frac{\lambda_1}{2} \sqrt{\frac{\pi}{\omega_+}} e^{\frac{v}{\varepsilon}} (1 + \mathcal{O}(\sqrt{\varepsilon})).
\end{aligned}$$

One can notice that all terms having $|a|$ as pre-factor are of the order $o(e^{\frac{V}{\varepsilon}})$. The terms without pre-factor $|a|$ except $\sqrt{\frac{\pi\varepsilon}{\omega_+}} e^{\frac{v}{\varepsilon}} (1 + \mathcal{O}(\varepsilon))$ are $o(e^{\frac{v}{\varepsilon}})$. This results in the inequality

$$a \geq -\sqrt{\frac{\omega_-}{\omega_+}} e^{-\frac{V-v}{\varepsilon}} (1 + \mathcal{O}(\varepsilon)).$$

Applying the arguments just given to $-\Phi_1$, one gets the converse inequality

$$a \leq -\sqrt{\frac{\omega_-}{\omega_+}} e^{-\frac{V-v}{\varepsilon}} (1 + \mathcal{O}(\varepsilon)).$$

This completes the proof. □

A good approximation of Φ_1 is described by the following theorem, which is a direct corollary of Lemmas 3.29 and 3.30.

THEOREM 3.31. *Let $0 < v - v' < \frac{1-\varkappa}{2} V$ and a be given according to Lemma 3.30. Let Φ_1 be defined according to the previous lemmas, and*

$$f_0(x) = a + (1-a) \frac{\int_{-1}^{x} e^{\frac{2U(y)}{\varepsilon}} \, dy}{\int_{-1}^{1} e^{\frac{2U(y)}{\varepsilon}} \, dy}, \quad x \in [-1, 1].$$

There exist $\varepsilon_0 > 0$ *such that for* $0 < \varepsilon \le \varepsilon_0$ *the following inequalities hold*

$$\max_{x \in (-\infty, x_-]} |\Phi_1(x) - a| \le e^{-\frac{1+\varkappa}{2\varepsilon}V}, \tag{3.87}$$

$$\max_{x \in [x_-, x_+]} |\Phi_1(x) - f_0(x)| \le e^{-\frac{v'}{\varepsilon}}, \tag{3.88}$$

$$\max_{x \in [x_+, \infty]} |\Phi_1(x) - 1| \le e^{-\frac{1+\varkappa}{2\varepsilon}v}. \tag{3.89}$$

PROOF. The statement of the theorem follows immediately from Lemmas 3.29 and 3.30. For example, recalling the assumption (**M**), for ε small enough

$$\max_{x \in (-\infty, -1]} |\Phi_1(x) - a| \le A|a| \frac{\lambda_1}{\sqrt{\varepsilon}} \le A_1 e^{-\frac{V-v}{\varepsilon}} e^{-\frac{v'}{\varepsilon}} \le e^{-\frac{1+\varkappa}{2\varepsilon}V}.$$

The latter inequality in the previous formula holds for small enough ε because $V - v + v' > \frac{1+\varkappa}{2}V$ for $0 < v - v' < \frac{1-\varkappa}{2}V$. Furthermore,

$$\max_{x \in (-1, x_-]} |\Phi_1(x) - a| \le A|1-a|\sqrt{\varepsilon} e^{-\frac{1+\varkappa}{2\varepsilon}V} \le e^{-\frac{1+\varkappa}{2\varepsilon}V}$$

for ε small enough. Combining these two inequalities gives (3.87). The inequalities (3.88) and (3.89) are proved analogously. □

COROLLARY 3.32. In the notation of Theorem 3.31 the following inequalities hold

$$a^2 - Be^{-\frac{(2+\varkappa)V - 2v}{2\varepsilon}} \le \max_{x \in (-\infty, x_-]} (\Phi_1(x))^2 \le a^2 + Be^{-\frac{(2+\varkappa)V - 2v}{2\varepsilon}}, \tag{3.90}$$

$$0 \le \max_{x \in [x_-, x_+]} (\Phi_1(x))^2 \le 2, \tag{3.91}$$

$$1 - Be^{-\frac{1+\varkappa}{2\varepsilon}v} \le \max_{x \in [x_+, \infty)} (\Phi_1(x))^2 \le 1 + Be^{-\frac{1+\varkappa}{2\varepsilon}v}. \tag{3.92}$$

PROOF. On the interval $(-\infty, x_-]$ we use (3.87) to get

$$a - e^{-\frac{1+\varkappa}{2\varepsilon}V} \le \Phi_1(x) \le a + e^{-\frac{1+\varkappa}{2\varepsilon}V}.$$

Hence, using that $a < 0$ one obtains

$$(a + e^{-\frac{1+\varkappa}{2\varepsilon}V})^2 \le (\Phi_1(x))^2 \le (a - e^{-\frac{1+\varkappa}{2\varepsilon}V})^2,$$

and therefore recalling the asymptotics of $a = \mathcal{O}(e^{-\frac{V-v}{\varepsilon}})$, for some $B > 0$

$$a^2 - Be^{-\frac{(2+\varkappa)V - 2v}{2\varepsilon}} \le (\Phi_1(x))^2 \le a^2 + Be^{-\frac{(2+\varkappa)V - 2v}{2\varepsilon}}.$$

The inequality (3.91) follows from the fact that Φ and f_0 are exponentially close, and $|f_0| \le |1 - a| \le \sqrt{2}$.

The inequality (3.92) is obtained analogously to (3.90). □

In order to be able to apply the variational principles for eigenvalues we have to study the derivative φ' of the solution of (3.74).

LEMMA 3.33. *Let φ be a solution of (3.74). Then there exist constants $\varepsilon_0 > 0$ and $A' > 0$ such that for $0 < \varepsilon \leq \varepsilon_0$*

$$\max_{x\in(-\infty,-1]} |\varphi'(x)| \leq A'|a|\frac{\lambda_1}{\varepsilon}, \tag{3.93}$$

$$\max_{x\in[-1,x_-]} |\varphi'(x)| \leq A'|b-a|\frac{e^{-\frac{1+\varkappa}{2\varepsilon}V}}{\varepsilon\sqrt{\varepsilon}}, \tag{3.94}$$

$$\max_{x\in[x_-,x_+]} |\varphi'(x) - f_0'(x)| \leq A'|b-a|\frac{\lambda_1}{\varepsilon\sqrt{\varepsilon}} \tag{3.95}$$

$$\max_{x\in[x_+,1]} |\varphi'(x)| \leq A'|b-a|\frac{e^{-\frac{1+\varkappa}{2\varepsilon}v}}{\varepsilon\sqrt{\varepsilon}}, \tag{3.96}$$

$$\max_{x\in[1,\infty)} |\varphi'(x)| \leq A'|b|\frac{\lambda_1}{\varepsilon}. \tag{3.97}$$

PROOF. On the interval $(-\infty,-1]$ the equation $\varphi = h = \sum_{k=0}^{\infty} h_k\lambda_1^k$ is valid. This yields, due to uniform convergence and (3.81) that

$$\begin{aligned} h'(x) = \sum_{k=0}^{\infty} h_k'\lambda_1^k &= -\sum_{k=1}^{\infty} \frac{2}{\varepsilon}\lambda_1^k e^{\frac{2U(x)}{\varepsilon}} \int_{-\infty}^{x} e^{-\frac{2U(y)}{\varepsilon}} h_{k-1}(y)\,dy \\ &= -\frac{2}{\varepsilon}\lambda_1 e^{\frac{2U(x)}{\varepsilon}} \int_{-\infty}^{x} e^{-\frac{2U(y)}{\varepsilon}} h(y)\,dy \end{aligned}$$

Using Lemma 3.29 implies in (3.93). Indeed for $x \leq -1$ and ε small enough

$$|h'(x)| \leq \frac{2}{\varepsilon}\lambda_1|a|(1 + A\frac{\lambda_1}{\sqrt{\varepsilon}})e^{\frac{2U(x)}{\varepsilon}} \int_{-\infty}^{x} e^{-\frac{2U(y)}{\varepsilon}}\,dy \leq c_1|a|\frac{\lambda_1}{\varepsilon},$$

for some $c_1 > 0$, since $e^{\frac{2U(x)}{\varepsilon}} \int_{-\infty}^{x} e^{-\frac{2U(y)}{\varepsilon}}\,dy$ is uniformly bounded.

On the interval $[-1,1]$ we have $\varphi = f = f_0 + \sum_{k=1}^{\infty} f_k\lambda_1^k$, hence

$$\begin{aligned} f'(x) = f_0'(x) + \sum_{k=1}^{\infty} f_k'\lambda_1^k &= (b-a)\frac{e^{\frac{2U(x)}{\varepsilon}}}{\int_{-1}^{1} e^{\frac{2U(y)}{\varepsilon}}\,dy} \\ &\quad - \frac{2}{\varepsilon}\lambda_1 e^{\frac{2U(x)}{\varepsilon}} \int_0^x e^{-\frac{2U(y)}{\varepsilon}} f(y)\,dy \\ &\quad + \frac{2}{\varepsilon}\lambda_1 e^{\frac{2U(x)}{\varepsilon}} \frac{\int_{-1}^{1} e^{\frac{2U(y)}{\varepsilon}} \int_0^y e^{-\frac{2U(z)}{\varepsilon}} f(z)\,dz\,dy}{\int_{-1}^{1} e^{\frac{2U(y)}{\varepsilon}}\,dy}, \end{aligned} \tag{3.98}$$

With the help of Laplace's method for small ε we obtain for some constant $c_2 > 0$ that the first summand in (3.98) can be estimated by

$$\max_{x\in[-1,x_-]} |f_0'(x)| \leq |b-a|\frac{e^{-\frac{1+\varkappa}{2\varepsilon}V}}{\sqrt{\frac{\pi\varepsilon}{\omega_0}}(1+\mathcal{O}(\varepsilon))} \leq c_2|b-a|\frac{e^{-\frac{1+\varkappa}{2\varepsilon}V}}{\sqrt{\varepsilon}}.$$

Due to the monotonicity of U on $[-1,0]$, the second summand in (3.98) can be estimated as

$$\max_{x\in[-1,x_-]} \left|\frac{2}{\varepsilon}\lambda_1 e^{\frac{2U(x)}{\varepsilon}} \int_0^x e^{-\frac{2U(y)}{\varepsilon}} f(y)\,dy\right| \leq c_3|b-a|\frac{\lambda_1}{\varepsilon}e^{-\frac{1+\varkappa}{2\varepsilon}V},$$

with a constant c_3 independent of ε. Analogously, for some $c_4 > 0$ the Laplace method and (3.79) imply

$$\max_{x\in[-1,x_-]}\left|\frac{2}{\varepsilon}\lambda_1 e^{\frac{2U(x)}{\varepsilon}}\frac{\int_{-1}^1 e^{\frac{2U(y)}{\varepsilon}}\int_0^y e^{\frac{2U(z)}{\varepsilon}} f(z)\,dz\,dy}{\int_{-1}^1 e^{\frac{2U(y)}{\varepsilon}}\,dy}\right| \le c_4|b-a|\frac{\lambda_1}{\varepsilon\sqrt{\varepsilon}}e^{-\frac{1+\varkappa}{2\varepsilon}V}.$$

Combining of the previous three inequalities results in (3.94).

The inequality (3.95) follows directly from (3.98). Inequalities (3.96) and (3.97) are proved analogously to (3.94) and (3.93). □

Recalling that the first eigenfunction Φ_1 is obtained as a solution of (3.74) for $b = 1$ and $a = -\sqrt{\frac{\omega_-}{\omega_+}}e^{-\frac{V-v}{\varepsilon}}(1+\mathcal{O}(\varepsilon))$ we obtain bounds for Φ_1'.

THEOREM 3.34. *Let $0 < v - v' < \frac{1-\varkappa}{2}V$ and a be given according to Lemma 3.30. Let Φ_1 be defined according to the previous lemmas, and*

$$f_0(x) = a + (1-a)\frac{\int_{-1}^x e^{\frac{2U(y)}{\varepsilon}}\,dy}{\int_{-1}^1 e^{\frac{2U(y)}{\varepsilon}}\,dy}, \qquad x \in [-1,1].$$

There exist $\varepsilon_0 > 0$ and $B' > 0$ such that for $0 < \varepsilon \le \varepsilon_0$ the following inequalities hold

$$\max_{x\in(-\infty,x_-]}|\Phi_1'(x)| \le B'\frac{e^{-\frac{1+\varkappa}{2\varepsilon}V}}{\varepsilon\sqrt{\varepsilon}}, \tag{3.99}$$

$$\max_{x\in[x_-,x_+]}|\Phi_1'(x) - f_0'(x)| \le e^{-\frac{v'}{\varepsilon}} \tag{3.100}$$

$$\max_{x\in[x_+,\infty)}|\Phi_1'(x)| \le B'\frac{e^{-\frac{1+\varkappa}{2\varepsilon}v}}{\varepsilon\sqrt{\varepsilon}}. \tag{3.101}$$

PROOF. Inequality (3.99) follows from (3.93) and (3.94). Indeed, for any $0 < v' < v$ there exists $\varepsilon_0 > 0$ such that for $0 < \varepsilon \le \varepsilon_0$ we have

$$A'|a|\frac{\lambda_1}{\varepsilon} \le e^{-\frac{V-v}{\varepsilon}}e^{-\frac{v'}{\varepsilon}} \le e^{-\frac{1+\varkappa}{2\varepsilon}V}.$$

For the second inequality we have to choose v' such that $0 < v - v' < \frac{1-\varkappa}{2}V$. Inequality (3.101) is proved analogously. Inequality (3.100) follows directly from (3.95). □

COROLLARY 3.35. In the notation of Theorem 3.34 there exists $B_1 > 0$ such that the following holds:

$$\max_{x\in(-\infty,x_-]}(\Phi_1'(x))^2 \le B_1\frac{e^{-\frac{1+\varkappa}{\varepsilon}V}}{\varepsilon^3},$$

$$(f_0'(x))^2 - B_1\frac{e^{\frac{2U}{\varepsilon}}e^{-\frac{v'}{\varepsilon}}}{\sqrt{\varepsilon}} \le (\Phi_1'(x))^2 \le (f_0'(x))^2 + B_1\frac{e^{\frac{2U}{\varepsilon}}e^{-\frac{v'}{\varepsilon}}}{\sqrt{\varepsilon}}, \quad x\in[x_-,x_+],$$

$$\max_{x\in[x_+,\infty]}(\Phi_1'(x))^2 \le B_1\frac{e^{-\frac{1+\varkappa}{\varepsilon}v}}{\varepsilon^3}.$$

PROOF. The inequalities follow directly from Theorem 3.34. The formula

$$f_0'(x) = (1-a)\frac{e^{\frac{2U(x)}{\varepsilon}}}{\int_{-1}^{1} e^{\frac{2U(y)}{\varepsilon}}\,dy}, \qquad x \in [-1,1]$$

and Laplace's method applied to the denominator in the latter formula. □

3.3.4.2. *Useful integrals of eigenfunctions.* For purposes of the next chapter we now calculate some integrals of Φ_0, Φ_1 and Φ_1'. In what follows, for any function $f : \mathbb{R} \to \mathbb{R}$ we denote $\overline{f}(x) = f(-x)$, $x \in \mathbb{R}$.

THEOREM 3.36. *In the small noise limit $\varepsilon \to 0$ we have*

$$\|\Phi_0\|^2_{\rho^{-1}} = \int_{\mathbb{R}} e^{-\frac{2U(y)}{\varepsilon}}\,dy = \sqrt{\frac{\pi\varepsilon}{\omega_-}}\,e^{\frac{V}{\varepsilon}}(1+\mathcal{O}(\varepsilon)), \tag{3.102}$$

$$\|\Phi_1\|^2_{\rho^{-1}} = \int_{\mathbb{R}} (\Phi_1(y))^2 e^{-\frac{2U(y)}{\varepsilon}}\,dy = \sqrt{\frac{\pi\varepsilon}{\omega_+}}\,e^{\frac{v}{\varepsilon}}(1+\mathcal{O}(\varepsilon)), \tag{3.103}$$

$$\|\Phi_1'\|^2_{\rho^{-1}} = \int_{\mathbb{R}} (\Phi_1'(y))^2 e^{-\frac{2U(y)}{\varepsilon}}\,dy = \sqrt{\frac{\omega_0}{\pi\varepsilon}}(1+\mathcal{O}(\varepsilon)), \tag{3.104}$$

$$\|\overline{\Phi}_1\|^2_{\rho^{-1}} = \int_{\mathbb{R}} (\Phi_1(-y))^2 e^{-\frac{2U(y)}{\varepsilon}}\,dy = \sqrt{\frac{\pi\varepsilon}{\omega_-}}\,e^{\frac{V}{\varepsilon}}(1+\mathcal{O}(\varepsilon)),$$

$$\langle \Phi_1, \overline{\Phi}_1\rangle_{\rho^{-1}} = \int_{\mathbb{R}} \Phi_1(y)\Phi_1(-y) e^{-\frac{2U(y)}{\varepsilon}}\,dy = -\sqrt{\frac{\pi\varepsilon}{\omega_+}}\,e^{\frac{v}{\varepsilon}}(1+\mathcal{O}(\varepsilon)),$$

$$\langle \Phi_0, \overline{\Phi}_1\rangle_{\rho^{-1}} = \int_{\mathbb{R}} \Phi_1(-y) e^{-\frac{2U(y)}{\varepsilon}}\,dy = \sqrt{\frac{\pi\varepsilon}{\omega_-}}\,e^{\frac{V}{\varepsilon}}(1+\mathcal{O}(\varepsilon)),$$

$$\langle x, \Phi_1\rangle_{\rho^{-1}} = \int_{\mathbb{R}} y\Phi_1(y) e^{-\frac{2U(y)}{\varepsilon}}\,dy = 2\sqrt{\frac{\pi\varepsilon}{\omega_+}}\,e^{\frac{v}{\varepsilon}}(1+\mathcal{O}(\varepsilon)).$$

PROOF. The evaluation of these integrals consists in combining Laplace's method, Theorem 3.23 concerning the zeroth eigenfunction, and Corollaries 3.32 and 3.35 which provide bounds for the first eigenfunction and its derivative. We give the complete set of arguments for the first three integrals, the remaining ones being treated similarly.

The norm (3.102) of Φ_0 is calculated by a direct application of Laplace's method. The global minimum of $U(x)$ is attained at $x = -1$ and $U(-1) = -\frac{V}{2}$. Formula (3.102) follows from (B.9) in the Appendix.

To obtain the norm $\|\Phi_1\|^2_{\rho^{-1}}$ we recall Corollary 3.32 to get with a suitable constant c_1

$$\begin{aligned}
\|\Phi_1\|^2_{\rho^{-1}} &= \int_{\mathbb{R}} (\Phi_1(y))^2 e^{-\frac{2U(y)}{\varepsilon}}\,dy \\
&\leq \left(\frac{\omega_-}{\omega_+} e^{-2\frac{V-v}{\varepsilon}}(1+\mathcal{O}(\varepsilon)) + Be^{-\frac{(2+\varkappa)V-2v}{2\varepsilon}}\right)\int_{-\infty}^{x_-} e^{-\frac{2U(y)}{\varepsilon}}\,dy \\
&+ 2\int_{x_-}^{x_+} e^{-\frac{2U(y)}{\varepsilon}}\,dy + (1+Be^{-\frac{(1+\varkappa)v}{v\varepsilon}})\int_{x_+}^{\infty} e^{-\frac{2U(y)}{\varepsilon}}\,dy \\
&\leq \sqrt{\frac{\pi\varepsilon}{\omega_+}}\,e^{\frac{v}{\varepsilon}}(1+\mathcal{O}(\varepsilon)) + c_1\left(e^{-\frac{V-2v}{\varepsilon}} + e^{-\frac{(1+\varkappa)V-2v}{2\varepsilon}} + e^{\frac{(1-\varkappa)V}{2\varepsilon}} + e^{\frac{(1-\varkappa)v}{2\varepsilon}}\right) \\
&= \sqrt{\frac{\pi\varepsilon}{\omega_+}}\,e^{\frac{v}{\varepsilon}}(1+\mathcal{O}(\varepsilon)).
\end{aligned}$$

The last line is justified, since the sum of the exponents in the previous line is $o(e^{\frac{v}{\varepsilon}})$. The converse inequality

$$\|\Phi_1\|^2_{\rho^{-1}} = \int_{\mathbb{R}} (\Phi_1(y))^2 e^{-\frac{2U(y)}{\varepsilon}}\, dy \geq \sqrt{\frac{\pi\varepsilon}{\omega_+}} e^{\frac{v}{\varepsilon}} (1 + \mathcal{O}(\varepsilon))$$

comes from the estimate of the integral on $[x_+, \infty)$. This proves (3.103)

To obtain the norm $\|\Phi_1'\|^2_{\rho^{-1}}$, we use Corollary 3.35, Laplace's method and the fact that $a = -\sqrt{\frac{\omega_-}{\omega_+}} e^{-\frac{V-v}{\varepsilon}} (1 + \mathcal{O}(\varepsilon))$. Indeed, with a suitable constant c_2

$$\begin{aligned}
\|\Phi_1'\|^2_{\rho^{-1}} &= \int_{\mathbb{R}} (\Phi_1'(y))^2 e^{-\frac{2U(y)}{\varepsilon}}\, dy \leq \frac{B_1}{\varepsilon^3} e^{-\frac{(1+\varkappa)V}{\varepsilon}} \int_{-\infty}^{x_-} e^{-\frac{2U(y)}{\varepsilon}}\, dy \\
&+ \frac{(1-a)^2}{\left(\int_{-1}^{1} e^{\frac{2U(y)}{\varepsilon}}\, dy\right)^2} \int_{x_-}^{x_+} e^{\frac{2U(y)}{\varepsilon}}\, dy + \frac{B_1}{\sqrt{\varepsilon}} e^{-\frac{v'}{\varepsilon}} \int_{x_-}^{x_+} dy \\
&+ \frac{B_1}{\varepsilon^3} e^{-\frac{(1+\varkappa)v}{\varepsilon}} \int_{x_+}^{\infty} e^{-\frac{2U(y)}{\varepsilon}}\, dy \\
&\leq \sqrt{\frac{\omega_0}{\pi\varepsilon}} (1 + \mathcal{O}(\varepsilon)) + \frac{c_2}{\varepsilon^3} \left(e^{-\frac{\varkappa V}{\varepsilon}} + e^{-\frac{(1-\varkappa)V}{\varepsilon}} + e^{-\frac{v'}{\varepsilon}} + e^{-\frac{(1+\varkappa)v}{2\varepsilon}} \right) \\
&= \sqrt{\frac{\omega_0}{\pi\varepsilon}} (1 + \mathcal{O}(\varepsilon)).
\end{aligned}$$

The converse inequality is more easily obtained by evaluating the integral only on $[x_-, x_+]$, and (3.104) follows. □

3.3.4.3. *The accurate asymptotics for* λ_1. In this section we refine the result of Theorem 3.24 for the case of a double-well potential.

THEOREM 3.37. *In the small noise limit* $\varepsilon \to 0$ *the first eigenvalue of* $-L$ *satisfies*

$$\lambda_1 = \frac{\sqrt{\omega_0 \omega_+}}{2\pi} e^{-\frac{v}{\varepsilon}} (1 + \mathcal{O}(\varepsilon)). \tag{3.105}$$

PROOF. According to the variational principle formulated in Theorem 3.20

$$\lambda_1 = \frac{\langle -L\Phi_1, \Phi_1 \rangle_{\rho^{-1}}}{\langle \Phi_1, \Phi_1 \rangle_{\rho^{-1}}} = \frac{\varepsilon}{2} \frac{\|\Phi_1'\|^2_{\rho^{-1}}}{\|\Phi_1\|^2_{\rho^{-1}}}.$$

The norms of Φ_1 and Φ_1' were obtained in Theorem 3.36. Applying (3.103) and (3.104) results in

$$\lambda_1 = \frac{\varepsilon}{2} \frac{\|\Phi_1'\|^2_{\rho^{-1}}}{\|\Phi_1\|^2_{\rho^{-1}}} = \frac{\varepsilon}{2} \sqrt{\frac{\omega_0}{\pi\varepsilon}} \sqrt{\frac{\omega_+}{\pi\varepsilon}} e^{-\frac{v}{\varepsilon}} (1 + \mathcal{O}(\varepsilon)) = \frac{\sqrt{\omega_0\omega_+}}{2\pi} e^{-\frac{v}{\varepsilon}} (1 + \mathcal{O}(\varepsilon)).$$

□

3.3.5. The spectral gap between first and second eigenvalues of L.

THEOREM 3.38. *There exists a constant* $M > 0$ *and* $\varepsilon_0 > 0$ *such that for* $0 < \varepsilon \leq \varepsilon_0$ *the second eigenvalue of the operator* $-L$ *satisfies*

$$\lambda_2 \geq M. \tag{3.106}$$

PROOF. Let $\varepsilon > 0$. To prove the theorem we combine Courant's minimax principle and the method used in Buslov and Makarov [**19**] in the proof of the analogous result for the first eigenvalue for the corresponding problem on a closed interval. Indeed, it follows from Theorem 3.21 that for any two functions $v, w \in L^2(\mathbb{R}, \mathbb{R}; \rho^{-1}dx)$ (sometimes called *constraints*) the following inequality holds:

$$\lambda_2 \geq \inf_{\substack{f \in \mathcal{D}_L \\ f \perp v, w \\ f \neq 0}} \frac{\varepsilon}{2} \frac{\|f'\|^2_{\rho^{-1}}}{\|f\|^2_{\rho^{-1}}}, \tag{3.107}$$

We choose v and w in $C_0^\infty(\mathbb{R}, \mathbb{R})$ as approximations of the Dirac δ-distributions δ_{-1} and δ_1 in the distributional sense. Then (3.107) implies the following inequality in which $f \in C_0^\infty(\mathbb{R}, \mathbb{R})$ appear with the constraint $f(\pm 1) = 0$ (see Remark 3.22).

$$\lambda_2 \geq \inf_{\substack{f \in \mathcal{D}_L \\ f(-1)=0, f(1)=0 \\ f \neq 0}} \frac{\varepsilon}{2} \frac{\|f'\|^2_{\rho^{-1}}}{\|f\|^2_{\rho^{-1}}} = \inf_{\substack{f \in C_0^\infty(\mathbb{R},\mathbb{R}) \\ f(-1)=0, f(1)=0 \\ f \neq 0}} \frac{\varepsilon}{2} \frac{\|f'\|^2_{\rho^{-1}}}{\|f\|^2_{\rho^{-1}}}.$$

Take an arbitrary function $f \in C_0^\infty(\mathbb{R}, \mathbb{R})$, such that $f(\pm 1) = 0$ and estimate its norm $\|f\|^2_{\rho^{-1}}$.

Consider the half-line $[1, \infty)$. The potential U increases on this interval, and $U' \geq 0$ there. Then for $x \geq 1$

$$\begin{aligned} 0 \leq (f(x))^2 e^{-\frac{2U(x)}{\varepsilon}} &= 2 \int_1^x (f e^{-\frac{U}{\varepsilon}})(f e^{-\frac{U}{\varepsilon}})' \, dy \\ &= 2 \int_1^x f f' e^{-\frac{2U}{\varepsilon}} \, dy - \frac{2}{\varepsilon} \int_1^x f^2 U' e^{-\frac{2U}{\varepsilon}} \, dy. \end{aligned} \tag{3.108}$$

The Cauchy–Schwarz inequality implies that for $x \geq 1$

$$2 \int_1^x f f' e^{-\frac{2U}{\varepsilon}} \, dy \leq 2 \left[\int_1^x f^2 e^{-\frac{2U}{\varepsilon}} \, dy \int_1^x (f')^2 e^{-\frac{2U}{\varepsilon}} \, dy \right]^{\frac{1}{2}} \leq 2 \|f\|_{\rho^{-1}} \cdot \|f'\|_{\rho^{-1}}.$$

From (3.108) we get for $x \geq 1$

$$\frac{2}{\varepsilon} \int_1^x f^2 U' e^{-\frac{2U}{\varepsilon}} \, dy \leq 2 \int_1^x f f' e^{-\frac{2U}{\varepsilon}} \, dy \leq 2 \|f\|_{\rho^{-1}} \cdot \|f'\|_{\rho^{-1}},$$

and

$$(f(x))^2 e^{-\frac{2U(x)}{\varepsilon}} \leq 2 \int_1^x f f' e^{-\frac{2U}{\varepsilon}} \, dy \leq 2 \|f\|_{\rho^{-1}} \cdot \|f'\|_{\rho^{-1}}. \tag{3.109}$$

Let us fix some $\delta > 0$ to be specified later. Then we estimate

$$\begin{aligned} &\frac{2\delta}{\sqrt{\varepsilon}} \int_{\{U' \geq \delta \sqrt{\varepsilon}\}} f^2 e^{-\frac{2U}{\varepsilon}} \, dy = \frac{2}{\varepsilon} \int_{\{U' \geq \delta \sqrt{\varepsilon}\}} f^2 \delta \sqrt{\varepsilon} e^{-\frac{2U}{\varepsilon}} \, dy \\ &\leq \frac{2}{\varepsilon} \int_{\{U' \geq \delta \sqrt{\varepsilon}\}} f^2 U' e^{-\frac{2U}{\varepsilon}} \, dy \leq \frac{2}{\varepsilon} \int_1^\infty f^2 U' e^{-\frac{2U}{\varepsilon}} \, dy \leq 2 \|f\|_{\rho^{-1}} \cdot \|f'\|_{\rho^{-1}}. \end{aligned} \tag{3.110}$$

On the other hand, using (3.109) and $U'(1) = 0$, $U''(1) = \omega_+ > 0$, we obtain

$$\int_{\{U' \leq \delta \sqrt{\varepsilon}\} \cap [1, \infty)} f^2 e^{-\frac{2U}{\varepsilon}} \, dy \leq 2 \|f\|_{\rho^{-1}} \cdot \|f'\|_{\rho^{-1}} \cdot \frac{\delta \sqrt{\varepsilon}}{\omega_+} (1 + \mathcal{O}(\sqrt{\varepsilon})). \tag{3.111}$$

Combining (3.110) and (3.111) results in

$$\int_1^\infty f^2 e^{-\frac{2U}{\varepsilon}}\,dy \le 2\sqrt{\varepsilon}\|f\|_{\rho^{-1}}\cdot\|f'\|_{\rho^{-1}}\left(\frac{1}{2\delta}+\frac{\delta(1+\mathcal{O}(\sqrt{\varepsilon})}{\omega_+}\right).$$

Acting analogously on the intervals $(-\infty,-1]$, $[-1,0]$, and $[0,1]$ we obtain

$$\begin{aligned}\|f\|^2_{\rho^{-1}} &= \int_{-\infty}^\infty f^2 e^{-\frac{2U}{\varepsilon}}\,dy\\ &\le 2\sqrt{\varepsilon}\|f\|_{\rho^{-1}}\cdot\|f'\|_{\rho^{-1}}\left(\frac{1}{\delta}+\delta\Big(\frac{1}{\omega_+}+\frac{1}{\omega_0}+\frac{1}{\omega_-}\Big)(1+\mathcal{O}(\sqrt{\varepsilon}))\right).\end{aligned}$$

Minimizing the expression in parenthesises in the last formula in δ for ε small enough we find that it is bounded below by some $A>0$. Hence finally, there exists $\varepsilon_0>0$ such that for $0<\varepsilon\le\varepsilon_0$

$$\lambda_2 \ge \frac{\varepsilon}{2}\frac{\|f'\|^2_{\rho^{-1}}}{(2A\sqrt{\varepsilon}\|f'\|_{\rho^{-1}})^2} = M > 0,$$

where the constant M does not depend on ε. □

3.3.6. Some remarks abiut multi-dimensional case. We have studied the spectral properties of the generator of a small-noise diffusion in $\mathbb{R}$. In this section we briefly discuss the possibility to extend the results to the d-dimensional setting, $d\in\mathbb{N}$.

Denote by

$$Hf(x) = \frac{\varepsilon}{2}\Delta f(x) - \nabla U(x)\nabla f(x),\quad x\in\mathbb{R}^d,\ f\in C_0^\infty(\mathbb{R}^d,\mathbb{R}),$$

the generator of the d-dimensional diffusion induced by the SDE

$$dX_t = -\nabla U(X_t)\,dt + \sqrt{\varepsilon}\,dW_t,$$

where W is a standard d-dimensional Wiener process. The potential $U\colon\mathbb{R}^d\to\mathbb{R}$ is an infinitely differentiable function, and $U(x)\to+\infty$ as $|x|\to\infty$ with polynomial growth.

We are interested in the eigenvalues and eigenfunctions of H in the small noise limit $\varepsilon\to0$, especially in the existence of a spectral gap between its eigenvalues.

To get criteria for the discreteness of the spectrum of H let us consider the d-dimensional Schrödinger operator

$$hf(x) = \Delta f(x) - v(x),\quad x\in\mathbb{R}^d,\ f\in C_0^\infty(\mathbb{R}^d,\mathbb{R}),$$

with the Schrödinger potential

$$v(x) = \frac{(\nabla U(x))^2}{\varepsilon^2} - \frac{\Delta U(x)}{\varepsilon},\quad x\in\mathbb{R}^d. \tag{3.112}$$

The diffusion and Schrödinger operators are related by the formula

$$hf = \frac{2}{\varepsilon}e^{-\frac{U}{\varepsilon}}H(e^{\frac{U}{\varepsilon}}f)$$

Then the operator h defined on $C_0^\infty(\mathbb{R}^d,\mathbb{R})$ is essentially-self adjoint if a condition similar to the corresponding condition in the one-dimensional case holds (see Theorem 3.17).

THEOREM 3.39 ([**7**]). *Let the Schrödinger potential v satisfy the condition*

$$v(x) \geq -Q(|x|), \quad x \in \mathbb{R}^d,$$

where $Q(r)$ is an increasing positive continuous function on $[0, +\infty)$ such that

$$\int_0^\infty \frac{dr}{\sqrt{Q(2r)}} = \infty.$$

Then $(h, C_0^\infty(\mathbb{R}^d, \mathbb{R}))$ is essentially self-adjoint.

The condition of the theorem is obviously satisfied for v defined by (3.112).

The diffusion generator $(H, C_0^\infty(\mathbb{R}^d, \mathbb{R}))$ is then an essentially self-adjoint and non-positive operator in $L^2(\mathbb{R}^d, \mathbb{R}; e^{-\frac{2U}{\varepsilon}} dx)$. The increase of the potential U at infinity implies discreteness of the spectrum and the existence of an orthonormal system of eigenfunctions in $L^2(\mathbb{R}^d, \mathbb{R}; e^{-\frac{2U}{\varepsilon}} dx)$. Moreover, the variational principles entail that the zeroth eigenvalue of H is zero, and the corresponding eigenfunction is constant. The spectrum of H is non-positive.

We have seen that in the one-dimensional case of a double-well potential the first eigenvalue is exponentially small in ε, and the second eigenvalue is bounded away from zero by a constant which does not depend on ε. It turns out that a similar result holds in $\mathbb{R}^d$.

THEOREM 3.40 (Kolokoltsov [**64**]). *Let the potential U have exactly n non-degenerate minima in $\mathbb{R}^d$. Then the first n eigenvalues $\mu_0(\varepsilon), \mu_1(\varepsilon), \ldots, \mu_{n-1}(\varepsilon)$ of H satisfy $\mu_0(\varepsilon) = 0$, $\mu_1(\varepsilon), \ldots, \mu_{n-1}(\varepsilon)$ are exponentially small in the small noise limit $\varepsilon \to 0$, and $|\mu_n(\varepsilon)|$ is bounded below by a positive constant not depending on ε.*

In particular in a double-well situation if we denote by Ω_i, $i = 1, 2$, the domains of attraction of the dynamical system $\dot{x} = -\nabla U(x)$, then it turns out that the eigenfunctions corresponding to the exponentially small eigenvalues of H are exponentially close in the $L^2(\mathbb{R}^d, \mathbb{R}; e^{-\frac{2U}{\varepsilon}} dx)$-norm to constants on Ω_i, see [**64**, Proposition 8.2.2 and Corollary]

The precise asymptotics of the exponentially small eigenvalues including the pre-factors has been obtained by Bovier *et al.* in [**15**] and [**16**] with help of the method of Newtonian capacities (see Theorem 1.1 in [**16**]). There they also gave another precise approximations of the corresponding normalized eigenfunctions.

3.4. Semiclassical approach to stochastic resonance

In this Section we return to the diffusion in a time-periodic double-well potential. Our aim is to compare the spectral power amplification coefficient of the diffusion with its counterpart for the dynamically adapted continuous-time Markov chain on the two-point space composed of the metastable states of the diffusion. The invariant density of the diffusion satisfies the forward Kolmogorov (Fokker–Planck) equation, which is in this case a parabolic partial differential equation with antisymmetric boundary conditions. The invariant density is described by some type of Fourier expansion along the discrete spectrum of the diffusion's infinitesimal generator. It is analysed with the fine asymptotic results of Section 3.3. The noise-independent spectral gap between the first and the second eigenvalues of the infinitesimal generator implies that only the terms of the invariant density expansion which correspond to the first two eigenvalues play a significant role in the induced expansion of the SPA coefficient. Theorem 3.50 contains the asymptotics of the SPA coefficient, if the noise parameter ε runs through the intervals $[\frac{v+\delta}{\ln T}, \frac{2V}{\ln T}]$, $\delta > 0$. In the large period limit $T \to \infty$ these intervals shrink in the natural scale on the one hand. Freidlin [**39**] on the other hand suggests that these are the relevant resonance intervals to look for. This impression is in particular supported by the observation made in Section 3.2.2 where we prove that the Markov chain SPA coefficient has a local maximum at $\varepsilon \sim \frac{v+V}{2\ln T}$. Surprisingly, it turns out that for the diffusion the SPA tuning curve is either decreasing or increasing on the resonance intervals. This means that the reduction to a Markov chain on the metastable states, however naturally it main retain the dynamical properties of the diffusion, does not preserve optimal tuning effects, at least not for the physicists' favourite measure of quality. The reason for this is hidden in the significance of many small random oscillations of the diffusion in the potential valley bottoms where it spends most of the time. If we cut off these fluctuations by identifying the valley bottoms with the minima themselves, we obtain a modified SPA coefficient which exactly shows the same resonance effects as the Markov chain in the large period limit (small noise limit). The exposition follows Pavlyukevich [**86**] and Imkeller and Pavlyukevich [**59**].

3.4.1. Diffusion with time-periodic drift and its invariant density. We return now to our principal subject of investigation and consider a family of diffusion processes $X^{\varepsilon,T} = (X_t^{\varepsilon,T})_{t\geq 0}$ given by the real-valued stochastic differential equation

$$dX_t^{\varepsilon,T} = -U'(X_t^{\varepsilon,T}, \frac{1}{T})\,dt + \sqrt{\varepsilon}\,dW_t, \quad \varepsilon, T > 0, \tag{3.113}$$

where W is a standard 1-dimensional Wiener process on some complete probability space $(\Omega, \mathcal{F}, \mathbb{P})$.

We have already introduced the potential function $U(\cdot,\cdot)$ in Chapter 1. It is periodic in time, i.e. $U(\cdot, t) = U(\cdot, t+1)$. We also assume that it is a step-function in time alternating between two spatially antisymmetric states, i.e.

$$U(x,t) = \begin{cases} U(x), & t \in [k, k+\frac{1}{2}), \\ U(-x), & t \in [k+\frac{1}{2}, k+1), \quad k = 0, 1, \ldots \end{cases} \tag{3.114}$$

We assume that the potential function U satisfies the assumptions (**S**), (**G**), and (**M**) formulated on the page 92.

The process $X^{\varepsilon,T}$ has continuous trajectories. The fast increase of the potential at infinity guarantees that the process is positively recurrent. This means that for any initial point, the mean time to enter any interval on $\mathbb{R}$ is finite. For details see [**48**], [**40**].

The results of Section 3.3 yield that under conditions formulated above the infinitesimal generator L associated with time-invariant potential function U considered as an operator in $L^2(\mathbb{R}, \mathbb{R}; e^{-\frac{2U}{\varepsilon}}\,dx)$ has a discrete spectrum and an orthogonal system of eigenfunctions.

Note that $X^{\varepsilon,T}$ is time-homogeneous during the half-periods $[\frac{T}{2}k, \frac{T}{2}(k+1))$, $k \geq 0$. Moreover, during the intervals $[Tk, Tk+\frac{T}{2})$, $k \geq 0$, it is a solution of

$$dX_t^{\varepsilon,T} = -U'(X_t^{\varepsilon,T})\,dt + \sqrt{\varepsilon}\,dW_t,$$

and during the intervals $[Tk + \frac{T}{2}, T(k+1))$, $k \geq 0$, a solution of

$$dX_t^{\varepsilon,T} = -\overline{U}'(X_t^{\varepsilon,T})\, dt + \sqrt{\varepsilon}\, dW_t,$$

where $\overline{U}(x) = U(-x)$, $x \in \mathbb{R}$.

Consider the time-homogeneous diffusion given by the SDE

$$dX_t^\varepsilon = -U'(X_t^\varepsilon)\, dt + \sqrt{\varepsilon}\, dW_t. \tag{3.115}$$

The process X^ε has a unique invariant measure m_0^ε, on the Borel σ-algebra $\mathcal{B}(\mathbb{R})$, i.e. the probability measure such that

$$m_0^\varepsilon(A) = \int_{\mathbb{R}} P^\varepsilon(t, x, A) m_0^\varepsilon(dx),$$

where $P^\varepsilon(t, x, A)$, $t \geq 0$, $x \in \mathbb{R}$, $A \in \mathcal{B}(\mathbb{R})$ is the transition function of X^ε. The measure m_0^ε has a density μ_0^ε, which is the unique positive solution of the forward Kolmogorov (Fokker–Planck) equation

$$\frac{\varepsilon}{2}\frac{d^2}{dx^2}\mu_0^\varepsilon(x) + \frac{d}{dx}(U'(x)\mu_0^\varepsilon(x)) = 0,$$
$$\mu_0^\varepsilon(x) > 0, \quad \int_{\mathbb{R}} \mu_0^\varepsilon(x)\, dx = 1, \quad x \in \mathbb{R}.$$

In case when the diffusion drift is a gradient of some potential, as for example in

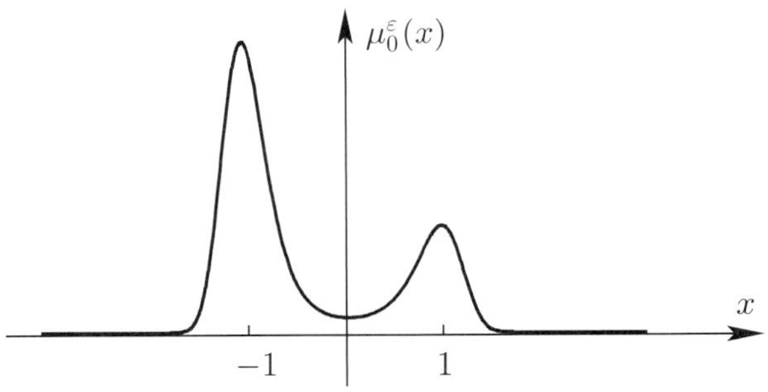

FIGURE 3.3. The invariant density $\mu_0^\varepsilon(x)$ of the time homogeneous diffusion X^ε.

(3.115), μ_0^ε is given explicitly by the formula

$$\mu_0^\varepsilon(x) = c_\varepsilon e^{-\frac{2U(x)}{\varepsilon}}, \quad x \in \mathbb{R},$$

with $c_\varepsilon = \int_{\mathbb{R}} e^{-\frac{2U(y)}{\varepsilon}}\, dy$. A typical form of the invariant density μ_0^ε of a time-homogeneous diffusion X^ε in a double-well potential is shown in Figure 3.3.

Moreover, for any initial conditions, the law of X_t^ε converges to an invariant measure m_0^ε as $t \to \infty$. This means that asymptotically the process ‘forgets’ about its initial conditions.

In case the process is defined by the SDE (3.113) there is no invariant measure in the usual sense as the process in not time-homogeneous. On the other hand, the time-periodicity of the drift $-U'(\cdot, t)$ suggests that the law of $X_t^{\varepsilon,T}$ should converge to some time-periodic law.

For convenience we rescale time $t \mapsto \frac{t}{T}$. To give a rigorous mathematical meaning to this we consider a new two-dimensional process

$$\mathbf{X}_t^{\varepsilon,T} = (X_{Tt}^{\varepsilon,T}, t \bmod 1), \quad t \geq 0. \tag{3.116}$$

The process $\mathbf{X}^{\varepsilon,T}$ takes values on the cylinder $\mathbb{R} \times S^1$, and is a *time-homogeneous* Markov process. This means that it has an invariant measure $\mathbf{m}^{\varepsilon,T}$ on $\mathcal{B}(\mathbb{R} \times S^1)$.

The measure $\mathbf{m}^{\varepsilon,T}$ has a density $\mu^{\varepsilon,T}$, which is a unique positive solution of the forward Kolmogorov equation for the process $\mathbf{X}^{\varepsilon,T}$, namely,

$$A_{\varepsilon,T}^* \mu^{\varepsilon,T}(x,\theta) = 0, \quad (x,\theta) \in \mathbb{R} \times (0,1), \tag{3.117}$$

with a continuity condition

$$\mu^{\varepsilon,T}(\cdot,0) = \mu^{\varepsilon,T}(\cdot,1) \tag{3.118}$$

and such that $\int_0^1 \int_{\mathbb{R}} \mu^{\varepsilon,T}(x,\theta)\, dx\, d\theta = 1$. Note that $\mu^{\varepsilon,T}(\cdot,\theta)$ determines the law of the r.v. $X_\theta^{\varepsilon,T}$, and therefore $\int_{\mathbb{R}} \mu^{\varepsilon,T}(x,\theta)\, dx = 1$ for any $\theta \in [0,1]$.

The operator

$$A_{\varepsilon,T}^* f = \frac{\varepsilon}{2} \frac{\partial^2}{\partial x^2} f + \frac{\partial}{\partial x}\left(f \frac{\partial}{\partial x} U \right) - \frac{1}{T} \frac{\partial}{\partial \theta} f, \quad f \in C_0^\infty(\mathbb{R} \times S^1, \mathbb{R}),$$

is the formal ajoint of the infinitesimal generator of $\mathbf{X}^{\varepsilon,T}$

$$A_{\varepsilon,T} f = \frac{\varepsilon}{2} \frac{\partial^2}{\partial x^2} f - \frac{\partial}{\partial x} f \frac{\partial}{\partial x} U + \frac{1}{T} \frac{\partial}{\partial \theta} f, \quad f \in C_0^\infty(\mathbb{R} \times S^1, \mathbb{R}).$$

Taking (3.114) into account we rewrite (3.117) and (3.118) in the form

$$\begin{cases} \frac{\varepsilon}{2} \frac{\partial^2}{\partial x^2} \mu^{\varepsilon,T} + \frac{\partial}{\partial x}\left(\mu^{\varepsilon,T} U' \right) = \frac{1}{T} \frac{\partial}{\partial \theta} \mu^{\varepsilon,T}, \text{ on } \mathbb{R} \times (0, \frac{1}{2}), \\ \frac{\varepsilon}{2} \frac{\partial^2}{\partial x^2} \mu^{\varepsilon,T}(x,\theta) + \frac{\partial}{\partial x}\left(\mu^{\varepsilon,T} \overline{U}' \right) = \frac{1}{T} \frac{\partial}{\partial \theta} \mu^{\varepsilon,T}, \text{ on } \mathbb{R} \times (\frac{1}{2}, 1), \\ \mu^{\varepsilon,T} \text{ is positive and continuous on } \mathbb{R} \times [0,1], \\ \int_{\mathbb{R}} \mu^{\varepsilon,T}(x,\theta)\, dx = 1, \;\; \theta \in [0,1]. \end{cases} \tag{3.119}$$

From the spatial antisymmetry (3.114) we can immediately deduce a similar antisymmetry property for $\mu^{\varepsilon,T}$.

PROPOSITION 3.41. For $x \in \mathbb{R}$ and $\theta \in (0, \frac{1}{2})$ we have

$$\mu^{\varepsilon,T}(x,\theta) = \mu^{\varepsilon,T}(-x, \theta + \tfrac{1}{2}).$$

This proposition is analogous to Proposition 3.7 concerning the invariant law of the Markov chain.

Proposition 3.41 together with (3.119) show that it is enough to find the invariant density on the first half-period, i.e. in the strip $\mathbb{R} \times [0, \frac{1}{2}]$, and in this strip $\mu^{\varepsilon,T}$ is a solution of the following boundary-value problem

$$\begin{cases} \frac{\varepsilon}{2} \frac{\partial^2}{\partial x^2} \mu^{\varepsilon,T} + \frac{\partial}{\partial x}(\mu^{\varepsilon,T} U') = \frac{1}{T} \frac{\partial}{\partial \theta} \mu^{\varepsilon,T} \text{ on } \mathbb{R} \times (0, \frac{1}{2}), \\ \mu^{\varepsilon,T}(\cdot, 0) = \mu^{\varepsilon,T}(-\cdot, \frac{1}{2}), \\ \mu^{\varepsilon,T} > 0, \quad \text{on } \mathbb{R} \times [0, \frac{1}{2}], \\ \int_{\mathbb{R}} \mu^{\varepsilon,T}(x,\theta)\, dx = 1, \;\; \theta \in [0, \frac{1}{2}]. \end{cases} \tag{3.120}$$

A typical form of the invariant density $\mu^{\varepsilon,T}$ as a function of x and θ is shown in Figure 3.4. In the following section we describe $\mu^{\varepsilon,T}$ in the small noise limit $\varepsilon \to 0$ in a Fourier type expansion.

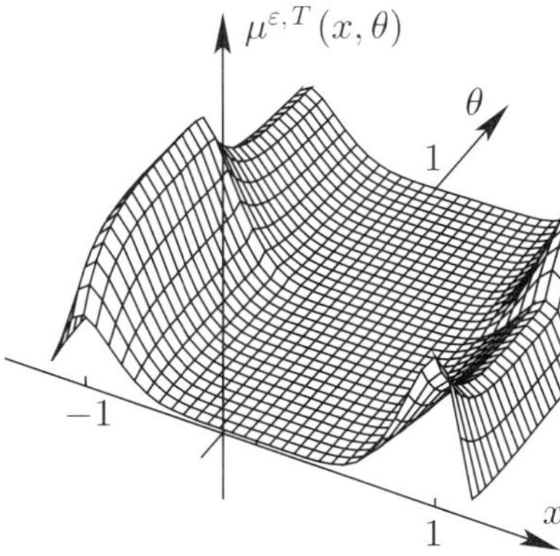

FIGURE 3.4. The invariant density $\mu^{\varepsilon,T}(x,\theta)$ of the time-inhomogeneous diffusion $X^{\varepsilon,T}$.

3.4.2. Asymptotic expansion of the invariant density. To solve (3.120), we proceed by separation of variables to an eigenvalue problem for the infinitesimal generator of a time-homogeneous diffusion. Assume that a solution $\mu^{\varepsilon,T}$ of the partial differential equation in (3.120) allows a factorization

$$\mu^{\varepsilon,T}(x,\theta) = \Psi^{\varepsilon,T}(x)\Theta^{\varepsilon,T}(\theta), \quad x \in \mathbb{R}, \theta \in (0, \tfrac{1}{2})$$

Then it follows from (3.120) that

$$\frac{\varepsilon}{2}\frac{(\Psi^{\varepsilon,T})''}{\Psi^{\varepsilon,T}} + U'\frac{(\Psi^{\varepsilon,T})'}{\Psi^{\varepsilon,T}} + U'' = \frac{1}{T}\frac{\dot{\Theta}^{\varepsilon,T}}{\Theta^{\varepsilon,T}} = -\lambda,$$

where $f' = \frac{df}{dx}$ and $\dot{g} = \frac{dg}{d\theta}$. The constant λ does not depend on x and θ and is an eigenvalue of the differential operator $-L^*$,

$$L^* f = \frac{\varepsilon}{2} f'' + U' f' + U'' f, \quad f \in C_0^\infty(\mathbb{R}, \mathbb{R}). \tag{3.121}$$

In the previous Section we have shown that the operator L^* is the formal adjoint of the infinitesimal generator of the time-homogeneous diffusion with potential U. It has a discrete spectrum $\{-\lambda_k\}_{k\geq 0}$ such that $0 = \lambda_0 < \lambda_1 < M < \lambda_2 < \cdots$. The first eigenvalue of L^* is exponentially small in ε and is equal to $\lambda_1 = \frac{\sqrt{\omega_0\omega_+}}{2\pi} e^{-\frac{v}{\varepsilon}}(1 + \mathcal{O}(\varepsilon))$. The ε-independent constant M determines the spectral gap between the first and the second eigenvalues of L^*.

The normalized eigenfunctions $\{\frac{\Psi_k}{\|\Psi_k\|_\rho}\}_{k\geq 0}$ provide a complete orthonormal system in $L^2(\mathbb{R}, \mathbb{R}; \rho dx)$ with $\rho = e^{\frac{2U}{\varepsilon}}$, i.e.

$$\int_{\mathbb{R}} \frac{\Psi_k(y)}{\|\Psi_k\|_\rho}\frac{\Psi_j(y)}{\|\Psi_j\|_\rho} e^{\frac{2U(y)}{\varepsilon}}\, dy = \delta_{kj}, \quad k, j = 0, 1, \ldots \tag{3.122}$$

Recall that the operator L has the same spectrum $\{-\lambda_k\}_{k\geq 0}$, and its eigenfunctions $\{\Phi_k\}_{k\geq 0}$ are related to those of L^* by $\Psi_k = e^{-\frac{2U}{\varepsilon}}\Phi_k$, $k \geq 0$.

Let $\mu^{\varepsilon,T}$ be the unique solution of (3.120). Consider $\mu^{\varepsilon,T}(\cdot,0)$ and expand it into the Fourier series with respect to the system $\{\Psi_k\}_{k\geq 0}$

$$\mu^{\varepsilon,T}(\cdot,0)=\sum_{k=0}^{\infty}a_k^{\varepsilon,T}\frac{\Psi_k}{\|\Psi_k\|_\rho},\tag{3.123}$$

where the Fourier coefficients are determined by the inner products

$$a_k^{\varepsilon,T}=\left\langle\frac{\Psi_k}{\|\Psi_k\|_\rho},\mu^{\varepsilon,T}(\cdot,0)\right\rangle_\rho,\quad k\geq 0.$$

The Parseval equality states that

$$\|\mu^{\varepsilon,T}(\cdot,0)\|_\rho^2=\sum_{k=0}^{\infty}(a_k^{\varepsilon,T})^2.$$

From the existence and uniqueness theorem for parabolic partial differential equations [**34, 92**] we conclude the following

PROPOSITION 3.42. $\mu^{\varepsilon,T}$, $x\in\mathbb{R}$, $s\in[0,1/2]$, can be represented by the series

$$\mu^{\varepsilon,T}(x,s)=\sum_{k=0}^{\infty}a_k^{\varepsilon,T}\frac{\Psi_k(x)}{\|\Psi_k\|_\rho}\exp\left(-T\lambda_k\, s\right),\quad x\in\mathbb{R}, s\in[0,\tfrac{1}{2}].\tag{3.124}$$

The coefficients $a_k^{\varepsilon,T}$ in the representation (3.123) can be expressed in terms of the function $\mu^{\varepsilon,T}(\cdot,0)$. Next we find the first coefficient in closed form.

PROPOSITION 3.43. For $\varepsilon>0$ and $T>0$

$$a_0^{\varepsilon,T}=a_0^{\varepsilon}=\|\Psi_0\|_\rho^{-1}=\left(\int_{\mathbb{R}}e^{-\frac{2U(y)}{\varepsilon}}\,dy\right)^{-1/2}.\tag{3.125}$$

PROOF. Using the condition that $\int_{\mathbb{R}}\mu^{\varepsilon,T}(x,s)\,dx=1$ for any s, and $\Psi_0=e^{-\frac{2U}{\varepsilon}}$ we find

$$\begin{aligned}1&=\int_{\mathbb{R}}\mu^{\varepsilon,T}(x,0)e^{\frac{2U(x)}{\varepsilon}}\frac{\Psi_0(x)}{\|\Psi_0\|_\rho}\|\Psi_0\|_\rho\,dx=\|\Psi_0\|_\rho\left\langle\mu^{\varepsilon,T}(\cdot,0),\frac{\Psi_0}{\|\Psi_0\|_\rho}\right\rangle_\rho\\&=a_0^{\varepsilon,T}\|\Psi_0\|_\rho=a_0^{\varepsilon,T}\left(\int_{\mathbb{R}}e^{-\frac{2U(y)}{\varepsilon}}\,dy\right)^{1/2}.\end{aligned}$$

□

COROLLARY 3.44. Let $T>0$ and $s\geq 0$. There exists $\varepsilon_0>0$ such that for $0<\varepsilon\leq\varepsilon_0$

$$\|\mu^{\varepsilon,T}(\cdot,s)\|_\rho\geq a_0^{\varepsilon}\geq C\varepsilon^{-1/4}e^{-\frac{V}{2\varepsilon}}$$

PROOF. The proof consists in evaluating the integral from (3.125) by Laplace's method, see (B.9). □

The second coefficient in the expansion (3.124) cannot be described explicitly. But the following proposition gives an approximation the goodness of which is guaranteed by the spectral gap.

PROPOSITION 3.45. Let M be the constant according to Theorem 3.38 marking the spectral gap. There exists $\varepsilon_0 > 0$ such that for $0 < \varepsilon \leq \varepsilon_0$ and $T > 0$ we have

$$a_1^{\varepsilon,T} = \frac{1}{\|\Psi_0\|_\rho} \frac{\left\langle \frac{\overline{\Psi}_0}{\|\Psi_0\|_\rho}, \frac{\Psi_1}{\|\Psi_1\|_\rho} \right\rangle_\rho + r(\varepsilon, T)}{1 - e^{-\lambda_1 T/2} \left\langle \frac{\overline{\Psi}_1}{\|\Psi_1\|_\rho}, \frac{\Psi_1}{\|\Psi_1\|_\rho} \right\rangle_\rho}, \tag{3.126}$$

where

$$|r(\varepsilon, T)| \leq e^{-MT/2} \|\mu^{\varepsilon,T}(\cdot, 0)\|_\rho \cdot \left\| \frac{\overline{\Psi}_1}{\|\Psi_1\|_\rho} e^{\frac{2(\overline{U} - U)}{\varepsilon}} \right\|_\rho. \tag{3.127}$$

PROOF. From (3.124) and the boundary condition $\mu^{\varepsilon,T}(x, 0) = \mu^{\varepsilon,T}(-x, \frac{1}{2})$, $x \in \mathbb{R}$, we obtain

$$\sum_{k=0}^{\infty} a_k^{\varepsilon,T} \frac{\Psi_k}{\|\Psi_k\|_\rho} = \sum_{k=0}^{\infty} a_k^{\varepsilon,T} e^{-\lambda_k T/2} \frac{\overline{\Psi}_k}{\|\Psi_k\|_\rho}. \tag{3.128}$$

Multiplying both sides of (3.128) by $e^{\frac{2U}{\varepsilon}} \frac{\Psi_1}{\|\Psi_1\|_\rho}$, integrating, and using (3.122) gives

$$a_1^{\varepsilon,T} = a_0^{\varepsilon} \left\langle \frac{\overline{\Psi}_0}{\|\Psi_0\|_\rho}, \frac{\Psi_1}{\|\Psi_1\|_\rho} \right\rangle_\rho + a_1^{\varepsilon,T} e^{-\lambda_1 T/2} \left\langle \frac{\overline{\Psi}_1}{\|\Psi_1\|_\rho}, \frac{\Psi_1}{\|\Psi_1\|_\rho} \right\rangle_\rho + r(\varepsilon, T),$$

where

$$r(\varepsilon, T) = \sum_{k=2}^{\infty} a_k^{\varepsilon,T} e^{-\lambda_k T/2} \left\langle \frac{\overline{\Psi}_k}{\|\Psi_k\|_\rho}, \frac{\Psi_1}{\|\Psi_1\|_\rho} \right\rangle_\rho.$$

To estimate the remainder term r, we now use the spectral gap result from Theorem 3.38. We may choose ε_0 such that for $\varepsilon \leq \varepsilon_0$ the third eigenvalue $\lambda_2 \geq M > 0$. Hence

$$\begin{aligned} |r(\varepsilon, T)| &\leq e^{-MT/2} \sum_{k=2}^{\infty} |a_k^{\varepsilon,T}| \left| \left\langle \frac{\overline{\Psi}_k}{\|\Psi_k\|_\rho}, \frac{\Psi_1}{\|\Psi_1\|_\rho} \right\rangle_\rho \right| \\ &\leq e^{-MT/2} \left[\sum_{k=2}^{\infty} (a_k^{\varepsilon,T})^2 \right]^{1/2} \left[\sum_{k=2}^{\infty} \left\langle \frac{\overline{\Psi}_k}{\|\Psi_k\|_\rho}, \frac{\Psi_1}{\|\Psi_1\|_\rho} \right\rangle_\rho^2 \right]^{1/2}. \end{aligned}$$

The inner products can be rewritten in the following form:

$$\begin{aligned} \left\langle \frac{\overline{\Psi}_k}{\|\Psi_k\|_\rho}, \frac{\Psi_1}{\|\Psi_1\|_\rho} \right\rangle_\rho &= \int_{\mathbb{R}} \frac{\overline{\Psi}_k}{\|\Psi_k\|_\rho} \frac{\Psi_1}{\|\Psi_1\|_\rho} e^{\frac{2U}{\varepsilon}} \, dx \\ &= \int_{\mathbb{R}} \frac{\Psi_k}{\|\Psi_k\|_\rho} \frac{\overline{\Psi}_1}{\|\Psi_1\|_\rho} e^{\frac{2\overline{U}}{\varepsilon}} e^{-\frac{2U}{\varepsilon}} e^{\frac{2U}{\varepsilon}} \, dx \\ &= \left\langle \frac{\overline{\Psi}_1}{\|\Psi_1\|_\rho} e^{\frac{2(\overline{U} - U)}{\varepsilon}}, \frac{\Psi_k}{\|\Psi_k\|_\rho} \right\rangle_\rho. \end{aligned}$$

Two applications of Parseval's equality complete the estimate (3.127). □

The only function not even approximately known which appears in the estimate (3.127) is $\mu^{\varepsilon,T}(\cdot, 0)$. The following proposition provides an upper bound for its ρ-norm in terms of the ρ-norm of the explicit function $\overline{\rho}^{-1} = e^{-\frac{2\overline{U}}{\varepsilon}}$. Denote $c = \int_{\mathbb{R}} \rho^{-1}(y) \, dy = (a_0^{\varepsilon})^{-2} \approx \sqrt{\frac{\pi\varepsilon}{\omega_-}} e^{\frac{V}{\varepsilon}}$ (see Theorem 3.36). Then, the function $\overline{\rho}^{-1}/c$ is the invariant density of the time-homogeneous diffusion in the potential

$\overline{U}(x) = U(-x)$, $x \in \mathbb{R}$. The spectral gap between $\lambda_0 = 0$ and λ_1 of the corresponding infinitesimal generator implies that the law of the diffusion converges to the invariant law exponentially fast and determines the rate of convergence. This indicates that the density $\mu^{\varepsilon,T}(\cdot,0) = \mu^{\varepsilon,T}(\cdot,1)$ should be close to $\overline{\rho}^{-1}/c$ if T is large enough. One can obtain the following estimate for the norm.

PROPOSITION 3.46. For any $\delta > 0$ there exist $T_0, \varepsilon_0 > 0$ such that for $\varepsilon \in [\frac{v+\delta}{\ln T}, \varepsilon_0]$

$$\left\|\mu^{\varepsilon,T}(\cdot,0)\right\|_\rho \le 6\left\|\overline{\rho}^{-1}/c\right\|_\rho. \tag{3.129}$$

PROOF. Combining the triangle inequality with the inequality $(a+b)^2 \le 2(a^2+b^2)$, $a, b \in \mathbb{R}$, gives

$$\left\|\mu^{\varepsilon,T}(\cdot,0)\right\|_\rho^2 \le 2\left\|\mu^{\varepsilon,T}(\cdot,0) - \frac{\overline{\rho}^{-1}}{c}\right\|_\rho^2 + 2\left\|\frac{\overline{\rho}^{-1}}{c}\right\|_\rho^2. \tag{3.130}$$

For $n \in \mathbb{N}$, denote by P_n^Ψ the orthogonal projector on the orthogonal complement of the span of the first n eigenfunctions $\Psi_0, \ldots, \Psi_{n-1}$. Let us estimate the first summand in (3.130). Using the boundary condition in (3.120) we obtain

$$\begin{aligned}
&\left\|\mu^{\varepsilon,T}(\cdot,0) - \frac{\overline{\rho}^{-1}}{c}\right\|_\rho^2 = \left\|\overline{\mu}^{\varepsilon,T}(\cdot,\tfrac{1}{2}) - \frac{\overline{\rho}^{-1}}{c}\right\|_\rho^2 = \left\|\mu^{\varepsilon,T}(\cdot,\tfrac{1}{2}) - \frac{\rho^{-1}}{c}\right\|_{\overline{\rho}}^2 \\
&\quad = \int_{\mathbb{R}} \left(P_1^\Psi \mu^{\varepsilon,T}(\cdot,\tfrac{1}{2})\right)^2 \overline{\rho}\, dy = \int_{\mathbb{R}} \left|P_1^\Psi \mu^{\varepsilon,T}(\cdot,\tfrac{1}{2})\right| \cdot \left|\mu^{\varepsilon,T}(\cdot,\tfrac{1}{2}) - \frac{\rho^{-1}}{c}\right| \overline{\rho}\, dy \\
&\quad \le \int_{\mathbb{R}} \left|P_1^\Psi \mu^{\varepsilon,T}(\cdot,\tfrac{1}{2})\right| \mu^{\varepsilon,T}(\cdot,\tfrac{1}{2})\overline{\rho}\, dy + \int_{\mathbb{R}} \left|P_1^\Psi \mu^{\varepsilon,T}(\cdot,\tfrac{1}{2})\right| \frac{\rho^{-1}}{c}\overline{\rho}\, dy \\
&\quad \le \int_{\mathbb{R}} \left|P_1^\Psi \mu^{\varepsilon,T}(\cdot,\tfrac{1}{2})\right| \mu^{\varepsilon,T}(\cdot,\tfrac{1}{2})\overline{\rho}\rho^{-1}\, \rho\, dy + \int_{\mathbb{R}} \left|P_1^\Psi \mu^{\varepsilon,T}(\cdot,\tfrac{1}{2})\right| \frac{\rho^{-2}}{c}\overline{\rho}\,\rho\, dy \\
&\quad \le \left\|P_1^\Psi \mu^{\varepsilon,T}(\cdot,\tfrac{1}{2})\right\|_\rho \left[\int_{\mathbb{R}} \left(\mu^{\varepsilon,T}(\cdot,\tfrac{1}{2})\right)^2 (\overline{\rho}\rho^{-1})^2 \rho\, dy\right]^{\frac{1}{2}} + \\
&\quad + \left\|P_1^\Psi \mu^{\varepsilon,T}(\cdot,\tfrac{1}{2})\right\|_\rho \left[\int_{\mathbb{R}} \left(\frac{\overline{\rho}\rho^{-2}}{c}\right)^2 \rho\, dy\right]^{\frac{1}{2}} \\
&\quad \le e^{-\lambda_1 T/2} \max\{\overline{\rho}\rho^{-1}\} \cdot \|\mu^{\varepsilon,T}(\cdot,0)\|_\rho^2 + \frac{e^{-\lambda_1 T/2}}{c} \left[\int_{\mathbb{R}} \overline{\rho}^2 \rho^{-3}\, dy\right]^{\frac{1}{2}} \|\mu^{\varepsilon,T}(\cdot,0)\|_\rho.
\end{aligned}$$

Taking into account the latter inequality and (3.130) we obtain a quadratic inequality for $\|\mu^{\varepsilon,T}(\cdot,0)\|_\rho$

$$\begin{aligned}
&\left(1 - 2e^{-\lambda_1 T} \max\{\overline{\rho}\rho^{-1}\}\right)\left\|\mu^{\varepsilon,T}(\cdot,0)\right\|_\rho^2 \\
&\qquad - 2\frac{e^{-\lambda_1 T}}{c}\left[\int_{\mathbb{R}} \overline{\rho}^2 \rho^{-3}\, dy\right]^{\frac{1}{2}} \left\|\mu^{\varepsilon,T}(\cdot,0)\right\|_\rho - 2\left\|\frac{\overline{\rho}^{-1}}{c}\right\|_\rho^2 \le 0.
\end{aligned} \tag{3.131}$$

Let us estimate the coefficients of (3.131) and thus find an upper bound for $\|\mu^{\varepsilon,T}(\cdot,0)\|_\rho$.

For any $\delta > 0$, let $\frac{v+\delta}{\ln T} \leq \varepsilon \leq \varepsilon_0$, where ε_0 is is given by Theorem 3.38. Then for some $C > 0$, recalling Theorem 3.37

$$e^{-\lambda_1 T/2} \leq \exp\Big\{ -\frac{\sqrt{\omega_0\omega_+}}{4\pi} T^{-\frac{v}{v+\delta}} T\Big(1+\mathcal{O}\Big(\frac{1}{\ln T}\Big)\Big)\Big\} \leq \exp\Big\{ -CT^{\frac{\delta}{v+\delta}}\Big\}. \tag{3.132}$$

This expression tends to 0 exponentially fast as $T \to \infty$. Moreover,

$$\max\{\overline{\rho}\rho^{-1}\} = \exp\Big\{\frac{2}{\varepsilon}\max_{x\in\mathbb{R}}(U(-x)-U(x))\Big\} = e^{\frac{\alpha}{\varepsilon}},$$

where $\alpha = 2\max_{x\in\mathbb{R}}\{U(-x)-U(x)\} \geq V - v > 0$ is a finite number, defined by the potential U. For $\varepsilon \geq \frac{v+\delta}{\ln T}$ we obtain

$$\max\{\overline{\rho}\rho^{-1}\} \leq T^{\frac{\alpha}{v+\delta}}. \tag{3.133}$$

Similarly, let $\beta' = 2\max\{2U(-x) - 3U(x)\} \geq 2(2U(1) - 3U(-1)) = 3V - 2v > 0$. Then, using Laplace's method, see (B.9), and recalling (**G**) we can estimate for some $C_1 > 0$

$$\frac{1}{c}\Big[\int_{\mathbb{R}} \overline{\rho}^2\rho^{-3}\,dy\Big]^{\frac{1}{2}} \leq C_1\sqrt{\varepsilon}e^{\frac{\beta'-V}{\varepsilon}} \leq T^{\frac{\beta}{v+\delta}} \tag{3.134}$$

for $\varepsilon \geq \frac{v+\delta}{\ln T}$ and some $\beta > \beta' - V \geq 2(V-v) > 0$.

Since $\max\{U(x) - 2U(-x)\} \geq U(1) - 2U(-1) = V - \frac{v}{2} > 0$, the free term of (3.131) is estimated in $\frac{v+\delta}{\ln T} \leq \varepsilon \leq \varepsilon_0$ by

$$\Big\|\frac{\overline{\rho}^{-1}}{c}\Big\|_\rho^2 = \frac{1}{c^2}\int_{\mathbb{R}} e^{\frac{2}{\varepsilon}(U(y)-2U(-y))}\,dy \geq C_2(\ln T)T^{-\frac{v}{v+\delta}} \tag{3.135}$$

for some positive constant C_2. This estimate means that the norm $\|\overline{\rho}^{-1}/c\|_\rho$ decays in T not faster than algebraically.

Using (3.130), (3.132), (3.133) and (3.134) implies that (3.131) holds if the following inequality holds:

$$\|\mu^{\varepsilon,T}(\cdot,0)\|_\rho^2(1-2e^{-CT^{\frac{\delta}{v+\delta}}}T^{\frac{\alpha}{v+\delta}}) - 2e^{-CT^{\frac{\delta}{v+\delta}}}T^{\frac{\beta}{v+\delta}}\|\mu^{\varepsilon,T}(\cdot,0)\|_\rho - 2\Big\|\frac{\overline{\rho}^{-1}}{c}\Big\|_\rho^2 \leq 0.$$

Consequently,

$$\begin{aligned}
\|\mu^{\varepsilon,T}(\cdot,0)\|_\rho \leq& \frac{e^{-CT^{\frac{\delta}{v+\delta}}}T^{\frac{\beta}{v+\delta}} + \sqrt{(e^{-CT^{\frac{\delta}{v+\delta}}}T^{\frac{\beta}{v+\delta}})^2 + 2\|\frac{\overline{\rho}^{-1}}{c}\|_\rho^2(1-2e^{-CT^{\frac{\delta}{v+\delta}}}T^{\frac{\alpha}{v+\delta}})}}{1-2e^{-CT^{\frac{\delta}{v+\delta}}}T^{\frac{\alpha}{v+\delta}}}\\
\leq& 2\left(\Big\|\frac{\overline{\rho}^{-1}}{c}\Big\|_\rho + \sqrt{\Big\|\frac{\overline{\rho}^{-1}}{c}\Big\|_\rho^2 + 2\Big\|\frac{\overline{\rho}^{-1}}{c}\Big\|_\rho^2}\right)\\
\leq& 6\Big\|\frac{\overline{\rho}^{-1}}{c}\Big\|_\rho
\end{aligned}$$

for $T \geq T_0$, where T_0 is the minimal value for which $\frac{v+\delta}{\ln T} \leq \varepsilon_0$, and

$$1 - 2e^{-CT^{\frac{\delta}{v+\delta}}}T^{\frac{\alpha}{v+\delta}} \geq \frac{1}{2}$$

and

$$e^{-CT^{\frac{\delta}{v+\delta}}}T^{\frac{\beta}{v+\delta}} \leq C_1T^{-\frac{v}{v+\delta}}\ln T \leq \Big\|\frac{\overline{\rho}^{-1}}{c}\Big\|_\rho.$$

□

3.4.3. Spectral power amplification. This and the following sections are devoted to the problem of stochastic resonance for the diffusion (3.113). As a measure of quality of tuning, we shall consider the spectral power amplification coefficient for $X^{\varepsilon,T}$, just as for the two-state reduction in Section 3.2. Of course, to make the two-state chain a consistent model of the reduced diffusion dynamics, we now have to adapt the pre-factors p and q to recover the true asymptotics of Kramers' times hidden in the precise description of λ_1. In this setting, our original plan was to prove that the resonance point obtained in Section 3.2 for the SPA coefficient of the Markov chain determines and thus characterizes a resonance point for the diffusion, in the small noise limit. To our surprise, this turns out not to be the case, as we shall now make precise.

The *spectral power amplification* coefficient is similarly to the two-state case defined by

$$\eta^X(\varepsilon, T) = \left| \int_0^1 \mathbb{E}_\mu X_{Ts}^{\varepsilon,T} e^{2\pi i s}\, ds \right|^2 \tag{3.136}$$

It will be compared with the analogous coefficient (3.21) of the Markov chain. Hereby we take the average with respect to the invariant law of $X^{\varepsilon,T}$ the density of which is $\mu^{\varepsilon,T}$.

The SPA coefficient describes the energy of the averaged trajectory $X^{\varepsilon,T}$ carried by the spectral component of period T, i.e. the period of the 'input signal'. For the sake of brevity, in this definition we omit the constant factor $\frac{\pi^2}{(V-v)^2}$ used in (3.21).

First, let us rewrite and simplify (3.136). We get

$$\begin{aligned}
\eta^X(\varepsilon, T) &= \left| \int_0^1 \mathbb{E}_\mu X_{Ts}^{\varepsilon,T} e^{2\pi i s}\, ds \right|^2 \\
&= \left| \int_0^{1/2} e^{2\pi i s} \int_{\mathbb{R}} x\mu^{\varepsilon,T}(x,s)\, dx\, ds + \int_{1/2}^1 e^{2\pi i s} \int_{\mathbb{R}} x\mu^{\varepsilon,T}(x,s)\, dx\, ds \right|^2 \\
&= \left| \int_0^{1/2} e^{2\pi i s} \int_{\mathbb{R}} x\mu^{\varepsilon,T}(x,s)\, dx\, ds + e^{\pi i} \int_0^{1/2} e^{2\pi i s} \int_{\mathbb{R}} x\mu^{\varepsilon,T}(x,s+\frac{1}{2})\, dx\, ds \right|^2 \\
&= \left| \int_0^{1/2} e^{2\pi i s} \int_{\mathbb{R}} x\mu^{\varepsilon,T}(x,s)\, dx\, ds - \int_0^{1/2} e^{2\pi i s} \int_{\mathbb{R}} x\mu^{\varepsilon,T}(-x,s)\, dx\, ds \right|^2 \\
&= 4 \left| \int_0^{1/2} e^{2\pi i s} \int_{\mathbb{R}} x\mu^{\varepsilon,T}(x,s)\, dx\, ds \right|^2 = 4 \left| S^X(\varepsilon, T) \right|^2.
\end{aligned}$$

Using (3.124) we find

$$\begin{aligned}
S^X(\varepsilon, T) &= \int_0^{1/2} e^{2\pi i s} \int_{\mathbb{R}} x\mu^{\varepsilon,T}(x,s)\, dx\, ds = \\
& a_0^\varepsilon \int_0^{1/2} e^{2\pi i s}\, ds \int_{\mathbb{R}} x \frac{\Psi_0(x)}{\|\Psi_0\|_\rho}\, dx + \\
& a_1^{\varepsilon,T} \int_0^{1/2} e^{2\pi i s} e^{-\lambda_1 T s}\, ds \int_{\mathbb{R}} x \frac{\Psi_1(x)}{\|\Psi_1\|_\rho}\, dx + r_1(\varepsilon, T),
\end{aligned} \tag{3.137}$$

where

$$r_1(\varepsilon, T) = \int_0^{1/2} e^{2\pi i s} \int_{\mathbb{R}} x \sum_{k=2}^{\infty} a_k^{\varepsilon,T} \frac{\Psi_k(x)}{\|\Psi_k\|_\rho} \exp\left(-T\lambda_k\, s\right) dx\, ds.$$

In the sequel, we shall occasionally use the symbol x to also denote the identity function on $\mathbb{R}$, $x \mapsto x$.

PROPOSITION 3.47. There exists $\varepsilon_0 > 0$ such that for $0 < \varepsilon \le \varepsilon_0$ and $T > 0$ we have

$$|r_1(\varepsilon, T)| \le \frac{1}{TM} \|\mu^{\varepsilon,T}(\cdot, 0)\|_\rho \cdot \|x e^{-\frac{2U}{\varepsilon}}\|_\rho, \tag{3.138}$$

with M from Theorem 3.38.

PROOF. Indeed, using Theorem 3.38 for the fourth inequality, we deduce

$$\begin{aligned}
|r_1(\varepsilon, T)| &\le \sum_{k=2}^{\infty} |a_k^{\varepsilon,T}| \int_0^{1/2} e^{-T\lambda_k s}\, ds \left| \int_{\mathbb{R}} x \frac{\Psi_k(x)}{\|\Psi_k\|_\rho}\, dx \right| \\
&\le \sum_{k=2}^{\infty} |a_k^{\varepsilon,T}| \left| \int_{\mathbb{R}} x \frac{\Psi_k(x)}{\|\Psi_k\|_\rho}\, dx \right| \frac{1 - \exp\left(-\lambda_k T/2\right)}{\lambda_k T} \\
&\le \frac{1}{\lambda_2 T} \sum_{k=2}^{\infty} |a_k^{\varepsilon,T}| \left| \int_{\mathbb{R}} x \frac{\Psi_k(x)}{\|\Psi_k\|_\rho}\, dx \right| \\
&\le \frac{1}{MT} \left[\sum_{k=2}^{\infty} (a_k^{\varepsilon,T})^2 \sum_{k=2}^{\infty} \left\langle x e^{-\frac{2U}{\varepsilon}}, \frac{\Psi_k}{\|\Psi_k\|_\rho} \right\rangle_\rho^2 \right]^{1/2}.
\end{aligned}$$

It remains to apply Parseval's equality. □

We next determine the leading term of $S^X(\varepsilon, T)$. We use formula (3.137), and recall that the coefficient a_0^ε was determined in Proposition 3.43, the coefficient $a_1^{\varepsilon,T}$ was found in Proposition 3.45.

LEMMA 3.48. *For any $\delta > 0$ there exist $T_0, \varepsilon_0 > 0$ such that for $\varepsilon \in [\frac{v+\delta}{\ln T}, \varepsilon_0]$ and $T \ge T_0$*

$$\begin{aligned}
S^X(\varepsilon, T) &= \frac{i}{\pi} \frac{\int_{\mathbb{R}} y \Psi_0(y)\, dy}{\|\Psi_0\|_\rho^2} \\
&\quad - \frac{1 + e^{-\lambda_1 T/2}}{2\pi i - \lambda_1 T} \frac{\int_{\mathbb{R}} y \Psi_1(y)\, dy}{\|\Psi_0\|_\rho^2} \frac{\langle \overline{\Psi}_0, \Psi_1 \rangle_\rho}{\|\Psi_1\|_\rho^2 - e^{-\lambda_1 T} \langle \overline{\Psi}_1, \Psi_1 \rangle_\rho} + r_2(\varepsilon, T),
\end{aligned} \tag{3.139}$$

where

$$|r_2(\varepsilon, T)| \le \frac{12}{MT} \left(\int_{\mathbb{R}} e^{-\frac{2U(y)}{\varepsilon}}\, dy \right)^{-1} \|e^{-\frac{2\overline{U}}{\varepsilon}}\|_\rho \cdot \|x e^{-\frac{2U}{\varepsilon}}\|_\rho, \tag{3.140}$$

and M is given by Theorem 3.38.

PROOF. The leading term of (3.139) is obtained from (3.137) by integration in $s \in [0, \frac{1}{2}]$. Let us estimate the error term $r_2(\varepsilon, T)$ which is composed of $r_1(\varepsilon, T)$ and $r(\varepsilon, T)$ from formula (3.126). Using Proposition 3.47 and Proposition 3.45, we

get
(3.141)

$$\begin{aligned}
|r_2(\varepsilon,T)| &\le |r_1(\varepsilon,T)| \\
&+ \frac{|r(\varepsilon,T)|}{\|\Psi_0\|_\rho |1-e^{-\lambda_1 T/2}\langle\frac{\overline{\Psi}_1}{\|\Psi_1\|_\rho},\frac{\Psi_1}{\|\Psi_1\|_\rho}\rangle_\rho|}\left|\frac{1+e^{-\lambda_1 T/2}}{2\pi i-\lambda_1 T}\int_{\mathbb{R}} y\frac{\Psi_1(y)}{\|\Psi_1\|_\rho}\,dy\right| \\
&\le \frac{1}{MT}\|\mu^{\varepsilon,T}(\cdot,0)\|_\rho\cdot\|xe^{-\frac{2U}{\varepsilon}}\|_\rho \\
&+ \frac{e^{-MT/2}}{\|\Psi_0\|_\rho}\frac{\|\mu^{\varepsilon,T}(\cdot,0)\|_\rho\cdot\|\overline{\Psi}_1 e^{2\frac{\overline{U}-U}{\varepsilon}}\|_\rho}{|\|\Psi_1\|_\rho^2-e^{-\lambda_1 T/2}\langle\overline{\Psi}_1,\Psi_1\rangle_\rho|}\left|\int_{\mathbb{R}} y\Psi_1(y)\,dy\right| \\
&= \frac{1}{MT}\|\mu^{\varepsilon,T}(\cdot,0)\|_\rho \\
&\times\left(\|xe^{-\frac{2U}{\varepsilon}}\|_\rho+\frac{MTe^{-MT/2}}{\|\Psi_0\|_\rho}\frac{\|\overline{\Psi}_1 e^{2\frac{\overline{U}-U}{\varepsilon}}\|_\rho\cdot|\int_{\mathbb{R}} y\Psi_1(y)\,dy|}{|\|\Psi_1\|_\rho^2-e^{-\lambda_1 T/2}\langle\overline{\Psi}_1,\Psi_1\rangle_\rho|}\right).
\end{aligned}$$

Note that for T large enough Laplace's method yields

$$C_1 \le \max_{\varepsilon\in[\frac{v+\delta}{\ln T},\varepsilon_0]}\|xe^{-\frac{2U}{\varepsilon}}\|_\rho = \max_{\varepsilon\in[\frac{v+\delta}{\ln T},\varepsilon_0]}\left(\int_{\mathbb{R}} y^2 e^{-\frac{2U(y)}{\varepsilon}}\,dy\right)^{\frac{1}{2}} \le C_2 T^{\frac{V}{2(v+\delta)}}$$

for some positive constants C_1 and C_2. This means that $\|xe^{-\frac{2U}{\varepsilon}}\|_\rho$ is bounded on $[\frac{v+\delta}{\ln T},\varepsilon_0]$ polynomially in T. Next we show that due to the factor $e^{-MT/2}$ the second summand in the parenthesis in (3.141) is exponentially small in T.

Recall that $\Psi_k = \Phi_k\rho^{-1}$, hence $\|\Psi_k\|_\rho = \|\Phi_k\|_{\rho^{-1}}$, $k\ge 0$, and note that $\|\overline{\Phi}_1\|_{\rho^{-1}} = \|\overline{\Psi}_1 e^{2\frac{\overline{U}-U}{\varepsilon}}\|_\rho$. Then using Theorem 3.36 we estimate in $\varepsilon\in[\frac{v+\delta}{\ln T},\varepsilon_0]$ for T large enough and with universal constant C changing from line to line

$$\begin{aligned}
&\frac{MTe^{-MT/2}}{\|\Psi_0\|_\rho}\frac{\|\overline{\Psi}_1 e^{2\frac{\overline{U}-U}{\varepsilon}}\|_\rho\cdot|\int_{\mathbb{R}} y\Psi_1(y)\,dy|}{|\|\Psi_1\|_\rho^2-e^{-\lambda_1 T/2}\langle\overline{\Psi}_1,\Psi\rangle_\rho|} \\
&\qquad= \frac{MTe^{-MT/2}}{\|\Phi_0\|_{\rho^{-1}}}\frac{\|\overline{\Phi}_1\|_{\rho^{-1}}\cdot|\int_{\mathbb{R}} y\Phi_1(y)e^{-\frac{2U(y)}{\varepsilon}}\,dy|}{|\|\Phi_1\|_{\rho^{-1}}^2-e^{-\lambda_1 T/2}\langle\overline{\Phi}_1,\Phi_1\rangle_{\rho^{-1}}|} \\
&\qquad\le CTe^{-MT/2}\frac{\sqrt[4]{\varepsilon}e^{\frac{V}{2\varepsilon}}\cdot\sqrt{\varepsilon}e^{\frac{v}{\varepsilon}}}{\sqrt[4]{\varepsilon}e^{\frac{V}{2\varepsilon}}\sqrt{\varepsilon}e^{\frac{v}{\varepsilon}}(1+e^{-\lambda_1 T/2})} \\
&\qquad\le CTe^{-MT/2}\le\|xe^{-\frac{2U}{\varepsilon}}\|_\rho.
\end{aligned}$$

Applying the previous formula and the inequality $\|\mu^{\varepsilon,T}(\cdot,0)\|_\rho\le 6\|\overline{\rho}^{-1}/c\|_\rho$ obtained in Proposition 3.46 to (3.141) completes the proof. □

Next we find the value of the spectral power amplification coefficient $\eta^X(\varepsilon,T)$ in the interval $[\frac{v+\delta}{\ln T},\varepsilon_0]$ for $\delta>0$ and large T. For abbreviation of the leading term, let us set

$$\begin{aligned}
b_0 = b_0^\varepsilon &= \frac{\int_{\mathbb{R}} y\Psi_0(y)\,dy}{\|\Psi_0\|_\rho^2} = \frac{\int_{\mathbb{R}} ye^{-\frac{2U(y)}{\varepsilon}}\,dy}{\int_{\mathbb{R}} e^{-\frac{2U(y)}{\varepsilon}}\,dy}, \\
b_1 = b_1^{\varepsilon,T} &= -\frac{1+e^{-\lambda_1 T/2}}{2}\frac{\int_{\mathbb{R}} y\Psi_1(y)\,dy}{\|\Psi_0\|_\rho^2}\frac{\langle\overline{\Psi}_0,\Psi_1\rangle_\rho}{\|\Psi_1\|_\rho^2-e^{-\lambda_1 T/2}\langle\overline{\Psi}_1,\Psi_1\rangle_\rho}.
\end{aligned}$$

Then (3.139) can be rewritten in the form

$$S^X(\varepsilon, T) = \frac{i}{\pi} b_0 + \frac{2}{2\pi i - \lambda_1 T} b_1 + r_2(\varepsilon, T),$$

and therefore

$$\eta^X(\varepsilon, T) = 4|S^X(\varepsilon, T)|^2 = \frac{4}{\pi^2} \frac{b_0^2(\lambda_1 T)^2}{4\pi^2 + (\lambda_1 T)^2} + 16\frac{(b_1 - b_0)^2}{4\pi^2 + (\lambda_1 T)^2} + r_3(\varepsilon, T), \tag{3.142}$$

where

$$r_3 = 4|r_2|^2 + 8\,\mathrm{Re}\,(s^X r_2^*), \tag{3.143}$$

'$*$' denotes the complex conjugate and $s^X = \frac{i}{\pi} b_0 + \frac{2}{2\pi i - \lambda_1 T} b_1$ is the leading term of S^X. We see that η^X is represented as a sum of three terms. Let us recall its Markov chain counterpart η^Y determined in Proposition 3.9. Up to the constant factor $\frac{\pi^2}{(V-v)^2}$ we have

$$\eta^Y(\varepsilon, T) = \frac{4}{\pi^2} \frac{T^2(\varphi - \psi)^2}{4\pi^2 + (\varphi + \psi)^2 T^2}. \tag{3.144}$$

It is clearly seen, that the first leading term of (3.142) is similar to (3.144). The correspondence were exact if $\lambda_1 \approx \psi \pm \varphi$ and $b_0 \approx 1$.

In the following Lemma we find asymptotic estimates for b_0 and b_1.

LEMMA 3.49. *There is $\varepsilon_0 > 0$ such that for $\varepsilon \leq \varepsilon_0$ we have*

$$b_0 = -1 - \frac{U^{(3)}(-1)}{4\omega_-^2}\varepsilon + \mathcal{O}(\varepsilon^2), \tag{3.145}$$

$$b_1 = -1 + \mathcal{O}(\varepsilon), \tag{3.146}$$

and consequently

$$b_0^2 = 1 + \frac{U^{(3)}(-1)}{2\omega_-^2}\varepsilon + \mathcal{O}(\varepsilon^2), \tag{3.147}$$

$$(b_1 - b_0)^2 = \mathcal{O}(\varepsilon^2). \tag{3.148}$$

PROOF. We use Laplace's method to obtain the asymptotic expansions for the integrals $\int_{\mathbb{R}} e^{-\frac{2U(y)}{\varepsilon}}\,dy$ and $\int_{\mathbb{R}} y e^{-\frac{2U(y)}{\varepsilon}}\,dy$; compare the formulae (B.8) and (B.10) in the Appendix B. We get

$$\int_{\mathbb{R}} e^{-\frac{2U(y)}{\varepsilon}}\,dy = \sqrt{\frac{\pi\varepsilon}{\omega_-}} e^{\frac{V}{\varepsilon}} \left[1 + \frac{\varepsilon}{16\omega_-^3}\left(\frac{5U^{(3)}(-1)^2}{3} - \omega_- U^{(4)}(-1)\right) + \mathcal{O}(\varepsilon^2)\right],$$

$$\int_{\mathbb{R}} y e^{-\frac{2U(y)}{\varepsilon}}\,dy = -\sqrt{\frac{\pi\varepsilon}{\omega_-}} e^{\frac{V}{\varepsilon}}$$
$$\times \left[1 + \frac{U^{(3)}(-1)}{4\omega_-^2}\varepsilon + \frac{\varepsilon}{16\omega_-^3}\left(\frac{5U^{(3)}(-1)^2}{3} - \omega_- U^{(4)}(-1)\right) + \mathcal{O}(\varepsilon^2)\right].$$

The relationships (3.145) and (3.147) follow from these formulae and the asymptotic expansion rule $\frac{1+a_1\varepsilon+\mathcal{O}(\varepsilon^2)}{1+a_2\varepsilon+\mathcal{O}(\varepsilon^2)} = 1 + (a_1 - a_2)\varepsilon + \mathcal{O}(\varepsilon^2)$, $a_1, a_2 \in \mathbb{R}$.

The estimate for b_1 is obtained analogously to Lemma 3.48 with the help of Theorem 3.36. More precisely, we get

$$\begin{aligned} b_1 &= -\frac{1+e^{-\lambda_1 T/2}}{2} \frac{\int_{\mathbb{R}} y\Psi_1(y)\,dy}{\|\Psi_0\|_\rho^2} \frac{\langle \overline{\Psi}_0, \Psi_1\rangle_\rho}{\|\Psi_1\|_\rho^2 - e^{-\lambda_1 T/2}\langle \overline{\Psi}_1, \Psi_1\rangle_\rho} \\ &= -\frac{1+e^{-\lambda_1 T/2}}{2} \frac{(x, \Phi_1)_{\rho^{-1}}}{\|\Phi_0\|_{\rho^{-1}}^2} \frac{\langle \Phi_0, \overline{\Phi}_1\rangle_{\rho^{-1}}}{\|\Phi_1\|_{\rho^{-1}}^2 - e^{-\lambda_1 T/2}\langle \overline{\Phi}_1, \Phi_1\rangle_{\rho^{-1}}} \\ &= -\frac{2\left(1+e^{-\lambda_1 T/2}\right)\cdot\sqrt{\frac{\pi\varepsilon}{\omega_+}}e^{\frac{v}{\varepsilon}}(1+\mathcal{O}(\varepsilon))\cdot\sqrt{\frac{\pi\varepsilon}{\omega_-}}e^{\frac{V}{\varepsilon}}(1+\mathcal{O}(\varepsilon))}{2\sqrt{\frac{\pi\varepsilon}{\omega_-}}e^{\frac{V}{\varepsilon}}(1+\mathcal{O}(\varepsilon))\cdot\left(\sqrt{\frac{\pi\varepsilon}{\omega_+}}e^{\frac{v}{\varepsilon}}(1+\mathcal{O}(\varepsilon)) + e^{-\lambda_1 T/2}\sqrt{\frac{\pi\varepsilon}{\omega_+}}e^{\frac{v}{\varepsilon}}(1+\mathcal{O}(\varepsilon))\right)} \\ &= -1+\mathcal{O}(\varepsilon). \end{aligned}$$

The combination of (3.145) and (3.146) leads to (3.148). □

Let us compare the first term in (3.142) with η^Y. Lemma 3.49 states that $b_0 \approx 1$ as $\varepsilon \to 0$. Recall that

$$\lambda_1 \approx \frac{\sqrt{\omega_0\omega_+}}{2\pi} e^{-\frac{v}{\varepsilon}},$$

and that the infinitesimal probabilities in Section 3.2 were defined by $\varphi = pe^{-\frac{V}{\varepsilon}}$ and $\psi = pe^{-\frac{v}{\varepsilon}}$, with $p, q > 0$.

We now choose p and q so that Kramers' times for the diffusion and the reducing Markov chain coincide not only to exponential order, but in addition up to the leading subexponential pre-factors. If we set

$$p = \frac{\sqrt{\omega_0\omega_-}}{2\pi}, \qquad q = \frac{\sqrt{\omega_0\omega_+}}{2\pi} \tag{3.149}$$

we also get

$$\psi + \varphi \approx \lambda_1, \qquad \psi - \varphi \approx \lambda_1$$

to the leading subexponential pre-factor. This moreover implies

$$b_0 \frac{(\lambda_1 T)^2}{4\pi^2 + (\lambda_1 T)^2} \approx \frac{(\psi-\varphi)^2 T^2}{4\pi^2 + (\psi+\varphi)^2 T^2}.$$

This correspondence is a sure temptation for suspecting that provided the remainder terms in (3.142) are small enough, the SPA coefficients of the diffusion and the Markov chain are close. This also strongly suggests that one could be able to relate asymptotically the optimal tuning rates for the diffusion and the reducing Markov chain.

This argument is in fact very common in the physics literature [**83, 43, 1**] and is used to pass to a simpler two-state framework for the investigation of various dynamucal properties of the diffusion in the small noise limit, especially in the context of stochastic resonance. This approach will now be shown to possibly have drastic side effects. In determining eigenvalues and Kramers' times we always had to take into account a multiplicative correction term of the type $1+\mathcal{O}(\varepsilon)$. The errors of order $\mathcal{O}(\varepsilon)$ in all the formulae we have derived reflect small random oscillations of the diffusion near the metastable states of the potential. Neglecting these terms means neglecting the 'diffusive' nature of the diffusion. Although these oscillations are small, they occur with a high probability. This leads very subtle drag effects in the potential wells' bottoms disturbing or even destroying the tuning picture the two-state reduction presents.

In Section 3.2 the optimal tuning rate for the Markov chain $Y^{\varepsilon,T}$ in the sense of the SPA coefficient was determined by

$$\varepsilon = \frac{V+v}{2\ln T}\left(1+\frac{\ln\left(\pi\sqrt{\frac{2}{pq}\frac{v}{V-v}}\right)}{\ln T}+\mathcal{O}\left(\frac{1}{\ln^2 T}\right)\right)$$

If the diffusion's tuning properties corresponding to the SPA coefficient were retained by the reducing Markov chain, we would expect a local maximum of the SPA coefficient $\eta^X(\cdot,T)$ at some point $\varepsilon_T \approx \frac{V+v}{2\ln T}$. Let us consider $\alpha \mapsto \eta^X(\frac{\alpha}{\ln T},T)$ for large T on the interval

$$[v+\delta,\Delta], \quad \text{for some} \quad \delta>0,\ \Delta>v+\delta.$$

On the ε-scale this corresponds to shrinking intervals $[\frac{v+\delta}{\ln T},\frac{\Delta}{\ln T}]$. We shall investigate, whether $\alpha\mapsto\eta^X(\frac{\alpha}{\ln T},T)$ posesses a local maximum for large T.

THEOREM 3.50. *Let $0<\delta<\frac{v}{3}$ and $\Delta>v+\delta$ be fixed. Let $\max_{x\in\mathbb{R}}\{U(x)-2U(-x)\}<V+v$. Then there exists $T_0>0$ such that for $T>T_0$, $\alpha\in[v+\delta,\Delta]$ we have*

$$\eta^X\left(\frac{\alpha}{\ln T},T\right)=\frac{4}{\pi^2}\left(1+\frac{U^{(3)}(-1)}{2\omega_-^2}\frac{\alpha}{\ln T}\right)+\mathcal{O}\left(\frac{1}{\ln^2 T}\right).$$

REMARK 3.51. Geometrically, the condition $\max_{x\in\mathbb{R}}\{U(x)-2U(-x)\}<V+v$ may be seen to express the fact that the potential is not too asymmetric outside of the wells.

PROOF. The proof consists in expanding of (3.142) as $T\to\infty$ and estimating the remainder terms.

First, we note that in order to apply Lemma 3.48 the noise parameter ε must satisfy $\varepsilon\in[\frac{v+\delta}{\ln T},\varepsilon_0]$ for some $\varepsilon_0>0$. It is clear that to verify this condition it is enough to take $T_0>e^{\Delta/\varepsilon_0}$.

Consider $\eta^X(\frac{\alpha}{\ln T},T)$ for $T>0$. The factor b_0 in the leading term of η^X is expanded with the help of Lemma 3.49.

On the interval $[\frac{v+\delta}{\ln T},\frac{\Delta}{\ln T}]$ with some constant $C>0$ we estimate

$$(\lambda_1 T)^2 \geq C(e^{-\frac{v}{\varepsilon}}T)^2 \geq C(T^{1-\frac{v}{v+\delta}})^2 = CT^{\frac{2\delta}{v+\delta}}. \tag{3.150}$$

This results in

$$\frac{(\lambda_1T)^2}{4\pi^2+(\lambda_1T)^2}=1-\frac{4\pi^2}{4\pi^2+(\lambda_1T)^2}=1+\mathcal{O}\left(\frac{1}{(\lambda_1T)^2}\right)=1+\mathcal{O}\left(\frac{1}{T^{\frac{2\delta}{v+\delta}}}\right).$$

Hence, we obtain the expansion of the leading term of (3.142)

$$\frac{4}{\pi^2}\frac{b_0^2(\lambda_1T)^2}{4\pi^2+(\lambda_1T)^2}=\frac{4}{\pi^2}\left(1+\frac{U^{(3)}(-1)}{2\omega_-^2}\frac{\alpha}{\ln T}\right)+\mathcal{O}\left(\frac{1}{\ln^2 T}\right).$$

It is left to estimate the terms $\frac{16(b_1-b_0)^2}{4\pi^2+(\lambda_1T)^2}$ and $|r_3|$. Analogously to (3.150), the first one is of the order $T^{-\frac{2\delta}{v+\delta}}\ln^{-2}T$ on $\alpha\in[v+\delta,\Delta]$ as $T\to\infty$ since $(b_1-b_0)^2=\mathcal{O}(\frac{1}{\ln^2 T})$.

Consider the third term given by (3.143):

$$|r_3(\varepsilon,T)|\leq 8|s^X(\varepsilon,T)||r_2(\varepsilon,T)|+4|r_2(\varepsilon,T)|^2,$$

where

$$|s^X(\varepsilon,T)| = \left(\frac{1}{\pi^2}\frac{b_0^2(\lambda_1 T)^2}{4\pi^2+(\lambda_1 T)^2} + \frac{16(b_1-b_0)^2}{4\pi^2+(\lambda_1 T)^2}\right)^{\frac{1}{2}} \le \frac{1}{2}$$

for $T \ge T_0$, $\frac{v+\delta}{\ln T} \le \varepsilon \le \frac{\Delta}{\ln T}$.

Let us estimate $|r_2(\varepsilon,T)|$. Lemma 3.48 states that

$$|r_2(\varepsilon,T)| \le \frac{12}{MT}\frac{\left[\int_{\mathbb{R}} e^{-\frac{2}{\varepsilon}(2U(-y)-U(y))}\,dy\right]^{\frac{1}{2}}\left[\int_{\mathbb{R}} ye^{-\frac{2U(y)}{\varepsilon}}\,dy\right]^{\frac{1}{2}}}{\int_{\mathbb{R}} e^{-\frac{2U(y)}{\varepsilon}}\,dy}.$$

Assume that $\max_{x\in\mathbb{R}}\{U(x)-2U(-x)\} < \kappa$, where κ is a positive number. Obviously, $\kappa > V - \frac{v}{2} > 0$, since

$$\max_{x\in\mathbb{R}}\{U(x)-2U(-x)\} \ge U(1)-2U(-1) = V - \frac{v}{2}.$$

Then, using Laplace's method we obtain for some $C>0$ independent of $\varepsilon \le \varepsilon_0$

$$|r_2(\varepsilon,T)| \le C\frac{e^{\frac{\kappa}{2\varepsilon}}e^{\frac{V}{2\varepsilon}}}{Te^{\frac{V}{\varepsilon}}} = C\frac{e^{\frac{\kappa-V}{2\varepsilon}}}{T}.$$

If we choose $\kappa = V+v$ and $\delta \in (0,\frac{v}{3})$, then the inequality $\frac{v+2\delta}{2(v+\delta)} > \frac{2\delta}{v+\delta}$ entails that

$$\max_{\alpha\in[v+\delta,\Delta]}\left|r_2(\frac{\alpha}{\ln T},T)\right| \le C\frac{1}{T^{1-\frac{v}{2(v+\delta)}}} = o\left(\frac{1}{T^{\frac{2\delta}{v+\delta}}}\right), \quad \delta\in(0,\tfrac{v}{3}).$$

Thus, the remainder terms in (3.142) are polynomially small and of the order $T^{-\frac{2\delta}{v+\delta}}\ln^{-2}T$. This completes the proof. □

As we see, the form of the tuning curve for the SPA coefficient crucially depends on the sign of $U^{(3)}(-1)$. If $U^{(3)}(-1)>0$, the tuning curve increases in α and does not have a local maximum on the interval $[\frac{v+\delta}{\ln T},\frac{\Delta}{\ln T}]$. If $U^{(3)}(-1)<0$ then η^X decreases and does not have a local maximum either.

Moreover, depending on the sign of $U^{(3)}(-1)$, the resonance curve is either greater or less than $\frac{4}{\pi^2}$. However, we can state, that the SPA coefficient $\eta^X(\varepsilon,T(\varepsilon)) = 1+\mathcal{O}(\varepsilon)$ for $T(\varepsilon)=e^{\frac{\alpha}{\varepsilon}}$, $\alpha\in[v+\delta,\Delta]$, which is near the maximal value of η^Y. This means, that amplification occurs, but an optimal tuning rate cannot be determined. Especially, the Markov chain behaviour is not shared by the diffusion, and the optimal tuning rate by the chain is not asymptotically equal to an optimal tuning rate for the diffusion, in which sense ever the latter may exist.

The observed subtle drag effect may be interpreted in the following way. If $U^{(3)}(-1)<0$, the potential in the deep well slightly leans towards the shallow one. Therefore in the ε-window considered increasing the noise intensity tends to reduce the averaged overall amplitude of the motion.

In case $U^{(3)}(-1)>0$, the outward leaning of the potential in its global maximum increases the averaged overall amplitude of the random motion with increasing noise intensity.

3.4.4. The small oscillations defect. Modified SPA coefficient. Stochastic resonance is an inter-well and not an intra-well effect. Given our experience gained in the previous section, we now suppress oscillations near the potential minima and take into account only big hoppings between the wells. In this modified setting, we shall now show that the behaviour of the reducing Markov chain is correctly retained asymptotically in the small noise limit.

In order to *cut off* small random fluctuations near -1 and 1 we define a function

$$g(x) = \begin{cases} x, & x \in (-\infty, x_1] \cup [x_2, y_1] \cup [y_2, \infty), \\ -1, & x \in [x_1, x_2], \\ 1, & x \in [y_1, y_2], \end{cases}$$

where $x_1 < -1 < x_2 < 0$ are such that $U(x_1) = U(x_2) = -\frac{V}{4}$, and $0 < y_1 < 1 < y_2$ are such that $U(y_1) = U(y_2) = -\frac{v}{4}$.

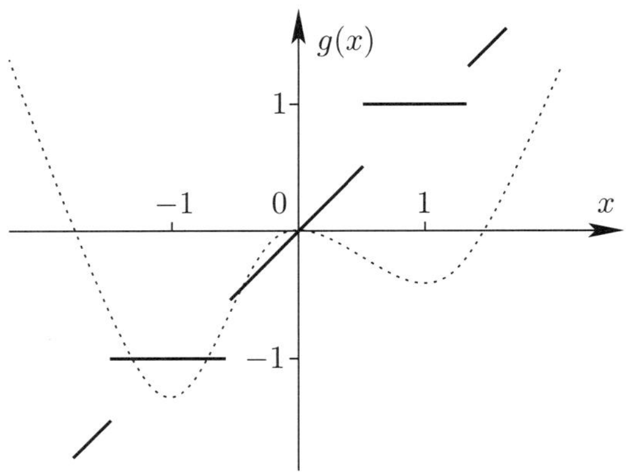

FIGURE 3.5. Function g designed to cut off small fluctuations.

For $\varepsilon, T > 0$ we consider the modified SPA coefficient

$$\widetilde{\eta}^X(\varepsilon, T) = \left| \int_0^1 \mathbb{E}_\mu g(X_{Ts}^{\varepsilon,T}) e^{2\pi i s}\, ds \right|^2 .$$

By inspection of the steps in the calculations of Section 3.4.3, replacing x with $g(x)$ if necessary, we obtain a formula for $\widetilde{\eta}^X$ which is analogous to (3.142):

$$\widetilde{\eta}^X(\varepsilon, T) = \frac{4}{\pi^2} \frac{\widetilde{b}_0^2(\lambda_1 T)^2}{4\pi^2 + (\lambda_1 T)^2} + 16 \frac{(\widetilde{b}_1 - \widetilde{b}_0)^2}{4\pi^2 + (\lambda_1 T)^2} + \widetilde{r}_3(\varepsilon, T), \tag{3.151}$$

where

$$\begin{aligned}
\widetilde{b}_0 &= \frac{\int_{\mathbb{R}} g(y) e^{-\frac{2U(y)}{\varepsilon}}\, dy}{\int_{\mathbb{R}} e^{-\frac{2U(y)}{\varepsilon}}\, dy}, \\
\widetilde{b}_1 &= -\frac{1 + e^{-\lambda_1 T/2}}{2} \frac{\int_{\mathbb{R}} g(y)\Psi_1(y)\, dy}{\|\Psi_0\|_\rho^2} \frac{\langle \overline{\Psi}_0, \Psi_1 \rangle_\rho}{\|\Psi_1\|_\rho^2 - e^{-\lambda_1 T/2} \langle \overline{\Psi}_1, \Psi_1 \rangle_\rho}, \\
\widetilde{r}_3(\varepsilon, T) &= 4|\widetilde{r}_2(\varepsilon, T)|^2 + 8 \operatorname{Re}(\widetilde{s}^X(\varepsilon, T) \widetilde{r}_2^*(\varepsilon, T)), \\
\widetilde{s}^X(\varepsilon, T) &= \frac{i}{\pi} \widetilde{b}_0 + \frac{2}{2\pi i - \lambda_1 T} \widetilde{b}_1, \\
|\widetilde{r}_2(\varepsilon, T)| &\le \frac{12}{MT} \left(\int_{\mathbb{R}} e^{-\frac{2U(y)}{\varepsilon}}\, dy \right)^{-1} \|e^{-\frac{2U}{\varepsilon}}\|_\rho \cdot \|g e^{-\frac{2U}{\varepsilon}}\|_\rho .
\end{aligned}$$

It turns out that due to different behaviour of the factor $\widetilde{b}_0$ has quite different asymptotics from b_0. Hence the modified SPA coefficient $\widetilde{\eta}^X(\varepsilon, T)$ has a local maximum close to the corresponding one for the Markov chain.

LEMMA 3.52. *There is $\varepsilon_0 > 0$ such that for $\varepsilon \le \varepsilon_0$, $T > 0$ we have*

$$\widetilde{b}_0 = -1 + 2\sqrt{\frac{\omega_-}{\omega_+}} e^{-\frac{V-v}{\varepsilon}} (1 + \mathcal{O}(\varepsilon)) \ge -1, \tag{3.152}$$

$$\widetilde{b}_1 = -1 + \mathcal{O}(\varepsilon), \tag{3.153}$$

and, consequently,

$$\widetilde{b}_0^2 = 1 - 4\sqrt{\frac{\omega_-}{\omega_+}} e^{-\frac{V-v}{\varepsilon}} (1 + \mathcal{O}(\varepsilon)) \le 1 \tag{3.154}$$

$$(\widetilde{b}_1 - \widetilde{b}_0)^2 = \mathcal{O}(\varepsilon^2). \tag{3.155}$$

PROOF. Consider $\widetilde{b}_0$ for small ε and use Laplace's method to obtain (3.152):

$$\begin{aligned}
\widetilde{b}_0 &= \frac{\int_{\mathbb{R}} g(y) e^{-\frac{2U(y)}{\varepsilon}} dy}{\int_{\mathbb{R}} e^{-\frac{2U(y)}{\varepsilon}} dy} \\
&= \frac{\left(-\int_{x_1}^{x_2} + \int_{y_1}^{y_2}\right) e^{-\frac{2U(y)}{\varepsilon}} dy + \left(\int_{-\infty}^{x_1} + \int_{x_2}^{y_1} + \int_{y_2}^{\infty}\right) y e^{-\frac{2U(y)}{\varepsilon}} dy}{\left(\int_{x_1}^{x_2} + \int_{y_1}^{y_2}\right) e^{-\frac{2U(y)}{\varepsilon}} dy + \left(\int_{-\infty}^{x_1} + \int_{x_2}^{y_1} + \int_{y_2}^{\infty}\right) e^{-\frac{2U(y)}{\varepsilon}} dy} \\
&= -\frac{\left(\int_{x_1}^{x_2} + \int_{y_1}^{y_2}\right) e^{-\frac{2U(y)}{\varepsilon}} dy - 2\int_{y_1}^{y_2} e^{-\frac{2U(y)}{\varepsilon}} dy + \mathcal{O}(e^{\frac{V}{2\varepsilon}})}{\left(\int_{x_1}^{x_2} + \int_{y_1}^{y_2}\right) e^{-\frac{2U(y)}{\varepsilon}} dy + \mathcal{O}(e^{\frac{V}{2\varepsilon}})} \\
&= -1 + \frac{2\sqrt{\frac{\pi\varepsilon}{\omega_+}} e^{\frac{v}{\varepsilon}} (1 + \mathcal{O}(\varepsilon))}{\sqrt{\frac{\pi\varepsilon}{\omega_-}} e^{\frac{V}{\varepsilon}} (1 + \mathcal{O}(\varepsilon))} = -1 + 2\sqrt{\frac{\omega_-}{\omega_+}} e^{-\frac{V-v}{\varepsilon}} (1 + \mathcal{O}(\varepsilon)).
\end{aligned}$$

For the third equation we used our hypothesis concerning the cutoff levels the inequality $V - v < \frac{V}{2}$ which follows from (**M**).

Formulae (3.153) and (3.155) are obtained analogously to Lemma 3.49. Expression (3.154) follows directly from (3.152). □

THEOREM 3.53. *Let $\max_{x\in\mathbb{R}}\{U(x) - 2U(-x)\} < V + v$ and $0 < \delta < \frac{v}{3}$. Then for any $1 < \gamma < \frac{V+v}{2(v+\delta)}$ there exists $T(\gamma)$ such that for $T > T(\gamma)$ the modified SPA coefficient $\varepsilon \mapsto \widetilde{\eta}^X(\varepsilon, T)$ has a local maximum on $[\gamma^{-1}\frac{V+v}{2\ln T}, \gamma\frac{V+v}{2\ln T}]$. The optimal tuning rate $\varepsilon(T)$ is exponentially equivalent to $\frac{V+v}{2\ln T}$ in the limit $T \to \infty$.*

PROOF. To show that $\varepsilon \mapsto \widetilde{\eta}^X(\varepsilon, T)$ given by (3.151) has a local maximum we consider it at the three points

$$\varepsilon_1(T) = \frac{V+v}{2\ln T}, \qquad \varepsilon_2(T) = \gamma\frac{V+v}{2\ln T}, \text{ and } \varepsilon_3(T) = \gamma^{-1}\frac{V+v}{2\ln T}. \tag{3.156}$$

Since $\gamma > 1$, we have $\varepsilon_3(T) < \varepsilon_1(T) < \varepsilon_2(T)$.

Consider the leading term of $\widetilde{\eta}^X(\varepsilon_1(T),T)$ for $T\to\infty$. The factor $\widetilde{b}_0^2$ is given by Lemma 3.52. Expanding the factor $\frac{(\lambda_1 T)^2}{\pi^2+(\lambda_1 T)^2}$ as $T\to\infty$ gives:

$$\begin{aligned}\frac{4}{\pi^2}\frac{\widetilde{b}_0^2(\lambda_1 T)^2}{4\pi^2+(\lambda_1 T)^2} &= \frac{4}{\pi^2}\left(1-4\sqrt{\frac{\omega_-}{\omega_+}}T^{-2\frac{V-v}{V+v}}(1+\mathcal{O}(\ln^{-1}T))\right)\\ &\quad\times\left(1-\frac{4\pi^4}{\omega_0\omega_+}T^{-2\frac{V-v}{V+v}}(1+\mathcal{O}(\ln^{-1}T))\right)\\ &= \frac{4}{\pi^2}-\frac{16}{\pi^2}\left(\sqrt{\frac{\omega_-}{\omega_+}}+\frac{\pi^4}{\omega_0\omega_+}\right)T^{-2\frac{V-v}{V+v}}\left(1+\mathcal{O}(\ln^{-1}T)\right).\end{aligned}$$

Analogously to Theorem 3.50 one shows that the remainder terms in (3.151) are of the order $T^{-2\frac{V-v}{V+v}}\ln^{-2}T$. This yields that

$$\widetilde{\eta}^X(\varepsilon_1(T),T)=\frac{4}{\pi^2}-\frac{16}{\pi^2}\left(\sqrt{\frac{\omega_-}{\omega_+}}+\frac{\pi^4}{\omega_0\omega_+}\right)\frac{1}{T^{2\frac{V-v}{V+v}}}\left(1+\mathcal{O}\left(\frac{1}{\ln T}\right)\right).$$

Analogously, we consider the modified SPA coefficient at $\varepsilon_2(T)$ and $\varepsilon_3(T)$ to obtain

$$\widetilde{\eta}^X(\varepsilon_2(T),T)=\frac{4}{\pi^2}-\frac{16}{\pi^2}\sqrt{\frac{\omega_-}{\omega_+}}\frac{1}{T^{2\frac{V-v}{\gamma(V+v)}}}\left(1+\mathcal{O}\left(\frac{1}{\ln T}\right)\right),$$

and

$$\widetilde{\eta}^X(\varepsilon_3(T),T)=\frac{4}{\pi^2}-\frac{16\pi^2}{\omega_0\omega_+}\frac{1}{T^{2\frac{V+v-2\gamma v}{V+v}}}\left(1+\mathcal{O}\left(\frac{1}{\ln T}\right)\right).$$

This entails that for $T(\gamma)$ large enough and $T>T(\gamma)$ we have

$$\begin{aligned}&\widetilde{\eta}^X(\varepsilon_1(T),T)>\widetilde{\eta}^X(\varepsilon_2(T),T) \qquad \text{and}\\ &\widetilde{\eta}^X(\varepsilon_1(T),T)>\widetilde{\eta}^X(\varepsilon_3(T),T)\end{aligned}$$

since $V-v>\frac{1}{\gamma}(V-v)$ and $V-v>V+v-2\gamma v$ for $\gamma>1$.

Since the differences between the values taken are of higher order than the bounds for the remainder terms, this completes the proof. □

Let us finish with some remarks concerning the dependence of the optimal tuning rate on the geometry of the potential U. We have seen in Chapter 3.2 that for some values of the pre-factors p and q and half-period T the tuning curve vanishes at certain noise levels or is monotonically increasing, see Proposition 3.9. We do not observe such a phenomenon in the present setting the SPA coefficient of the diffusion in the small noise and large period limits. Recall that the Markov chain SPA coefficient vanishes at $\widehat{\varepsilon}=(V-v)/\ln\left(\frac{p}{q}\right)$ which is a positive number independent of ε and T. Of course, it can happen that η^X or $\widetilde{\eta}^X$ vanishes for some noise intensity. However, our approach describes neither this effect nor monotonicity of $\widetilde{\eta}^X$. The reason is this: in the small noise limit considered here we are outside of the domains of parameter space for which this behaviour is exhibited.

CHAPTER 4

Large deviations and transitions between meta-stable states of dynamical systems with small noise and weak inhomogeneity

Motivated by the small oscillations defect exhibited in Chapter 3 the aim of this chapter is to investigate notions of periodic tuning for the solution trajectories of diffusions with weakly periodic drifts and additive small noise which are robust when passing to the reduced dynamics Markov chain. The main object of our interest are weakly time inhomogeneous diffusion processes given as solutions of the SDE

$$(4.1) \qquad dX_t^{\varepsilon,x} = b\Big(\frac{t}{T}, X_t^{\varepsilon,x}\Big)\, dt + \sqrt{\varepsilon} dW_t, \quad t \geq 0, \quad X_0^{\varepsilon,x} = x \in \mathbb{R}^d,$$

where b is assumed to be one-periodic in time such that in the re-scaled version given T is the period of the deterministic input, and $W = (W_t)_{t\geq 0}$ is a standard d-dimensional Brownian motion. In Chapter 3 the time-dependence of the drift term was reduced to a periodic switching between two gradients of (one-dimensional) double-well potentials: $-U_1'$ for the first half-period and $-U_2'$ for the second half-period (see Figure 1.12). In this Chapter, the drift b, still having two stable states, will change continuously with respect to the time variable. In the *gradient* case, $b = -\nabla_x U$, one may think of the periodicity of U as resulting from a periodic modulation of a symmetric double-well potential, for instance in the one-dimensional case $U(t,x) = U_0(x) - Ax\sin(\frac{2\pi t}{T})$ with $U_0(x) = \frac{1}{4}x^4 - \frac{1}{2}x^2$. In this case (see Chapter 1), equation (4.1) roughly describes the motion of a Brownian particle in a time inhomogeneous double-well potential or more precisely in a homogeneous double-well potential perturbed by a slow external periodic forcing.

The notions of quality of periodic tuning we consider are related to the maximal probabilities that the random exit or transition times between the two domains of attraction of the underlying dynamical systems $\dot{x}_t = b(t, x_t)$ happen in time windows parametrized by free energy parameters μ on an exponential scale where time period and noise strength are related by $T(\varepsilon) = \exp(\frac{\mu}{\varepsilon})$. For the two-state Markov chains describing the effective dynamics of the diffusions with slow and weak time inhomogeneity this optimal transition rate is readily calculated. Our concept moreover has the advantage that the related transition times allow a treatment by methods of large deviations in the small noise limit as well as the diffusions with a weak noise dependent time inhomogeneity.

We therefore start with a careful extension of large deviations theory to diffusions with slow time inhomogeneity in Section 4.1. The central result for the subsequent analysis of their exit times is contained in a large deviations principle, uniform with respect to the energy parameter (see Propositions 4.6 and 4.8). It allows us in Section 4.3 to derive upper (Proposition 4.22) and lower bounds (Proposition 4.24) for the asymptotic exponential exit rate from domains of attraction for slowly time dependent diffusions. They combine to the main large deviations result (Theorem 4.19) describing the exact asymptotic exponential exit rates for slowly and weakly time inhomogeneous diffusions in the small noise limit. This central result is tailor made for providing in Section 4.4 the optimal tuning rate related to maximal probability of transition during an exponential time window (see Theorems 4.29 and 4.31). We finally compare the resulting stochastic resonance point to the ones obtained in Section 4.2 for the Markov chains of the reduced dynamics (Theorem 4.15) on the basis of a related large deviation principle for exit and transition times (Theorem 4.13), and conclude that they agree in the small noise limit, thus establishing robustness (Theorem 4.32). The exposition of this Chapter follows Herrmann and Imkeller [**49, 50**] and Herrmann et al. [**51**].

4.1. Large deviations for diffusions with weakly inhomogeneous coefficients

In this section we shall extend the large deviations result by Freidlin and Wentzell to time inhomogeneous diffusions. This extension concerns only diffusions for which slowly time dependent drift coefficients $b^\varepsilon = b^\varepsilon(\cdot,\cdot)$ converge to a homogeneous drift coefficient $b = b(\cdot,\cdot)$ in the small noise limit. Since the aim is the study of stochastic resonance in a double-well potential, we cannot assume that the drift coefficient is globally Lipschitz. Therefore we shall use the extension of the theory of Freidlin–Wentzell to diffusion with coefficients that possess local Lipschitz properties, and present a generalization to slowly time inhomogeneous diffusions. It is noteworthy that all these extensions are not sufficient for a direct treatment of stochastic resonance. They just serve to prepare the ground for a subsequent study of large deviations for exit and transition times.

4.1.1. A general result on weakly time inhomogeneous diffusions. For $x \in \mathbb{R}^d$ consider the family $X^{\varepsilon,x}$, $\varepsilon > 0$, of solutions of the SDE

$$dX_t^{\varepsilon,x} = b^\varepsilon(t, X_t^\varepsilon)\,dt + \sqrt{\varepsilon}\,dW_t, \qquad t \geq 0,\ X_0^{\varepsilon,x} = x \in \mathbb{R}^d. \tag{4.2}$$

Here b^ε is Lipschitz continuous with respect to the time variable, and locally Lipschitz with respect to x. Our main large deviations result for diffusions for which time inhomogeneity fades out in the small noise limit is summarized in the following Proposition. We just recall that $|\cdot|$ is the Riemann norm in $\mathbb{R}^d$.

PROPOSITION 4.1 (Large deviations principle). *Assume that the drift of the SDE (4.2) satisfies*

$$\lim_{\varepsilon\to 0} b^\varepsilon(t,x) = b(x) \tag{4.3}$$

for all $t \geq 0$, uniformly w.r.t. x on compact subsets of $\mathbb{R}^d$, for some locally Lipschitz function $b\colon \mathbb{R}^d \to \mathbb{R}^d$. Moreover we assume the following dissipativity condition: there exist $R_0, \eta > 0$ such that

$$\langle x, b^\varepsilon(t,x)\rangle < -\eta|x| \quad \text{for all } \varepsilon > 0,\ t \geq 0,\ |x| \geq R_0. \tag{4.4}$$

Then for any $T \geq 0$, the laws of $(X^{\varepsilon,x})_{\varepsilon>0}$ satisfy a large deviations principle on $C([0,T],\mathbb{R}^d)$ with good rate function $I_{x,T}$ given by

$$I_{x,T}(\varphi) = \begin{cases} \dfrac{1}{2}\displaystyle\int_0^T |\dot{\varphi}_t - b(\varphi_t)|^2\,dt, & \text{if } \varphi - x \in \mathcal{H}_1([0,T],\mathbb{R}^d), \\ +\infty, & \text{otherwise.} \end{cases} \tag{4.5}$$

PROOF. Let $T \geq 0$ be fixed. First we introduce $Y^{\varepsilon,x}$ as the diffusion associated to the limiting drift term b. It is governed by the SDE

$$dY_t^{\varepsilon,x} = b(Y_t^{\varepsilon,x})\,dt + \sqrt{\varepsilon}\,dW_t, \qquad t \geq 0,\ Y_0^{\varepsilon,x} = x \in \mathbb{R}^d.$$

For notational convenience, we drop the ε, x-dependence of X and Y. As explained in Section 2.2, it suffices to establish exponential equivalence between the laws of X and Y. The corresponding LDPs are then governed by the same rate function.

We show that for any $\delta > 0$, we have

$$\limsup_{\varepsilon\to 0} \varepsilon \ln \mathbb{P}\Big(\|X - Y\| > \delta\Big) = -\infty, \tag{4.6}$$

where $\|\cdot\|$ is the norm of uniform convergence on $[0,T]$. In order to verify this, fix $\delta > 0$, and observe that

$$\begin{aligned} |X_t - Y_t| &= \Big|\int_0^t b^\varepsilon(u, X_u)\,du - \int_0^t b(Y_u)\,du\Big| \\ &\leq \int_0^t |b^\varepsilon(u,X_u) - b(X_u)|\,du + \int_0^t |b(X_u) - b(Y_u)|\,du. \end{aligned}$$

For $R > 0$, let $\tau_R := \inf\{t \geq 0\colon X_t \notin B_R(x)\}$, let $\tilde{\tau}_R$ be defined similarly with X replaced by Y, and $\sigma_R := \tau_R \wedge \tilde{\tau}_R$. The local Lipschitz continuity of b implies the existence of a constant L_R such that $|b(y_1) - b(y_2)| \leq L_R|y_1 - y_2|$ for $y_1, y_2 \in B_R(x)$.

$$\|X - Y\| \leq e^{L_R T}\int_0^T |b^\varepsilon(u, X_u) - b(X_u)|\,du \quad \text{ on } \{\sigma_R > T\}.$$

Due to uniform convergence, for any $\eta > 0$ we can find $\varepsilon_0 > 0$ such that

$$\sup_{y\in B_R(x)} |b^\varepsilon(t,y) - b(y)| \leq \eta \quad \text{ for } t \in [0,T],\ \varepsilon < \varepsilon_0.$$

This implies

$$\|X - Y\| \leq \eta T e^{L_R T} \quad \text{ for } \varepsilon < \varepsilon_0 \text{ on } \{\sigma_R > T\}. \tag{4.7}$$

If $\sigma_R > T$ then by choosing η (and ε_0) small enough $\|X - Y\| \leq \delta/2$, i.e. X and Y are very close together before they exit from $B_R(x)$. It remains to deal with the event $\{\sigma_R \leq T\}$ when one of the paths X or Y exits from the bounded domain $B_R(x)$. We see that for $\varepsilon < \varepsilon_0$

$$\mathbb{P}\Big(\|X - Y\| > \delta\Big) \leq \mathbb{P}(\tau_R \leq T) + \mathbb{P}(\tilde{\tau}_R \leq T).$$

Since X and Y are close within the ball $B_R(x)$, we deduce that if X escapes from $B_R(x)$ before time T, then Y must at least escape from $B_{R/2}(x)$ before time T (if $R > \delta$). So we have

$$\mathbb{P}\Big(\|X - Y\| > \delta\Big) \leq 2\mathbb{P}(\tilde{\tau}_{R/2} \leq T)$$

for $\varepsilon < \varepsilon_0$. Hence the exponential tightness of the law of Y (see Corollary 2.35) implies that for $R \geq R_0$

$$\limsup_{\varepsilon \to 0} \varepsilon \ln \mathbb{P}\Big(\|X - Y\| > \delta\Big) \leq -\eta R/4$$

Therefore we can send $R \to \infty$ in order to prove the exponential equivalence for the laws of X and Y. □

The large deviations result of Proposition 4.1 can be extended in a classical way to a uniform large deviation principle with respect to the initial condition. The arguments based on Gronwall's Lemma are similar to those used in the proof of Corollary 2.38. We omit the details.

4.1.2. Weak inhomogeneity through slow periodic variation. In this subsection we shall deal with diffusion processes given as solutions of (4.1) for which the drift b is subject to a very slow periodic time inhomogeneity. More precisely, we shall be concerned with solutions of the following stochastic differential equation taking their values in d-dimensional Euclidean space, driven by a d-dimensional Brownian motion W of intensity ε:

$$dX_t^{\varepsilon,x} = b\Big(\frac{t}{T(\varepsilon)}, X_t^{\varepsilon,x}\Big)\, dt + \sqrt{\varepsilon}\, dW_t, \quad t \geq 0, \; X_0^{\varepsilon,x} = x \in \mathbb{R}^d. \tag{4.8}$$

In accordance with the heuristics of the Kramers–Eyring law (see Chapter 1), $T(\varepsilon)$ is taken to be an exponentially large time scale

$$T(\varepsilon) = e^{\frac{\mu}{\varepsilon}} \quad \text{with } \mu > 0. \tag{4.9}$$

The drift $b(t,x)$ of (4.8) is supposed to be a time-periodic function of period 1. Concerning its regularity properties, we suppose it to be locally Lipschitz in both variables, i.e. for $R > 0$, $x \in \mathbb{R}^d$ there are constants $L_R(x)$ and $\kappa_R(x)$ such that

$$|b(t,y_1) - b(t,y_2)| \leq L_R(x)|y_1 - y_2|, \tag{4.10}$$

$$|b(t,y) - b(s,y)| \leq \kappa_R(x)|t - s| \tag{4.11}$$

for all $y, y_1, y_2 \in B_R(x)$ and $s, t \geq 0$. Since stochastic resonance will be typically studied for Brownian particles moving in double-well potential landscapes, we shall assume that the drift term forces the diffusion to stay in compact sets for long times in order to get sufficiently small level sets. We suppose that there exist constants η, $R_0 > 0$ such that

$$\langle x, b(t,x) \rangle < -\eta |x| \tag{4.12}$$

for $t \geq 0$ and $|x| \geq R_0$. This growth condition implies exponential tightness of the diffusion (see Remark 2.34 for the precise asymptotics).

4.1.2.1. *Properties of the quasi-potential.* Since the diffusion $X^{\varepsilon,x}$ is subject to very slow periodic variations, we are in a situation in which on small time intervals the behavior of the diffusion does not change drastically if we freeze the time parameter in the drift term. For a time parameter $s \geq 0$ we are therefore motivated to analyze the solution Y^ε of the SDE

$$dY_t^{\varepsilon,x} = b(s, Y_t^{\varepsilon,x})\, dt + \sqrt{\varepsilon}\, dW_t, \quad t \geq 0, \; Y_0^{\varepsilon,x} = x.$$

Applying Theorem 2.36 we obtain a LDP for $Y^{\varepsilon,x}$ on $C([0,T],\mathbb{R}^d)$, $T>0$, with good rate function

$$I_T^s(\varphi)=\begin{cases}\frac{1}{2}\int_0^T|\dot{\varphi}_t-b(s,\varphi_t)|^2\,dt, & \text{if } \varphi-x\in\mathcal{H}_1([0,T],\mathbb{R}^d),\\ +\infty, & \text{otherwise.}\end{cases}\tag{4.13}$$

The corresponding rate functions are given a separate treatment in this subsection. We shall briefly discuss their regularity properties. Let us define the *cost* functions. For $s\geq 0$, $x,y\in\mathbb{R}^d$ they are given by

$$V^s(x,y,t)=\inf\{I_t^s(\varphi)\colon \varphi\in C([0,t],\mathbb{R}^d),\varphi_0=x,\varphi_t=y\}.\tag{4.14}$$

The value $V^s(x,y,t)$ is the cost of forcing the *frozen* system Y^ε to be at the point y at time t when starting at x. The corresponding *quasi-potential*

$$V^s(x,y)=\inf_{t>0}V^s(x,y,t)\tag{4.15}$$

describes the cost for the frozen system to go from x to y eventually. The regularity of the drift term is transmitted to both the quasi-potential and the cost function. Moreover, since (4.13) is a good rate function, the level sets of the quasi-potential are compact. Trajectories of the diffusion that reach a point far away from the origin are really costly. The following properties of the cost functions are immediate.

LEMMA 4.2. *For any $x,y,z\in\mathbb{R}^d$ and $s,t,u\geq 0$ we have*

a) $V^s(x,y,t+u)\leq V^s(x,z,t)+V^s(z,y,u)$,
b) $(s,y)\mapsto V^s(x,y,t)$ *is continuous on* $\mathbb{R}_+\times\mathbb{R}^d$,
c) $\inf_{|y|\geq R}V^s(x,y,t)\to\infty$ *as* $R\to\infty$ *uniformly w.r.t.* $s\geq 0$.

The regularity of the drift term implies local Lipschitz continuity for the quasi-potential.

LEMMA 4.3. *For any compact subset K of $\mathbb{R}^d$ there exists $\Gamma_K\geq 0$ such that*

$$\sup_{s\geq 0}V^s(x,y)\leq\Gamma_K d(x,y)$$

for all $x,y\in K$.

PROOF. Let x and y belong to K. There exists $R>0$ such that $K\subset B_R(0)$. Set $T=d(x,y)$. We construct a path $\varphi\in C([0,T],\mathbb{R}^d)$ by setting

$$\varphi_t=x+\frac{y-x}{d(x,y)}t\quad\text{for } t\in[0,T].$$

Hence $\varphi_0=x$ and $\varphi_T=y$. Since b is locally Lipschitz uniformly w.r.t. $s\geq 0$, we obtain an upper bound for the energy of φ by

$$I_T^s(\varphi)\leq\frac{1}{2}\sup_{s\geq 0}\int_0^T|\dot{\varphi}_t-b(s,\varphi_t)|^2dt\leq\frac{1}{2}\int_0^T\left(\frac{|y-x|}{d(x,y)}+\sup_{0\leq u\leq 1}|b(u,\varphi_t)|\right)^2dt.$$

Since b is locally Lipschitz uniformly w.r.t. $s\geq 0$ and x we further get

$$\begin{aligned}I_T^s(\varphi)&\leq\frac{1}{2}\int_0^T\Big(1+\kappa_R(0)+|b(0,\varphi_t)|\Big)^2dt\\&\leq\frac{1}{2}\int_0^T\Big(1+\kappa_R(0)+L_R(0)|\varphi_t|+|b(0,0)|\Big)^2dt\\&\leq\frac{T}{2}\Big(1+\kappa_R(0)+RL_R(0)+|b(0,0)|\Big)^2.\end{aligned}$$

For $\Gamma_K := \frac{1}{2}(1+\kappa_R(0)+RL_R(0)+|b(0,0)|)^2$ and by the definition of T, we obtain

$$\sup_{s\geq 0} V^s(x,y) \leq \sup_{s\geq 0} I_T^s(\varphi) \leq \Gamma_K d(x,y).$$

□

4.1.2.2. *Large deviations.* Let us now study the family $(X^{\varepsilon,x})_{\varepsilon>0}$ of solutions of the SDE

$$dX_t^{\varepsilon,x} = b\Big(\frac{t}{T(\varepsilon)}, X_t^{\varepsilon,x}\Big)\,dt + \sqrt{\varepsilon}\,dW_t, \quad t\geq 0,\ X_0^{\varepsilon,x} = x \in \mathbb{R}^d.$$

We shall first derive a large deviations principle for the laws $\mu_\varepsilon = \mathbb{P}\circ(X^{\varepsilon,x})^{-1}$, $\varepsilon>0$. In a second step this allows us to describe precisely the typical trajectories of the system in order to obtain information about the transitions between the meta-stable states given by the equilibria of b.

We will present a detailed analysis on the large deviations of μ_ε, $\varepsilon>0$, exhibiting exponential rates that are uniform w.r.t. to some of the system's parameters: the free energy parameter μ, the starting time, and the initial condition. These improvements are essential for the treatment of the exit problem in the sequel.

First let us note that $(\mu_\varepsilon)_{\varepsilon>0}$ satisfies a large deviations principle on $C([0,T],\mathbb{R}^d)$ endowed with the norm of uniform convergence and with good rate function I_T^0, the rate function of a homogeneous diffusion governed by the frozen drift $b(0,\cdot)$ and given by

$$I_T^0(\varphi) = \begin{cases} \frac{1}{2}\int_0^T |\dot\varphi_t - b(0,\varphi_t)|^2\,dt, & \text{if } \varphi - x \in \mathcal{H}_1([0,T],\mathbb{R}^d),\\ +\infty, & \text{otherwise.}\end{cases}$$

This is an immediate consequence of Proposition 4.1. One only has to mention that

$$\lim_{\varepsilon\to 0} b\Big(\frac{t}{T(\varepsilon)}, x\Big) = b(0,x)$$

locally uniformly w.r.t. x due to the Lipschitz assumptions on b.

To prove uniformity w.r.t. the above mentioned system parameters, we have to refine the comparison arguments involved in the proof of Proposition 4.1. Moreover we shall consider exponentially large starting times, which — through the scaling with $T(\varepsilon)$ — result in arbitrary autonomous drift terms $b(s,\cdot)$ with $s\geq 0$.

The diffusion (4.8) is a time inhomogeneous Markov process. The solution starting at time $r\geq 0$ with initial condition $x\in\mathbb{R}^d$ has the same law as the solution $X^{r,x}$ of the SDE

$$dX_t^{\varepsilon,r,x} = b\Big(\frac{r+t}{T(\varepsilon)}, X_t^{\varepsilon,r,x}\Big)\,dt + \sqrt{\varepsilon}\,dW_t,\ t\geq 0,\ X_0^{\varepsilon,r,x} = x\in\mathbb{R}^d. \tag{4.16}$$

We denote its law by $\mathbb{P}_{x,r}(\cdot)$, assume from now on that $T(\varepsilon)=\exp(\frac{\mu}{\varepsilon})$ for some $\mu>0$, and fix $T\geq 0$.

Let us first specify the exponentially large starting times. We introduce a neighborhood S^r of the time scale $rT(\varepsilon)$ and the diffusion with frozen drift $Y^{\varepsilon,r,x}$:

$$dY_t^{\varepsilon,r,x} = b(r, Y_t^{\varepsilon,r,x})\,dt + \sqrt{\varepsilon}\,dW_t, \quad t\geq 0,\ Y_0^{\varepsilon,r,x} = x. \tag{4.17}$$

In order to use localization arguments, for $R>0$ we define the following first exit time

$$\sigma_R^{u,r,x} = \inf\{t\geq 0\colon X_t^{\varepsilon,u,x}\notin B_R(0) \text{ or } Y_t^{\varepsilon,r,x}\notin B_R(0)\} = \tau_R^{u,x}\wedge\tilde\tau_R^{r,x}, \tag{4.18}$$

where $\tau_R^{u,x} := \inf\{t \geq 0\colon X_t^{\varepsilon,u,x} \notin B_R(0)\}$, and $\tilde{\tau}_R^{\varepsilon,r,x}$ is defined similarly with $X^{\varepsilon,u,x}$ replaced by $Y^{\varepsilon,r,x}$.

LEMMA 4.4. *Let $x \in \mathbb{R}^d$, $T \geq 0$, $u \in S^r$. On $\{\sigma_R^{u,r,x} > T\}$ we have*

$$\|X^{\varepsilon,u,x} - Y^{\varepsilon,r,x}\| \leq \kappa_R(0) T e^{L_R(0)T} \frac{\operatorname{diam}(S^r) + T}{T(\varepsilon)},$$

where $\kappa_R(0)$, $L_R(0)$ are the Lipschitz constants associated to the drift term b and $\|\cdot\|$ represents the norm of uniform convergence on $C([0,T],\mathbb{R}^d)$.

PROOF. Since $X^{\varepsilon,u,x}$ and $Y^{\varepsilon,r,x}$ are defined with respect to the same Brownian motion, we get

$$\begin{aligned}\|X^{\varepsilon,u,x} - Y^{\varepsilon,r,x}\| &\leq \int_0^T \Big| b\Big(\frac{u+s}{T(\varepsilon)}, X_s^{\varepsilon,u,x}\Big) - b(r, Y_s^{\varepsilon,r,x})\Big|\, ds \\ &\leq \int_0^T \Big| b\Big(\frac{u+s}{T(\varepsilon)}, X_s^{u,x}\Big) - b(r, X_s^{u,x})\Big|\, ds + \int_0^T |b(r, X_s^{u,x}) - b(r, Y_s^{r,x})|\, ds.\end{aligned}$$

Let $R > 0$ and $\sigma_R^{u,r,x}$ the exit time of $B_R(0)$ defined by (4.18). The local Lipschitz continuity of b implies the existence of a constant L_R such that $|b(r,x) - b(r,y)| \leq L_R|x-y|$ for $x, y \in B_R(x_0)$. An application of Gronwall's Lemma (see A.1) yields

$$\|X^{u,x} - Y^{r,x}\| \leq e^{L_R T} \int_0^T \Big| b\Big(\frac{u+s}{T(\varepsilon)}, X_s^{u,x}\Big) - b(r, X_s^{u,x})\Big|\, du \quad \text{on } \{\sigma_R^{u,x,r} > T\}.$$

The Lipschitz continuity of b with respect to the time variable implies

$$\|X^{u,x} - Y^{r,x}\| \leq \kappa_R(0) T e^{L_R T} \cdot \sup_{s \in [0,T]} \Big| \frac{u+s}{T(\varepsilon)} - r \Big| \quad \text{on } \{\sigma_R^{u,x,r} > T\}.$$

Since $u \in S^r$ we obtain the announced inequality. □

The neighborhood S_r of $rT(\varepsilon)$ appearing in Lemma 4.4 can in practice depend on several parameters, namely

- the noise intensity ε;
- the period of the slow perturbation of the dynamical system $T(\varepsilon)$; since $T(\varepsilon) = \exp(\frac{\mu}{\varepsilon})$ the dependence will only be represented by the logarithmic free energy parameter μ;
- the "center"of the neighborhood r;
- some additional parameter β which captures the growth of the neighborhood w.r.t. ε. The notion becomes precise in the following definition.

DEFINITION 4.5. Let $\mathcal{V} \subset (0,\infty)$. For $\mu \in \mathcal{V}$, $0 \leq r \leq 1$ and $\beta \geq 0$, we consider a neighborhood of the starting time $rT(\varepsilon)$ denoted by $S^{r,\beta}(\varepsilon,\mu)$. The family $(S^{r,\beta}(\varepsilon,\mu))_{\mu \in \mathcal{V}, 0 \leq r \leq 1, \beta \geq 0}$ satisfies the **condition (H)** if

$$\limsup_{\varepsilon \to 0} \sup_{\mu \in \mathcal{V}, r \in [0,1]} \frac{\operatorname{diam}(S^{r,\beta}(\varepsilon,\mu))}{T(\varepsilon)} \leq \beta.$$

For example, the condition **(H)** is satisfied for

$$S^{r,\beta}(\varepsilon,\mu) = \Big[\Big(r - \frac{\beta}{2}\Big)T(\varepsilon), \Big(r + \frac{\beta}{2}\Big)T(\varepsilon)\Big].$$

Condition **(H)** is crucial. If the neighborhoods $S^{r,\beta}(\varepsilon,\mu)$ do not grow too fast as $\varepsilon \to 0$, then by Lemma 4.4, $X^{\varepsilon,u,x}$ and $Y^{\varepsilon,r,x}$ remain close together as long as they stay in $B_R(0)$.

PROPOSITION 4.6. Let $K \subset \mathbb{R}^d$ be a compact set and $\mathcal{V} \subset (0, \infty)$. For $\mu \in \mathcal{V}$, $r \in [0,1]$ and $\beta \geq 0$ let $S^{r,\beta}(\varepsilon, \mu)$ be a neighborhood of $rT(\varepsilon)$, and assume that the family $(S^{r,\beta}(\varepsilon,\mu))_{\mu\in\mathcal{V},0\leq r\leq 1,\beta\geq 0}$ satisfies the condition **(H)**. Then for any closed $F \subset C([0,T], \mathbb{R}^d)$, there exists $\delta = \delta(F)$ such that

$$\limsup_{\varepsilon\to 0} \varepsilon \ln \sup_{\substack{y\in K,\mu\in\mathcal{V},\\ u\in S^{r,\beta}(\varepsilon,\mu)}} \mathbb{P}(X^{\varepsilon,u,y} \in F) \leq - \inf_{y\in K} \inf_{\substack{\varphi\in F^{\gamma_0},\\ \varphi_0=y}} I_T^r(\varphi),$$

where $\gamma_0 = \gamma_0(F) = \beta\delta(F)$ and F^{γ_0} is the closed γ_0-neighborhood of F i.e. $F^{\gamma_0} = F + \overline{B_{\gamma_0}(0)}$ in Minkowski's notation. This bound holds uniformly w.r.t. $r \in [0,1]$.

REMARK 4.7. The upper bound means that for any $\theta > 0$ we can find $\varepsilon_0 > 0$ such that for $\varepsilon \leq \varepsilon_0$ we have

$$\varepsilon \ln \sup_{\substack{y\in K,\mu\in\mathcal{V},\\ u\in S^{r,\beta}(\varepsilon,\mu)}} \mathbb{P}(X^{\varepsilon,u,y} \in F) \leq - \inf_{y\in K} \inf_{\substack{\varphi\in F^{\gamma_0},\\ \varphi_0=y}} I_T^r(\varphi) + \theta.$$

The uniformity in the statement means that ε_0 can be chosen independently of r.

Observe that the expression for the blowup-factor $\gamma_0(F)$ depends on the set F only through $\delta(F)$ which is independent of β, and that $\gamma_0(F) \to 0$ as $\beta \to 0$ for all F. In particular, if β is equal to zero, we recover the classical bound of the uniform LDP.

PROOF. 1. Fix a closed set $F \subset C([0,T], \mathbb{R}^d)$, and a compact set $\mathcal{W} \subseteq [0,1]$. Denoting as before by $\|\cdot\|$ the uniform norm in $C([0,T], \mathbb{R}^d)$, for all $y \in K$, $\gamma > 0$, $\mu \in \mathcal{V}$, $r, r_0 \in \mathcal{W}$, and $u \in S^{r,\beta}(\varepsilon, \mu)$ we have

$$\mathbb{P}(X^{\varepsilon,u,y} \in F) \leq \mathbb{P}(Y^{\varepsilon,r_0,y} \in F^\gamma) + \mathbb{P}(\|X^{\varepsilon,u,y} - Y^{\varepsilon,r_0,y}\| > \gamma).$$

This yields

$$\begin{aligned} &\limsup_{\varepsilon\to 0} \varepsilon \ln \sup_{\substack{y\in K,\mu\in\mathcal{V},r\in\mathcal{W},\\ u\in S^{r,\beta}(\varepsilon,\mu)}} \mathbb{P}(X^{\varepsilon,u,y} \in F) \\ &\leq \limsup_{\varepsilon\to 0} \varepsilon \ln \max\Big\{ \sup_{y\in K} \mathbb{P}(Y^{\varepsilon,r_0,y} \in F^\gamma), \\ &\qquad \sup_{\substack{y\in K,\mu\in\mathcal{V},r\in\mathcal{W},\\ u\in S^{r,\beta}(\varepsilon,\mu)}} \mathbb{P}(\|X^{\varepsilon,u,y} - Y^{\varepsilon,r_0,y}\| > \gamma)\Big\}. \end{aligned} \tag{4.19}$$

We show that there exists γ such that the asymptotics of the maximum in (4.19) is determined by the left hand term $\sup_{y\in K} \mathbb{P}(Y^{\varepsilon,r_0,y} \in F^\gamma)$.

In this case the uniform LDP for Y will provide the bound

$$\limsup_{\varepsilon\to 0} \varepsilon \ln \sup_{\substack{y\in K,\mu\in\mathcal{V},r\in\mathcal{W},\\ u\in S^{r,\beta}(\varepsilon,\mu)}} \mathbb{P}(X^{\varepsilon,u,y} \in F) \leq - \inf_{y\in K} \inf_{\substack{\varphi\in F^{\gamma},\\ \varphi_0=y}} I_T^{r_0}(\varphi), \tag{4.20}$$

where unfortunately γ will depend on F.

For this purpose, we estimate $\mathbb{P}(\|X^{\varepsilon,u,y} - Y^{\varepsilon,r_0,y}\| > \gamma)$. Due to condition **(H)**, for $R > 0$ Lemma 4.4 implies the following upper bound. On the event $\{\sigma_R^{u,r_0,y} > T\}$, we have

$$\|X^{\varepsilon,u,y} - Y^{\varepsilon,r_0,y}\|_T \leq \beta_1, \quad \text{as } \varepsilon \to 0,$$

where

$$\beta_1 = \beta_1(\mathcal{W}) = \beta_0(\mathcal{W})M(R), \tag{4.21}$$

with

$$\beta_0(\mathcal{W}) := \beta + \sup_{r\in\mathcal{W}} |r - r_0| \quad \text{and} \quad M(R) := T\kappa_R(0)e^{L_R(0)T}. \tag{4.22}$$

In particular, if $\gamma > \beta_1(\mathcal{W})$, we can find $\varepsilon_0 > 0$ such that for $\varepsilon \le \varepsilon_0$

$$\sup_{\substack{y\in K,\mu\in\mathcal{V},r\in\mathcal{W},\\ u\in S^{r,\beta}(\varepsilon,\mu)}} \mathbb{P}(\|X^{\varepsilon,u,y} - Y^{\varepsilon,r_0,y}\| > \gamma) \le \Theta(R,\varepsilon), \tag{4.23}$$

where $\Theta(R,\varepsilon)$ is defined by

$$\Theta(R,\varepsilon) := \sup_{\substack{r\in[0,1],y\in K,\mu\in\mathcal{V},\\ u\in S^{r,\beta}(\varepsilon,\mu)}} \mathbb{P}(\tau_R^{u,y} \le T) + \sup_{\substack{r\in[0,1],\\ y\in K}} \mathbb{P}(\tilde{\tau}_R^{r,y} \le T).$$

By Theorem 2.32 and Remark 2.33 we have

$$\limsup_{\varepsilon\to 0} \varepsilon \ln \Theta(R,\varepsilon) \le -\eta R$$

for all $R \ge R_0$. Hence we may find $R \ge R_0$ such that

$$\limsup_{\varepsilon\to 0} \varepsilon \ln \Theta(R,\varepsilon) \le -\sup_{r\in[0,1]} \inf_{y\in K} \inf_{\substack{\varphi\in F,\\ \varphi_0=y}} I_T^r(\varphi).$$

We summarize our findings so far. For $R \ge R_0$, let $\delta(F) = M(R)$. Then $\delta(F)$ is independent of β and $\mathcal{W}$. Choose $\gamma > \beta_1(\mathcal{W}) = \beta_0(\mathcal{W})\delta(F)$. Then the maximum in (4.19) is given by the left hand term, and we obtain

$$\begin{aligned}\limsup_{\varepsilon\to 0} \varepsilon \ln \sup_{\substack{r\in\mathcal{W},y\in K,\mu\in\mathcal{V},\\ u\in S^{r,\beta}(\varepsilon,\mu)}} \mathbb{P}(\|X^{\varepsilon,u,y} - Y^{\varepsilon,r_0,y}\| > \gamma) &\le \limsup_{\varepsilon\to 0} \varepsilon \ln \Theta(R,\varepsilon)\\ &\le -\sup_{r\in[0,1]} \inf_{y\in K} \inf_{\substack{\varphi\in F,\\ \varphi_0=y}} I_T^r(\varphi)\\ &\le -\inf_{y\in K} \inf_{\substack{\varphi\in F^\gamma,\\ \varphi_0=y}} I_T^{r_0}(\varphi).\end{aligned}$$

This implies the inequality (4.20). The particular choice $\mathcal{W} = \{r_0\}$ yields this bound for all $\gamma > \gamma_0(F) = \beta\delta(F)$ given in the statement and proves the claimed bound. By taking the limit $\gamma \to \gamma_0(F)$ we get the asserted upper bound since I is a good rate function (see Theorem 2.31).

2. We prove the uniformity of the LDP w.r.t. r. The proof is essentially based on the continuity of the rate function. For this purpose fix $\theta > 0$, and for $r_0 \in [0,1]$ let $\mathcal{W}_{r_0}$ be a neighborhood of r_0. Choose $\phi^* \in F^{\gamma_0}$ (here $\gamma_0 = \gamma_0(F) = \beta\delta(F)$) starting at some point $y_0 \in K$ with

$$\inf_{y\in K} \inf_{\substack{\varphi\in F^{\gamma_0},\\ \varphi_0=y}} I_T^{r_0}(\varphi) \ge I_T^{r_0}(\phi^*) - \frac{\theta}{8}.$$

By the proof of Lemma 2.43 and the continuity of $r \mapsto I_T^r(\phi^*)$ we can assume $\mathcal{W}_{r_0}$ to be small enough such that for $r \in \mathcal{W}_{r_0}$, denoting $\gamma^* = \beta_0(\mathcal{W}_{r_0})\delta(F)$,

$$\begin{aligned}\inf_{y\in K} \inf_{\substack{\varphi\in F^{\gamma^*},\\ \varphi_0=y}} I_T^{r_0}(\varphi) &\ge \inf_{y\in K} \inf_{\substack{\varphi\in F^{\gamma_0},\\ \varphi_0=y}} I_T^{r_0}(\varphi) - \frac{\theta}{8} \ge I_T^{r_0}(\phi^*) - \frac{\theta}{4}\\ &\ge I_T^r(\phi^*) - \frac{\theta}{2} \ge \inf_{y\in K} \inf_{\substack{\varphi\in F^{\gamma_0},\\ \varphi_0=y}} I_T^r(\varphi) - \frac{\theta}{2}.\end{aligned}$$

Due to compactness we can choose finitely many points $r_1, \dots, r_N$, such that their corresponding neighborhoods cover $[0,1]$. For $1 \le n \le N$ denote $\gamma_n^* := \beta_0(\mathcal{W}_{r_n})\delta(F)$. For each n, $1 \le n \le N$, there exists $\varepsilon_n > 0$ such that for $\varepsilon \le \varepsilon_n$ and $r \in \mathcal{W}_{r_n}$,

$$\begin{aligned} \varepsilon \ln \sup_{\substack{y \in K, \mu \in \mathcal{V}, \\ u \in S^{r,\beta}(\varepsilon,\mu)}} \mathbb{P}(X^{\varepsilon,u,y} \in F) &\le - \inf_{y \in K} \inf_{\substack{\varphi \in F^{\gamma_n^*}, \\ \varphi_0 = y}} I_T^{r_n}(\varphi) + \frac{\theta}{2} \\ &\le - \inf_{y \in K} \inf_{\substack{\varphi \in F^{\gamma_0}, \\ \varphi_0 = y}} I_T^{r}(\varphi) + \theta. \end{aligned}$$

Hence for $\varepsilon \le \min_{1 \le n \le N} \varepsilon_n$, the preceding inequality holds for all $r \in [0,1]$. □

PROPOSITION 4.8. Under the same conditions as stated in Proposition 4.6, for any open $G \subset C([0,T], \mathbb{R}^d)$, there exists $\delta = \delta(G)$ and $\beta_0 = \beta_0(G)$ such that, if $\beta \le \beta_0$,

$$\liminf_{\varepsilon \to 0} \varepsilon \ln \inf_{\substack{y \in K, \mu \in \mathcal{V}, \\ u \in S^{r,\beta}(\varepsilon,\mu)}} \mathbb{P}(X^{\varepsilon,u,y} \in G) \ \ge - \sup_{y \in K} \inf_{\substack{\varphi \in G^{\gamma_0}, \\ \varphi_0 = y}} I_T^{r}(\varphi),$$

where $\gamma_0 = \gamma_0(G) = \beta\delta(G)$ and G^{γ_0} is the complement of $(G^c)^{\gamma_0}$. This bound holds uniformly w.r.t. $r \in [0,1]$.

PROOF. Let $G \subset C([0,T], \mathbb{R}^d)$ be an open set, $\mathcal{W} \subset [0,1]$ compact. Consider the increasing function

$$f(l) := \frac{1}{\eta} \sup_{y \in K} \inf_{\phi \in G^l :\, \phi_0 = y} I_T^{r_0}(\phi),$$

let

$$l_0 = \inf\{l \ge 0 \colon f(l) = +\infty\},$$

and recall that η is the constant introduced in the growth condition for the drift (4.4).

Step 1: the bounded case. Assume first that $l_0 < \infty$ (this is guaranteed if G is bounded), and set

$$R := f\Big((l_0 - \beta_0(\mathcal{W})) \vee \frac{l_0}{2}\Big) \vee R_0 \quad \text{and} \quad \gamma := \beta_0(\mathcal{W})M(R),$$

where R_0 is given by Corollary 2.35 and both β_0 and M are defined by (4.22). Then for $y \in K$, $\gamma > 0$, $\mu \in \mathcal{V}$, $r, r_0 \in \mathcal{W}$, and $u \in S^{r,\beta}(\varepsilon,\mu)$

$$\mathbb{P}(Y^{\varepsilon,r_0,y} \in G^\gamma) \le \mathbb{P}(X^{\varepsilon,u,y} \in G) + \mathbb{P}(\|Y^{\varepsilon,r_0,y} - X^{\varepsilon,u,y}\| > \gamma).$$

The last term in the previous inequality was already studied in the proof of Proposition 4.8, namely it is related to $\Theta(R,\varepsilon)$ by (4.23). By the uniform LDP for $Y^{\varepsilon,r_0,y}$

we conclude that

$$
\begin{aligned}
-\eta f(\gamma) &= -\sup_{y\in K}\inf_{\substack{\varphi\in G^\gamma,\\ \varphi_0=y}} I_T^{r_0}(\varphi)\\
&\le \liminf_{\varepsilon\to 0}\varepsilon\ln\inf_{y\in K}\mathbb{P}(Y^{\varepsilon,r_0,y}\in G^\gamma)\\
&\le \max\Big\{\liminf_{\varepsilon\to 0}\varepsilon\ln\inf_{\substack{r\in\mathcal{W},y\in K,\mu\in\mathcal{V},\\ u\in S^{r,\beta}(\varepsilon,\mu)}}\mathbb{P}(X^{\varepsilon,u,y}\in G),\\
&\qquad \limsup_{\varepsilon\to 0}\varepsilon\ln\sup_{\substack{r\in\mathcal{W},y\in K,\mu\in\mathcal{V},\\ u\in S^{r,\beta}(\varepsilon,\mu)}}\mathbb{P}(\|Y^{\varepsilon,r_0,y}-X^{\varepsilon,u,y}\|>\gamma)\Big\}\\
&\le \max\Big\{\liminf_{\varepsilon\to 0}\varepsilon\ln\inf_{\substack{r\in\mathcal{W},y\in K,\mu\in\mathcal{V},\\ u\in S^{r,\beta}(\varepsilon,\mu)}}\mathbb{P}(X^{\varepsilon,u,y}\in G),\\
&\qquad \limsup_{\varepsilon\to 0}\varepsilon\ln\Theta(R,\varepsilon)\Big\}.
\end{aligned}
$$

Since f is increasing and $R\ge R_0$, we obtain by Corollary 2.35

$$
\begin{aligned}
-\eta f(\gamma+\beta_0(\mathcal{W})) &\le -\eta f(\gamma)\\
&\le \max\Big\{\liminf_{\varepsilon\to 0}\varepsilon\ln\inf_{\substack{r\in\mathcal{W},y\in K,\mu\in\mathcal{V},\\ u\in S^{r,\beta}(\varepsilon,\mu)}}\mathbb{P}(X^{\varepsilon,u,y}\in G),-\eta R\Big\}.
\end{aligned}
$$

Now we have to compare $f(\gamma)$ and R in order to see under which conditions the maximum in the preceding inequality is given by the left hand term.

If $f(\gamma)>R$, then $\gamma>(l_0-\beta_0(\mathcal{W}))\vee\frac{l_0}{2}\ge l_0-\beta_0(\mathcal{W})$ by monotonicity of f, hence $f(\gamma+\beta_0(\mathcal{W}))=+\infty$ by definition of l_0. Otherwise we have $f(\gamma)\le R$, which means that the left term dominates the maximum.

In both cases we get

$$
-\eta f(\gamma+\beta_0(\mathcal{W}))\le\liminf_{\varepsilon\to 0}\varepsilon\ln\inf_{\substack{r\in\mathcal{W},y\in K,\mu\in\mathcal{V},\\ u\in S^{r,\beta}(\varepsilon,\mu)}}\mathbb{P}(X^{\varepsilon,u,y}\in G).
$$

Step 2: the unbounded case. Now consider $l_0=+\infty$. Let $\beta_0(G):=\sup_{l\ge 0}\frac{l}{M(f(l))}$, the existence of which was claimed in the statement.

If $\beta_0(\mathcal{W})<\beta_0(G)$, we can choose l_1 such that $\frac{l_1}{M(f(l_1))}\ge\beta_0(\mathcal{W})$ and set $\gamma:=\beta_0(\mathcal{W})M(f(l_1))$. Using the same arguments as in Step 1, we deduce that

$$
-\eta f(\gamma)\le\max\Big\{\liminf_{\varepsilon\to 0}\varepsilon\ln\inf_{\substack{r\in\mathcal{W},y\in K,\mu\in\mathcal{V},\\ u\in S^{r,\beta}(\varepsilon,\mu)}}\mathbb{P}(X^{\varepsilon,u,y}\in G),-\eta f(l_1)\Big\}.
$$

Since f is increasing and $l_1\ge\gamma$, we obtain

$$
-\eta f(\gamma)\le\liminf_{\varepsilon\to 0}\varepsilon\ln\inf_{\substack{r\in\mathcal{W},y\in K,\mu\in\mathcal{V},\\ u\in S^{r,\beta}(\varepsilon,\mu)}}\mathbb{P}(X^{\varepsilon,u,y}\in G).
$$

Summarizing our findings in both the bounded and the unbounded case, we obtain $\gamma=\beta_0(\mathcal{W})\delta(G)$ such that the desired bound holds. Namely, we have $\delta(G)=M(R)+1$ in the bounded case and $\delta(G)=M(f(l_1))$ in the unbounded case. Furthermore, the choice $\mathcal{W}=\{r_0\}$ corresponds to $\beta_0(\mathcal{W})=\beta$ and yields $\gamma_0(G)=\beta\delta(G)$. Uniformity in r is proved in exactly the same way as for the upper bound in Proposition 4.6. □

4.2. A new measure of periodic tuning induced by Markov chains

In the small noise limit $\varepsilon \to 0$, it seems reasonable to assume that the periodicity properties of the diffusion trajectories caused by the periodic forcing appearing in the drift term b, are essentially captured by a simpler, reduced stochastic process: a continuous time Markov chain which just jumps between two states x_- and x_+ representing the meta-stable states at rates corresponding to the transition mechanism of the diffusion. In the potential diffusion case, x_- and x_+ are the bottoms of the wells of the double well potential.

This is just the reduction idea ubiquitous in the physics literature, and explained for example in McNamara, Wiesenfeld [**74**]. In Section 3.4.3 it is shown that this idea may conflict with the intra well fluctuations of the diffusion if the quality of periodic tuning is measured by concepts using spectral decompositions of the trajectories.
We shall now show that in the small noise limit both models, diffusion and Markov chain, produce the same resonance picture, if measuring quality of periodic tuning is based on transition rates.

The aim of this section is to give a precise account of the transitions in the reduced Markov chain model (see Figure 4.1 for the transitions in Markov models with different tunings).

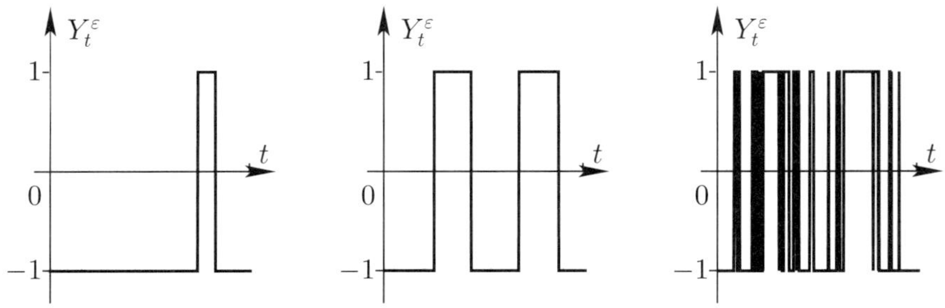

FIGURE 4.1. Different tunings for the reduced model.

This shall permit to provide the intuition for the construction of the new quality measure of periodic tuning. We first describe the reduced model. It consists in version of the Markov chain introduced in Section 3.2 with time continuous transition characteristics. The Kramers–Eyring law characterizes the exit law from a domain of attraction for classical homogeneous diffusions and by extension the transitions between its meta-stable states. The mean exit time is exponentially large and the exponential scale depends on the quasi-potential. If the homogeneous dynamical system has two domains of attraction A_- and A_+ ($x_\pm \in A_\pm$ and $\overline{A_- \cup A_+} = \mathbb{R}^d$) and if we define $\tau_\pm^\varepsilon$ the exit times, then we know that there exist $e_- > 0$ and $e_+ > 0$ such that

$$\lim_{\varepsilon \to 0} \varepsilon \ln \mathbb{E}_{x_\pm}[\tau_\pm^\varepsilon] = e_\pm.$$

In this case the continuous time Markov chain $\{Y_t,\, t \geq 0\}$ with state space $\{-1, 1\}$ that mimics the *inter well dynamics* of the diffusion has the infinitesimal generator

$$Q = \begin{pmatrix} -\varphi & \varphi \\ \psi & -\psi \end{pmatrix},$$

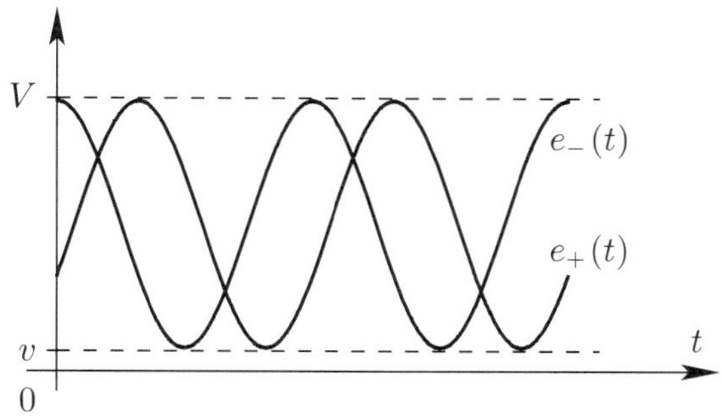

FIGURE 4.2. The periodic rates e_- and e_+.

with $\varphi = e^{-e_-/\varepsilon}$ and $\psi = e^{-e_+/\varepsilon}$. We have to adapt this correspondence between diffusions and Markov chains to diffusions the drift of which is dominated by slow periodic variations with period $T(\varepsilon) = \exp(\frac{\mu}{\varepsilon})$. In the previous section, it was proven that $X^{\varepsilon,x}$ defined by

$$dX_t^{\varepsilon,x} = b\Big(\frac{t}{T(\varepsilon)}, X_t^{\varepsilon,x}\Big)\,dt + \sqrt{\varepsilon}\,dW_t, \qquad t \geq 0,\ X_0^{\varepsilon,x} = x \in \mathbb{R}^d,$$

on time intervals of the form $[rT(\varepsilon), (r+1)T(\varepsilon)]$ has similar behavior as the solution $Z^{\varepsilon,r,x}$ of the SDE

$$dZ_t^{\varepsilon,r,x} = b(r, Z_t^{\varepsilon,r,x})\,dt + \sqrt{\varepsilon}\,dW_t, \quad t \geq 0,\ Z_0^{\varepsilon,r,x} = x.$$

It is therefore natural to conjecture that — locally around the time $rT(\varepsilon)$ — the Markov chain mimics a diffusion with frozen drift term. So we will let $e_\pm$ depend on time continuously in a slow periodic way.

Hence the reduced model of interest is given by the following settings. We consider a continuous time Markov chain $\{Y_t^\varepsilon,\, t \geq 0\}$ in the state space $\{-1, 1\}$ with initial data $Y_0^\varepsilon = -1$. The infinitesimal generator is given by

$$Q(t) = \begin{pmatrix} -\varphi(t) & \varphi(t) \\ \psi(t) & -\psi(t) \end{pmatrix}$$

with

$$\varphi(t) = \exp\Big(-\frac{e_-(t/T(\varepsilon))}{\varepsilon}\Big) \quad \text{and} \quad \psi(t) = \exp\Big(-\frac{e_+(t/T(\varepsilon))}{\varepsilon}\Big). \tag{4.24}$$

Here $e_\pm$ are continuous 1-periodic functions.

As our standard example (see Chapter 1) we can choose functions varying sinusoidally between the levels v and V $(v < V)$ and given by

$$e_-(t) = \frac{v+V}{2} + \frac{V-v}{2}\cos(2\pi t) \tag{4.25}$$

and $e_+(t) = e_-(t+\phi)$ where ϕ represents the phase of the system.

4.2.1. Law of the transitions. For $n \in \mathbb{N}$ let us denote by T_n the time of the nth jump from one state to the other (if n is odd, the Markov chain jumps from -1 to $+1$ and the opposite happens if n is even). Then we define the normalized time of the nth jump by

$$\tau_n := \frac{T_n}{T(\varepsilon)} \quad \text{for} \quad n \in \mathbb{N}. \tag{4.26}$$

Due to the periodic changes of the potential (matrix of infinitesimal probabilities) the law of the nth jump depends strongly on the position of the Markov chain and on the time of the last jump. The following result is obtained easily with the help of the Markov property.

LEMMA 4.9. *For $n \in \mathbb{N}$ the density $p_\pm$ of τ_n, given $\tau_{n-1} = u$ is equal to*

$$p_+(t) = T(\varepsilon)\psi(T(\varepsilon)t)\exp\left[-T(\varepsilon)\int_u^t \psi(T(\varepsilon)s)\,ds\right]\cdot \mathbb{I}_{[u,\infty)}(t), \tag{4.27}$$

$$p_-(t) = T(\varepsilon)\varphi(T(\varepsilon)t)\exp\left[-T(\varepsilon)\int_u^t \varphi(T(\varepsilon)s)\,ds\right]\cdot \mathbb{I}_{[u,\infty)}(t). \tag{4.28}$$

In the sequel we shall control laws of exit or transition times by the Laplace transforms of these variables.

LEMMA 4.10. *For $n \in \mathbb{N}$, the Laplace transform of τ_n, given $\tau_{n-1} = u$ and $(-1)^n = -1$, is equal to*

$$\mathcal{L}_-^{(u)}(\lambda) = \frac{\displaystyle\int_u^{u+1} T(\varepsilon)\varphi(tT(\varepsilon))\exp\left\{-T(\varepsilon)\int_u^t \varphi(T(\varepsilon)s)ds - \lambda t\right\}dt}{1-\exp\left\{-T(\varepsilon)\displaystyle\int_0^1 \varphi(T(\varepsilon)s)\,ds - \lambda\right\}}, \quad \lambda \geq 0. \tag{4.29}$$

A similar expression can be obtained for $(-1)^n = 1$, by just replacing φ by ψ in expression (4.29).

PROOF. By (4.28), for $\lambda \geq 0$,

$$\begin{aligned}
\mathcal{L}_-^{(u)}(\lambda) &= \mathbb{E}\left[e^{-\lambda\tau_n}\,\middle|\,\tau_{n-1} = u,\ (-1)^n = -1\right] \\
&= \int_u^\infty T(\varepsilon)\varphi(T(\varepsilon)t)\exp\left\{-T(\varepsilon)\int_u^t \varphi(T(\varepsilon)s)\,ds - \lambda t\right\}dt \\
&= \sum_{k\geq 0}\int_u^{u+1} T(\varepsilon)\varphi(T(\varepsilon)t)e^{-\lambda t-\lambda k} \\
&\qquad\times\exp\left\{-T(\varepsilon)\int_u^t \varphi(T(\varepsilon)s)\,ds - kT(\varepsilon)\int_0^1 \varphi(T(\varepsilon)s)ds\right\}dt.
\end{aligned}$$

Summing up the geometric series

$$\sum_{k\geq 0}\exp\left\{-kT(\varepsilon)\int_0^1 \varphi(T(\varepsilon)s)\,ds - \lambda k\right\} = \left(1-\exp\left\{-T(\varepsilon)\int_0^1 \varphi(T(\varepsilon)s)ds - \lambda\right\}\right)^{-1}$$

we obtain the expression (4.29). □

4.2.2. Asymptotic behaviour. Since the density of τ_n given τ_{n-1} depends on ε and $T(\varepsilon)$, we shall consider the limit as $\varepsilon \to 0$. Then, by the definition (4.24) of φ the Laplace transform $\mathcal{L}_-$ becomes

$$\mathcal{L}_-^{(u)}(\lambda) = \frac{\displaystyle\int_u^{u+1}\exp\left\{\frac{\mu - e_-(t)}{\varepsilon} - \int_u^t \exp\left\{\frac{\mu - e_-(s)}{\varepsilon}\right\}ds - \lambda t\right\}dt}{1-\exp\left\{-\displaystyle\int_0^1 \exp\left\{\frac{\mu - e_-(s)}{\varepsilon}\right\}ds - \lambda\right\}}. \tag{4.30}$$

As in the preceding section, for $u \in [0,1]$ we define S_u to be the normalized time of the first jump after the normalized time u. Since the considered process is Markovian, the Laplace transform of S_u given $Y^\varepsilon_{uT(\varepsilon)} = \pm 1$ is equal to $\mathcal{L}^{(u)}_\pm$.

THEOREM 4.11. *Let $u \in [0,1]$. Then in the limit as ε tend to zero the following holds true:*

(1) *if*

$$\mu > \sup_{t\geq 0} e_-(t)$$

then the law of $S_u - u$ tends to the Dirac measure in the origin;

(2) *if*

$$\mu \in \left(\inf_{t\geq 0} e_-(t), \sup_{t\geq 0} e_-(t)\right]$$

then the conditional law of S_u given $Y^\varepsilon_{uT(\varepsilon)} = -1$ tends to the Dirac measure in the point $a^-_\mu(u)$ defined by

$$a^\pm_\mu(u) = \inf\{t > u \colon \mu - e_\pm(t) \geq 0\}, \; u \geq 0; \tag{4.31}$$

if $Y^\varepsilon_{uT(\varepsilon)} = 1$, then the same result holds with the Dirac measure in the point $a^+_\mu(u)$;

(3) *if*

$$\mu \leq \inf_{t\geq 0} e_-(t),$$

then the probability measure of S_u tends weakly to the null measure.

PROOF. (1) Let us first study the case $\mu > \sup_{t\geq 0} e_-(t)$. By (4.30), the numerator of the Laplace transform is equal to

$$N_\varepsilon(\lambda) = \int_u^{u+1} W_\varepsilon(t,\lambda)\left(e^{(\mu-e_-(t))/\varepsilon} + \lambda\right) \exp\left\{-\int_u^t e^{(\mu-e_-(s))/\varepsilon}\, ds - \lambda t\right\} dt,$$

where

$$W_\varepsilon(t,\lambda) = \frac{e^{(\mu-e_-(t))/\varepsilon}}{e^{(\mu-e_-(t))/\varepsilon} + \lambda}.$$

Let us fix $\lambda \geq 0$. Since $\mu > \sup_{t\geq 0} e_-(t)$, we get, as $\varepsilon \to 0$, uniformly with respect to the variable t

$$1 - o(\varepsilon) \leq W_\varepsilon(t,\lambda) \leq 1.$$

Hence, asymptotically as $\varepsilon \to 0$, the numerator $N_\varepsilon(\lambda)$ is equivalent to

$$e^{-\lambda u} - \exp\left\{-\int_u^{u+1} e^{(\mu-e_-(s))/\varepsilon}\, ds - \lambda(u+1)\right\}$$

which tends to $e^{-\lambda u}$ as ε tends to zero. Moreover the denominator in the Laplace transform obviously tends to 1. So $\lim_{\varepsilon\to 0} \mathcal{L}^{(u)}_-(\lambda) = e^{-\lambda u}$. The same limit can be obtained for the Laplace transform $\mathcal{L}^{(u)}_+$. We deduce that the law of $S_u - u$ tends to the Dirac measure in the origin. This proof can be extended to the case $\mu = \sup_{t\geq 0} e_-(t)$.

(2) In the second case $\mu \in (\inf_{t\geq 0} e_-(t), \sup_{t\geq 0} e_-(t)]$, decomposing the integral $\int_0^1 e^{(\mu-e_-(s))/\varepsilon}\, ds$ into integrals on disjoint intervals, we obtain that the denominator of the Laplace transform tends to 1, exponentially fast if $\mu > \inf_{t\geq 0} e_-(t)$ and, otherwise, using Laplace's method (see Appendix B), with a speed of order $\sqrt{\varepsilon}$.

The numerator will be decomposed into the three parts

$$N_\varepsilon(\lambda) = R_1(\varepsilon,\lambda) + R_2(\varepsilon,\lambda) + R_3(\varepsilon,\lambda),$$

where

$$R_i(\varepsilon,\lambda) = \int_{\Delta_i} \mathcal{A}(\mu,t,\varepsilon)\,dt = \int_{\Delta_i} \exp\left\{\frac{\mu - e_-(t)}{\varepsilon} - \int_u^t e^{(\mu-e_-(s))/\varepsilon}\,ds - \lambda t\right\} dt,$$

$i = 1,2,3$, and where

$$\begin{aligned}\Delta_1 &= [u,u+1]\cap\{t\geq 0\colon \mu - e_-(t) < -\sqrt{\varepsilon}\},\\ \Delta_2 &= [u,u+1]\cap\{t\geq 0\colon \mu - e_-(t) > 0\},\\ \Delta_3 &= [u,u+1]\cap\{t\geq 0\colon -\sqrt{\varepsilon}\leq \mu - e_-(t) \leq 0\}.\end{aligned}$$

By the definition of Δ_1, we get $R_1(\varepsilon,\lambda)\leq e^{-1/\sqrt{\varepsilon}}$ which tends to zero as ε decreases. Otherwise $R_3(\varepsilon,\lambda)\leq |\Delta_3| = \mathcal{O}(\sqrt{\varepsilon})$ where $|\cdot|$ is the Euclidean length. It remains to determine the limit of the expression $R_2(\varepsilon,\lambda)$.

We notice that if $\mu = \inf_{t\geq 0} e_-(t)$ then $\Delta_2 = \emptyset$. This implies that the Laplace transform tends to 0: the measure of τ_n given τ_{n-1} tends to the null measure. Recalling from (4.31) the definition of a_μ^-, let us now assume that $\mu > \inf_{t\geq 0} e_-(t)$ and that $a_\mu^-(u) > u$. We postpone the study of the case $a_\mu^-(u) = u$. There exists $\delta > 0$ small enough, independent of ε, such that

$$R_2(\varepsilon,\lambda) = \int_{[a_\mu^-(u),a_\mu(u)+\delta]} \mathcal{A}(\mu,t,\varepsilon)\,dt + \int_{\Delta_2\cap[a_\mu(u),a_\mu(u)+\delta]^c} \mathcal{A}(\mu,t,\varepsilon)\,dt.$$

The second integral can be bounded above by

$$e^{-\lambda(a_\mu(u)+\delta)}\exp\Big(-\int_u^{a_\mu(u)+\delta} e^{(\mu-e_-(s))/\varepsilon}\,ds\Big).$$

Using Laplace's method, we get that this expression tends to zero, as $\varepsilon\to 0$. Furthermore, the first integral satisfies the inequalities

$$e^{-\lambda(a_\mu(u)+\delta)} I \leq \int_{[a_\mu(u),a_\mu(u)+\delta]} \mathcal{A}(\mu,t,\varepsilon)\,dt \leq e^{-\lambda a_\mu(u)} I,$$

with

$$I = \exp\left(-\int_u^{a_\mu(u)} e^{(\mu-e_-(s))/\varepsilon}\,ds\right) - \exp\left(-\int_u^{a_\mu(u)+\delta} e^{(\mu-e_-(s))/\varepsilon}\,ds\right).$$

For $\delta > 0$ fixed, I tends to 1 as $\varepsilon\to 0$. Hence $\lim_{\varepsilon\to 0}\mathcal{L}_-^{(u)}(\lambda) = e^{-\lambda a_\mu(u)}$ for $Y^\varepsilon_{uT(\varepsilon)} = -1$.

If $\mu > \inf_{t\geq 0} e_-(t)$ and $a_\mu(u) = u$, we denote by δ the first time greater than u such that $e_-(\delta) = \mu$. By definition (4.31), we obtain $\delta > u$. Let us decompose the expression $R_2(\varepsilon,\lambda)$ as follows:

$$R_2(\varepsilon,\lambda) = \int_{[u,\delta]} \mathcal{A}(\mu,t,\varepsilon)\,dt + \int_{\Delta_2\cap[u,\delta]^c} \mathcal{A}(\mu,t,\varepsilon)\,dt.$$

The second term is bounded above by

$$e^{-\lambda\delta}\exp\Big(-\int_u^{a_\mu(\delta)} e^{(\mu-e_-(t))/\varepsilon}\,dt\Big).$$

This term tends to zero as $\varepsilon\to 0$. Using the arguments presented in Step 1, we obtain that the first expression tends to $e^{-\lambda u}$.

All the results of Step 2 can be proved for $Y^{\varepsilon}_{uT(\varepsilon)} = 1$ using the function e_+ instead of e_-.

(3) In case $\mu \leq \inf_{t\geq 0} e_-(t)$, it is straightforward to prove that $\lim_{\varepsilon\to 0} \mathcal{L}^{(u)}_{\pm}(\lambda) = 0$. We deduce that the law of S_u converges weakly to the null measure. $\square$

The following definition singles out the window of scales for which the Markov chain can be said to show periodic behavior.

DEFINITION 4.12. For $\mu \in \mathbb{R}$ let $T(\varepsilon) = e^{\mu/\varepsilon}$, and recall that the dependence of the Laplace transforms on the parameter μ resides in the scaling of time. Further let

$$I_0 = \Big\{\mu \in \mathbb{R}\colon \lim_{\varepsilon\to 0} \mathcal{L}^{(u)}_{-}(\lambda) \neq 0 \text{ for all } \lambda \in \mathbb{R}_+,\ u \in \mathbb{R}_+\Big\},$$
$$I_1 = \Big\{\mu \in \mathbb{R}\colon \{u \in [0,1), \lim_{\varepsilon\to 0} \mathcal{L}^{(u)}_{-}(\lambda) = \lim_{\varepsilon\to 0} \mathcal{L}^{(u)}_{+}(\lambda) = e^{-\lambda u} \text{ for all } \lambda \in \mathbb{R}_+\} = \emptyset\Big\}.$$

Then the interval $I_{\mathcal{R}} := I_0 \cap I_1$ is called *interval of resonance.*

On this interval, in the small noise limit the process cannot remain in the same state with positive probability. Moreover, in the small noise limit the process cannot exhibit chaotic behavior, i.e. instantaneously jump between the states. Indeed, on the set

$$\Omega^{\mu} := \Big\{t \in [0,1)\colon \mu - e_-(t) < 0\Big\} \cap \Big\{t \in [0,1)\colon \mu - e_+(t) < 0\Big\}$$

the process will asymptotically show chaotic behavior: if it is in one state then it jumps instantaneously to the other one.

So if there is a phase ϕ between e_- and e_+ i.e. $e_+(t) = e_-(t+\phi)$, the interval of resonance should be equal to the set of all

$$\mu \in \Big(\inf_{t\geq 0} e_-(t), \sup_{t\geq 0} e_-(t)\Big]$$

such that, for this μ, the set Ω^{μ} is empty. In the example we have

$$e_-(t) = \frac{v+V}{2} + \frac{V-v}{2}\cos(2\pi t) \quad \text{and } e_+(t) = e_-(t+\phi),$$

Figure 4.3 presents a $\mu \in (v, V]$ such that Ω^{μ} has positive Lebesgue measure.

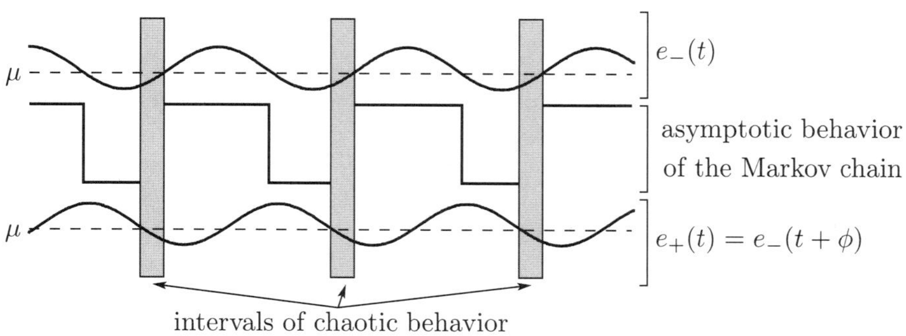

FIGURE 4.3. Chaotic behavior of the two-state Markov chain.

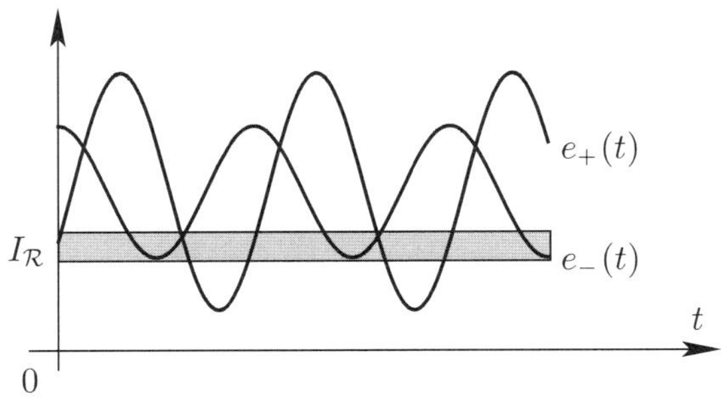

FIGURE 4.4. The resonance interval $I_{\mathcal{R}}$.

By (4.24), the set of all μ belonging to this particular interval satisfies the conditions

$$\text{(4.32)} \qquad \frac{\pi - \arccos(\frac{2\mu - V - v}{V - v})}{\pi} \leq \phi \leq \frac{\arccos(\frac{2\mu - V - v}{V - v})}{\pi} \quad \text{and } \mu > v.$$

Let us now assume that $\phi = 1/2$. The behavior of the two-state Markov chain then imitates the barrier crossings of a diffusion in a double-well potential. Here the potential is like $U(x) + x\cos(2\pi t/T(\varepsilon))$, where U is a symmetric double-well potential. Indeed, whether the diffusion starts in the deepest position of the left well at time t or starts in the right well at time $t + T(\varepsilon)/2$, it has to cross a barrier of the same hight to reach the other well. So the phase between the infinitesimal rate of jumps of the associated Markov chain has to be equal to $T(\varepsilon)/2$. In this particular case, we obtain the following interval of resonance:

$$I_{\mathcal{R}} = \left(v, \frac{v + V}{2}\right].$$

The time scale corresponding to the upper bound is of the same order as the mean exit time of one well (Kramers' rate), by the diffusion starting in the deepest position of this well, for the symmetric double-well potential U. This potential is, in fact, the average potential of $U(x) + x\cos(\frac{2\pi t}{T(\varepsilon)})$ over one period. Let us note that $\mu_0 = \frac{v+V}{2}$ is a bifurcation point: for $\mu \leq \mu_0$ the asymptotic behavior has no chaotic part, for $\mu > \mu_0$ chaoticity appears.

We claim that the resonance interval is determined in general by the interval

$$I_{\mathcal{R}} = \left(\max_{i=\pm} \inf_{t \geq 0} e_i(t), \inf_{t \geq 0} \max_{i=\pm} e_i(t)\right),$$

see Figure 4.4.

4.2.3. Optimal tuning using escape-time distribution and large deviations. In this Section, we shall concentrate on a measure of quality of periodic tuning which is based on the *interspike distributions*, describing the probabilities of the first exit as a function of time which possesses *spikes* at multiples of the (unscaled) period 1. We shall find the time scales for optimal tuning, in the small noise limit $\varepsilon \to 0$. In physical jargon, we consider the intensity of the first peak of the escape time distribution.

Mathematically, in our setting this intensity may be described by the probability to jump for the first time from the initial state to the other one. In the original scale, it means tuning the noise intensity to a value which maximizes the probability to observe the first jump in an exponentially wide time interval. The optimal time scale will be determined by a combination of a large deviations result concerning the first jump of the Markov chain parametrized by the logarithmic scale μ of time $T(\varepsilon) = \exp(\frac{\mu}{\varepsilon})$, and a maximization problem for the uniformly obtained large deviation rates.

By (4.31), a_μ^- is well defined on the interval $[\inf_{t\geq 0} e_-(t), \sup_{t\geq 0} e_-(t)]$. We extend this function continuously to $(-\infty, \sup_{t\geq 0} e_-(t)]$ by setting $a_\mu^- = a_{e_*}^-$ for $\mu \leq e_*$ where $e_* = \inf_{t\geq 0} e_-(t)$.

For $\mu \leq \sup_{t\geq 0} e_-(t)$ and $h > 0$, $\mathcal{A}^h$ will be the event defined by

$$\mathcal{A}^h := \Big\{\omega\colon |\tau_1(\omega) - (a_\mu(0) + k)| \geq h \text{ for all } k \in \mathbb{N}\Big\}. \tag{4.33}$$

The following theorem, proved by large deviations methods, essentially determines exponential rates of probabilities for $\mathcal{A}^h$, in particular if the exponential scale μ is in the resonance interval. Its primary result, formula (4.34), states that the main contribution to this probability of jumping comes from the instant during the time interval under consideration at which just the minimal barrier height is obtained. Its exponential rate is given by the remaining barrier height to overcome.

THEOREM 4.13. *Let $Y_0^\varepsilon = -1$. 1. If*

$$\mu \in \Big(\inf_{t\geq 0} e_-(t), \sup_{t\geq 0} e_-(t)\Big)$$

and

$$h < \min\Big\{a_\mu^-(0), 1 - a_\mu^-(0)\Big\}$$

then

$$\lim_{\varepsilon\to 0} \varepsilon \ln \mathbb{P}(\mathcal{A}^h) = \mu - e_-(a_\mu^-(0) - h). \tag{4.34}$$

2. If

$$\mu < \inf_{t\geq 0} e_-(t),$$

then, for h small enough,

$$\lim_{\varepsilon\to 0} \varepsilon \ln \mathbb{P}(\mathcal{A}^h) = e_-(a_\mu^-(0)) - e_-(a_\mu^-(0) - h). \tag{4.35}$$

3. If $Y_0^\varepsilon = +1$, it suffices to replace in 1. and 2. e_- and a_μ^- by e_+ and a_μ^+ respectively.

PROOF. Denote $\mathcal{D}_h := \{t \geq 0\colon |t - (a_\mu^-(0) + k)| \geq h \text{ for all } k \in \mathbb{N}\}$. Using the expression (4.28) for the density, we get

$$\mathbb{P}(\mathcal{A}^h) = \int_{\mathcal{D}_h} T(\varepsilon)\varphi(T(\varepsilon)t) \exp\Big(-T(\varepsilon)\int_0^t \varphi(T(\varepsilon)s)\, ds\Big)\, dt.$$

Then, for h small,

$$\begin{aligned}\mathbb{P}(\mathcal{A}^h) = \int_0^{a_\mu^-(0)-h} \Bigg[& T(\varepsilon)\varphi(T(\varepsilon)t) e^{-T(\varepsilon)\int_0^t \varphi(T(\varepsilon)s)\,ds} \\ & + \frac{e^{-T(\varepsilon)\int_0^{a_\mu^-(0)} \varphi(T(\varepsilon)s)\,ds}}{1 - e^{-T(\varepsilon)\int_0^1 \varphi(T(\varepsilon)s)\,ds}} \\ & \times \int_h^{1-h} T(\varepsilon)\varphi(T(\varepsilon)(a_\mu^-(0)+t)) e^{-T(\varepsilon)\int_0^t \varphi(T(\varepsilon)(a_\mu^-(0)+s))\,ds}\Bigg]\,dt.\end{aligned}$$

Then $\mathbb{P}(\mathcal{A}^h) = A + B \cdot C$, where

$$\begin{aligned} A &= 1 - \exp\Big(- \int_0^{a_\mu^-(0)-h} e^{(\mu - e_-(s))/\varepsilon}\,ds\Big), \\ B &= \frac{\exp\Big(- \int_0^{a_\mu^-(0)+h} e^{(\mu-e_-(s))/\varepsilon}\,ds\Big)}{1 - \exp\Big(- \int_0^1 e^{(\mu-e_-(s))/\varepsilon}\,ds\Big)}, \\ C &= 1 - \exp\Big(- \int_h^{1-h} e^{(\mu - e_-(s+a_\mu^-(0)))/\varepsilon}\,ds\Big). \end{aligned}$$

1. If $\mu \in (\inf_{t\geq 0} e_-(t), \sup_{t\geq 0} e_-(t))$, then, using Laplace's method to get equivalent expressions for the integrals as ε tends to zero, there exists some constant $K > 0$ such that

$$\begin{aligned}\mathbb{P}(\mathcal{A}^h) &\sim 1 - \exp\Big(- \int_0^{a_\mu^-(0)-h} e^{(\mu-e_-(s))/\varepsilon}\,ds\Big) \sim \int_0^{a_\mu^-(0)-h} e^{(\mu-e_-(s))/\varepsilon}\,ds \\ &\sim K e^{(\mu - e_-(a_\mu^-(0)-h))/\varepsilon}. \end{aligned} \tag{4.36}$$

2. If $\mu < \inf_{t\geq 0} e_-(t)$, then, again by Laplace's method, as $\varepsilon \to 0$, we get

$$\begin{aligned} \varepsilon \ln(A) &\sim \mu - e_-(a_\mu^-(0) - h), \\ \varepsilon \ln(B) &\sim e_-(a_\mu^-(0)) - \mu, \end{aligned}$$

and

$$\varepsilon \ln(C) \sim \mu - e_-(a_\mu^-(0) - h).$$

Hence

$$\lim_{\varepsilon\to 0} \varepsilon \ln \mathbb{P}(\mathcal{A}^h) = \max\Big\{ \mu - e_-(a_\mu^-(0) - h), e_-(a_\mu^-(0)) - e_-(a_\mu^-(0) - h)\Big\}.$$

We deduce (4.35) for h small enough. □

Let us now describe the optimal tuning rate corresponding to the *stronger first peak of the density of the first jump.* For μ belonging to the resonance interval, we will start the Markov chain at time t^* such that $a_\mu^{\pm}(t^*) > 0$; t^* becomes the new time origin.

DEFINITION 4.14. Assume that $a_\mu^\pm(0) > 0$, and let $h > 0$ be small enough. Then the *measure of quality of periodic tuning* is defined by

$$\mathcal{N}^h(\varepsilon, \mu) = \min_{i=\pm} \mathbb{P}_i\Big(\tau_{-i} \in [a_\mu^i(0) - h, a_\mu^i(0) + h]\Big), \quad \varepsilon > 0, \ \mu \in I_{\mathcal{R}}, \tag{4.37}$$

which is called *transition probability for a time window of width* h for the Markov chain.

Here is the asymptotic result obtained as a consequence of Theorem 4.13.

THEOREM 4.15. *Let* Γ *be a compact subset of* $I_{\mathcal{R}}$, $h_0 < \min\{a_\mu^\pm(0), 1 - a_\mu^\pm(0)\}$. *Then for* $0 < h \leq h_0$

$$\lim_{\varepsilon \to 0} \varepsilon \ln \Big(1 - \mathcal{N}^h(\varepsilon, \mu)\Big) = \max_{i=\pm 1} \Big\{\mu - e_i(a_\mu^i - h)\Big\} \tag{4.38}$$

uniformly for $\mu \in \Gamma$.

PROOF. Let us study the family of functions $\varepsilon \ln F_\varepsilon(\mu)$, $\varepsilon > 0$, where $F_\varepsilon(\mu)$ is defined as an integral such as one in $\mathbb{P}(\mathcal{A}^\delta)$ but on the domain $\{t \geq 0\colon |t - a_\mu^-(0)| \geq h\}$. We get

$$F_\varepsilon(\mu) = 1 - \exp\Big(-\int_0^{a_\mu^-(0)-h} e^{\frac{\mu - e_-(s)}{\varepsilon}}\, ds\Big) + \exp\Big(-\int_0^{a_\mu^-(0)+h} e^{\frac{\mu - e_-(s)}{\varepsilon}}\, ds\Big).$$

Theorem 4.13 then provides the simple convergence of the family of functions

$$\lim_{\varepsilon \to 0} \varepsilon \ln F_\varepsilon(\mu) = \mu - e_-(a_\mu^-(0) - \delta).$$

Let us prove uniform convergence. By Ascoli's theorem, it suffices to show that the derivative with respect to the variable μ is bounded as $\varepsilon \to 0$. There exists a constant $C > 0$, such that

$$\left|\frac{\varepsilon F_\varepsilon'(\mu)}{F_\varepsilon(\mu)}\right| \leq C\, F_\varepsilon^{-1}(\mu) \cdot \Big[\int_0^{a_\mu^-(0)-h} e^{\frac{\mu - e_-(s)}{\varepsilon}}\, ds + \\ + \varepsilon \frac{\partial a_\mu^-(0)}{\partial \mu} e^{\frac{\mu - e_-(a_\mu^+(0) - h)}{\varepsilon}} + o\left(\varepsilon e^{\frac{\mu - e_-(a_\mu^-(0) - h)}{\varepsilon}}\right)\Big].$$

We deduce that this derivative is bounded using the arguments related to (4.36). In order to finish the proof it suffices to use the same arguments for τ_+. $\square$

EXAMPLE 4.16. We choose the function varying between the levels v and V, $0 < v < V$, given by

$$e_-(t) = \frac{v + V}{2} + \frac{V - v}{2} \cdot \cos(2\pi t),$$
$$e_+(t) = e_-(t + \phi), \quad t \geq 0,$$

where ϕ represents the phase of the system.

We recall that the resonance interval is equal to $I_{\mathcal{R}} = (v, \frac{v+V}{2}]$ if $\phi = 1/2$, otherwise $I_{\mathcal{R}} \subset (v, \frac{v+V}{2}]$. Let $[a, b] \subset (v, \frac{v+V}{2})$. We need that $a_\mu^\pm(0) > 0$ for all $\mu \in [a, b]$. If this condition is not satisfied, it suffices to shift the time origin. In our particular case, the assumption is true for $\phi \leq 1/4$. Then we have

$$a_\mu^-(0) = \frac{1}{2\pi} \arccos\left(\frac{2\mu - v - V}{V - v}\right) > 1/4,$$

$$a_\mu^+(0) = \frac{1}{2\pi} \arccos\left(\frac{2\mu - v - V}{V - v}\right) - \phi.$$

For $0 < h$ small enough we want to maximize

$$\min_{i=\pm} \mathbb{P}_i\Big(\tau_i \in [a^i_\mu(0) - h, a^i_\mu(0) + h]\Big) \text{ for } \mu \in [a, b].$$

The optimal $\mu_0^\varepsilon(h)$ then corresponds to the optimal tuning between the small noise intensity and the exponentially large period of the perturbation. We shall find $\lim_{\varepsilon\to 0} \mu_0^\varepsilon(h)$. Since the convergence result (4.38) is uniform with respect to μ it suffices to minimize $\mu - e_\pm(a^\pm_\mu(0) - h)$. This function admits a minimum $\mu_0(h)$ on $[v, \frac{v+V}{2}]$. It is reached if and only if

$$\arccos\left(\frac{2\mu_0 - v - V}{V - v}\right) = \frac{\pi}{2}(1 + h),$$

i.e. when

$$\mu_0(h) = \frac{V - v}{2} \cdot \cos\left(\frac{\pi}{2} + \frac{\pi h}{2}\right) + \frac{v + V}{2}.$$

Let us just note that it does not depend on the phase between e_- and e_+.

Moreover the limit function $\mu - e_\pm(a^\pm_\mu(0) - \delta)$ is decreasing for $a \le \mu \le \mu_0$. Hence, if $\mu_0(h)$ does not belong to the interval of resonance, i.e. if there exists $r < \mu_0(h)$ such that $I_\mathcal{R} = (v, r]$, the optimal tuning in the resonance interval is r.

As a summary of our discussion of the optimal tuning measured by the probability for escape during an exponential time window we may state the following.

Let $I_\mathcal{R} = (v, r]$ for some $0 < r < \frac{v+V}{2}$, and $h > 0$ small enough. Then the optimal tuning corresponding to the quality measure $\mathcal{N}^h(\varepsilon, \mu)$ is given by

$$\mu_\mathcal{R}(h) = \min(r, \mu_0(h)) = \min\left\{r, \frac{v + V}{2} - \frac{V - v}{2}\sin\frac{\pi h}{2}\right\}.$$

We may call the critical free energy

$$\mu_\mathcal{R} = \lim_{h\to 0} \mu_0(h) = \frac{v + V}{2}$$

the *resonance point*.

The concepts developed in this subsection will be extended to the framework of diffusion processes in Section 4.4. This is why we need to study the first transition time between domains of attraction of meta-stable states (the stationary equilibria of the dynamical system determined by the drift) in the diffusion model.

4.3. Exit and entrance times of domains of attraction

The aim of this section is to determine the exact small noise asymptotics of exit times from the domains of attraction of the two stable states of the weakly and periodically perturbed system (4.8). This is achieved by using suitable splittings of exponentially long time intervals, which allow us to employ the large deviations results of Section 4.1. We obtain asymptotic exponential exit or transition rates between domains of attraction of the diffusion in exponential windows in units of the free energy, analogous to the ones described in Section 4.2. As for the associated Markov chains investigated in Section 4.2, they constitute the main ingredient for our probabilistic notion of stochastic resonance treated in the following section.

In Subsection 4.3.1 we discuss key properties of the geometry of the underlying deterministic system given by (4.8). They essentially state that geometrically

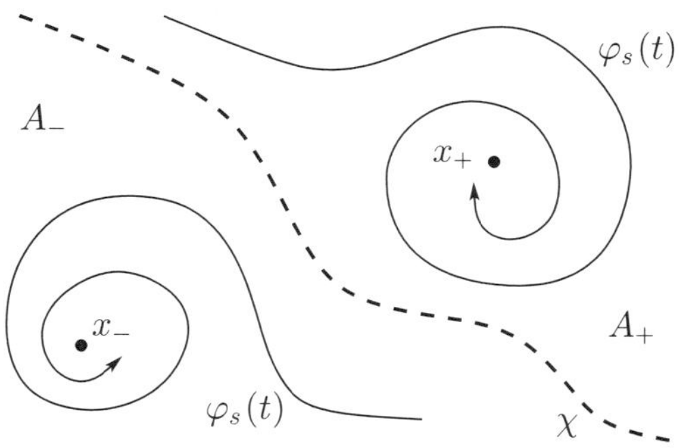

FIGURE 4.5. Domains of attraction

our system resembles a double well potential, although it does not need to be conservative. The two remaining subsections are devoted to establishing upper and lower bounds for the asymptotic exponential rates in the main result of this section (Theorem 4.19) on the transition time asymptotics.

4.3.1. Geometric preliminaries. We continue to study asymptotic properties of diffusions with weakly periodic drifts given by the SDE

$$dX_t^{\varepsilon,x} = b\Big(\frac{t}{T(\varepsilon)}, X_t^{\varepsilon,x}\Big)\, dt + \sqrt{\varepsilon}\, dW_t, \quad t \geq 0,\ X_0^{\varepsilon,x} = x \in \mathbb{R}^d,. \tag{4.39}$$

on exponential time scales $T(\varepsilon) = \exp(\frac{\mu}{\varepsilon})$. We assume as before that b satisfies the local Lipschitz conditions (4.10) and (4.11), and that the growth of the inward drift is sufficiently strong near infinity, as made precise by (4.4). The latter condition ensures that the diffusion will asymptotically be concentrated on a compact set, and this holds on an exponential scale (see Corollary 2.32).

To study the effects of weak periodicity of the drift on the asymptotic behavior of exit times from domains of attraction, we shall make the following simple assumptions on the geometry of b concerning its attraction and conservation properties. Essentially, we shall assume that $\mathbb{R}^d$ is split into two domains of attraction, separated by a simple geometric boundary which is invariant in time. The additional conditions concerning the geometry of b are as follows.

ASSUMPTION 4.17. For all $s \geq 0$ the d-dimensional ordinary differential equation

$$\dot{\varphi}_s(t) = b(s, \varphi_s(t)),\ t \geq 0, \tag{4.40}$$

admits two stable equilibria x_- and x_+ in $\mathbb{R}^d$ which do not depend on $s \geq 0$. Moreover, the domains of attraction defined by

$$A_\pm = \Big\{y \in \mathbb{R}^d \colon \varphi_s(0) = y \quad \text{implies} \quad \lim_{t\to\infty} \varphi_s(t) = x_\pm\Big\} \tag{4.41}$$

are independent of $s \geq 0$. They are supposed to satisfy $\overline{A_- \cup A_+} = \mathbb{R}^d$, and $\partial A_- = \partial A_+$. We denote by χ the common boundary.

This assumption can be weakened. Both the stable equilibria and the domains of attraction can depend on $s \geq 0$. The asymptotic results on the exit and entrance times remain true in this more general setting. We stick to Assumption 4.17 for reasons of notational simplicity. According to the Kramers–Eyring law, the mean

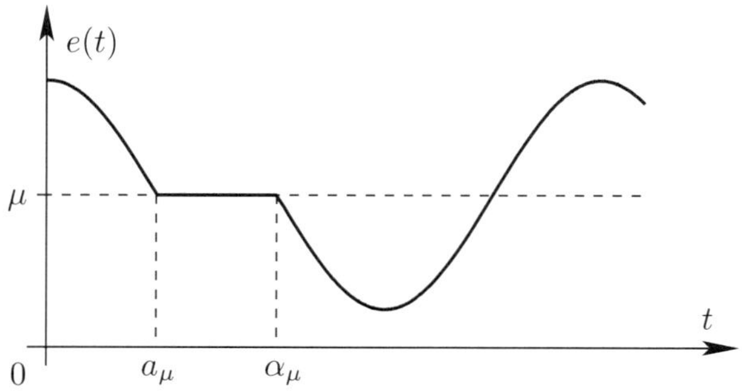

FIGURE 4.6. Definition of a_μ and α_μ.

time a homogeneous diffusion of noise intensity ε needs to leave a potential well of depth $\frac{v}{2}$ is of the order $\exp(\frac{v}{\varepsilon})$ (see Chapter 1 and Section 4.2). It is therefore natural to consider the time scales $T(\varepsilon) = \exp(\frac{\mu}{\varepsilon})$, $\mu > 0$, on which one can observe transitions between the domains of attraction. For simplicity we measure these scales in their corresponding free energy units μ.

Exit times of the domains of attraction $A_\pm$, provided that the weakly time inhomogeneous diffusion starts near the equilibrium points $x_\pm$, are the core subject of interest in this section. By obvious symmetry reasons, we may restrict our attention to the case of an exit from A_-. As we shall show, this exit time depends on the quasi-potential, that is on the minimal cost for the system described by (4.40) starting in a neighborhood of x_- to exit its domain of attraction through χ. For this reason we introduce the one-periodic *free energy function*

$$e(s) := \inf_{y \in \chi} V^s(x_-, y) < \infty \qquad \text{for } s \geq 0, \tag{4.42}$$

which is continuous on $\mathbb{R}_+$. Recall that V^s is the quasi-potential defined in (4.15). The function e corresponds in fact to the function e_- introduced in the Markov chain framework (see Section 4.2). We choose to simplify the notations since we shall only study the exit time from the domain A_-.

In the gradient case $b = -\nabla_x U$, this function coincides with twice the depth of the potential barrier to be overcome in order to exit from A_-, i.e. the energy the diffusion needs to leave A_-. Therefore scales μ — corresponding to the Kramers–Eyring times $T(\varepsilon) = \exp(\frac{\mu}{\varepsilon})$ according to the chosen parametrization — at which we expect transitions between the domains of attraction must be comprised between

$$\mu_* := \inf_{t \geq 0} e(t) \qquad \text{and} \qquad \mu^* := \sup_{t \geq 0} e(t).$$

These two constants are finite and are reached at least once per period since e is continuous and periodic. Now fix a time scale parameter μ. This parameter serves as a threshold for the energy, and we expect to observe an exit from A_- at the first time t at which $e(t)$ falls below μ. For $\mu \in (\mu_*, \mu^*)$ we therefore define

$$a_\mu = \inf\{t \geq 0 \colon e(t) \leq \mu\}, \qquad \alpha_\mu = \inf\{t \geq 0 \colon e(t) < \mu\}. \tag{4.43}$$

The subtle difference between a_μ and α_μ may be important, but we shall rule it out for our considerations by making the following assumption.

ASSUMPTION 4.18. The energy function e is strictly monotone between its (discrete) extremes, and every local extremum is global.

Under this assumption we have $a_\mu = \alpha_\mu$. We are now in a position to state the main result of this section. Let $\varrho > 0$ be small enough such that the Euclidean ball $B_\varrho(x_+) \subset A_+$, and let us define the first entrance time into this ball by

$$\tau_\varrho^\varepsilon = \inf\{t \geq 0 : \; X_t^\varepsilon \in B_\varrho(x_+)\}. \tag{4.44}$$

In agreement with previous notation, the stopping time just defined should also carry a subscript '+' indicating the direction of transition. Not to overburden notation, we suppress this subscript as long as this direction is clear from the context.

THEOREM 4.19. *Let $\mu < e(0)$. There exist $\eta > 0$ and $h_0 > 0$ such that for $h \leq h_0$*

$$\lim_{\varepsilon \to 0} \varepsilon \ln \sup_{y \in B_\eta(x_-)} \mathbb{P}_y\Big(\tau_\varrho^\varepsilon \notin [(a_\mu - h)T(\varepsilon), (a_\mu + h)T(\varepsilon)]\Big) \;=\; \mu - e(a_\mu - h).$$

Moreover, this convergence is uniform w.r.t. μ on compact subsets of $(\mu_*, e(0))$.

Note that Assumption 4.18 implies the continuity of $\mu \mapsto \mu - e(a_\mu - h)$. The statement of the theorem may be paraphrased in the following way. It specifies time windows in which transitions between the domains of attraction will be observed with very high probability. In particular, if e is strictly monotonous between its extremes, on the free energy scale μ we prove that the entrance time into a neighborhood of x_+ will be located near $a_\mu T(\varepsilon)$ in the small noise limit. The assumption $\mu < e(0)$ is only a technical assumption in order to avoid instantaneous jumping of the diffusion to the opposite domain. It can always be achieved by simply starting the diffusion a little later. We could even assume that $e(0) = \mu^*$ which then would yield uniform convergence on compact subsets of (μ_*, μ^*).

The rest of this section is devoted to the proof of this main result and is subdivided into separate subsections in which lower and upper bounds are treated separately.

4.3.2. Lower bound for the exit rate. We have to establish upper and lower bounds on the transition time τ_ϱ^ε which should both be exceeded with an exponentially small probability that has to be determined exactly. It will turn out that — in the free energy scale μ and with $h > 0$ — the probability of exceeding the upper bound $(\alpha_\mu + h)T(\varepsilon)$ vanishes asymptotically to all exponential orders in the small ε limit. So the exact large deviations rate shall be determined only by the probability $\mathbb{P}_x(\tau_\varrho^\varepsilon \leq (a_\mu - h)T(\varepsilon))$ of falling below the lower bound $(a_\mu - h)T(\varepsilon)$.

Deriving a lower bound of the latter probability and an upper bound of the probability $\mathbb{P}_x(\tau_\varrho^\varepsilon \geq (\alpha_\mu + h)T(\varepsilon))$ essentially leads to the same problem. In both cases one has to prove large deviations type upper bounds for probabilities of the form $\mathbb{P}_x(\tau_\varrho^\varepsilon \geq s(\varepsilon))$ for suitably chosen $s(\varepsilon)$, $\varepsilon > 0$. This can be expressed in terms of the problem of *diffusion exit* from a carefully chosen bounded domain.

Recall the *first entrance time* τ_ϱ^ε of a small neighborhood $B_\varrho(x_+)$ of the equilibrium point x_+. For $R, \varrho > 0$ consider the bounded domain

$$D = D(R, \varrho) := \overline{B_R(0)} \backslash B_\varrho(x_+),$$

and let

$$\tau_D^\varepsilon := \inf\{t \geq 0 \colon X_t^{\varepsilon, x} \notin D\}$$

be the *first exit time* of $X^{\varepsilon,x}$ from D.

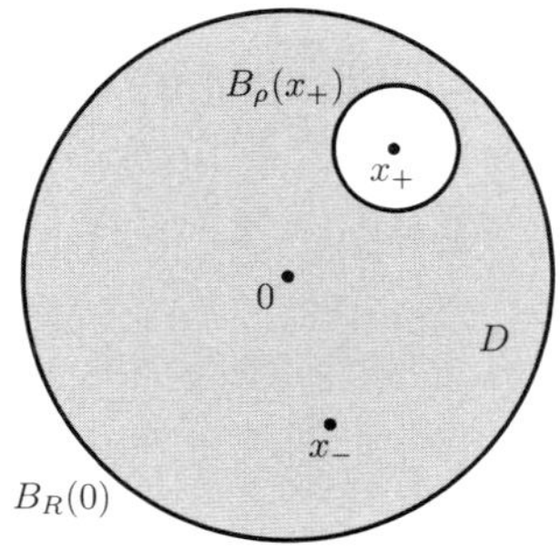

FIGURE 4.7. Domain D and the points x_- and x_+.

An exit from D means that either $X^{\varepsilon,x}$ enters $B_\varrho(x_+)$, i.e. we have a transition to the other equilibrium, or $X^{\varepsilon,x}$ leaves $B_R(0)$. But, as a consequence of our growth condition (2.44), the probability of the latter event does not contribute on the large deviations scale due to Theorem 2.32, as the following simple argument shows.

For $\varepsilon > 0, s > 0$ let $s(\varepsilon,\mu) = sT(\varepsilon)$. Since $\tau_D^\varepsilon = \tau_\varrho^\varepsilon \wedge \sigma_R$ where σ_R is the time of the diffusion's first exit from $B_R(0)$, Theorem 2.32 provides constants $R_1, \varepsilon_1 > 0$ such that for $R \geq R_1$, $\varepsilon \leq \varepsilon_1$, and $|x| \leq R/2$

(4.45)

$$\begin{aligned}\mathbb{P}_x\Big(\tau_\varrho^\varepsilon \geq s(\varepsilon,\mu)\Big) &\leq \mathbb{P}_x\Big(\{\tau_\varrho^\varepsilon \geq s(\varepsilon,\mu)\} \cap \{\sigma_R \geq s(\varepsilon,\mu)\}\Big) + \mathbb{P}_x\Big(\sigma_R < s(\varepsilon,\mu)\Big)\\ &\leq \mathbb{P}_x\Big(\tau_D^\varepsilon \geq s(\varepsilon,\mu)\Big) + C\eta^2\,\frac{s(\varepsilon,\mu)}{\varepsilon}\,e^{-\frac{\eta R}{\varepsilon}}.\end{aligned}$$

By the choice of $s(\varepsilon,\mu)$ and $T(\varepsilon) = \exp(\frac{\mu}{\varepsilon})$, the right term in the last sum is of the order $\frac{1}{\varepsilon}\exp(\frac{\mu-\eta R}{\varepsilon})$, i.e. it can be assumed to be exponentially small of any exponential order required by choosing R suitably large. Obviously, this holds uniformly with respect to μ on compact sets. This argument shows that the investigation of asymptotic properties of the laws of τ_ϱ^ε may be replaced by a study of similar properties of τ_D^ε, with an error that may be chosen arbitrarily small by increasing R.

Similarly to the time homogeneous exit problem, we need a lemma which shows how to approximate the energy of a transition by the cost along particular trajectories that leave a neighborhood of D. This is of central importance to the estimation of the asymptotic law of τ_D^ε.

LEMMA 4.20. *Let $\theta > 0$ and M be a compact interval of $\mathbb{R}_+$. Then there exist $T_0 > 0$ and $\delta > 0$ with the following property. For all $x \in D$ and $s \in M$, we can find a continuous path $\zeta^{x,s} \in C([0,T_0],\mathbb{R}^d)$ starting in $\zeta_0^{x,s} = x$ and ending at some point of distance $d(\zeta_{T_0}^{x,s}, D) \geq \delta$ away from D such that*

$$I_{T_0}^s(\zeta^{x,s}) \leq e(s) + \theta \quad \textit{for all} \quad s \in M.$$

PROOF. Fix $\theta > 0$, and let us decompose the domain D into three different ones. Fixing $l > 0$, define a domain β_l by

$$\beta_l = \{x \in D\colon d(x,\chi) < l\}.$$

We recall that χ separates A_- from A_+. Then we define two closed sets $D_- = (D \setminus \beta_l) \cap A_-$ and $D_+ = (D \setminus \beta_l) \cap A_+$. We shall construct appropriate paths from

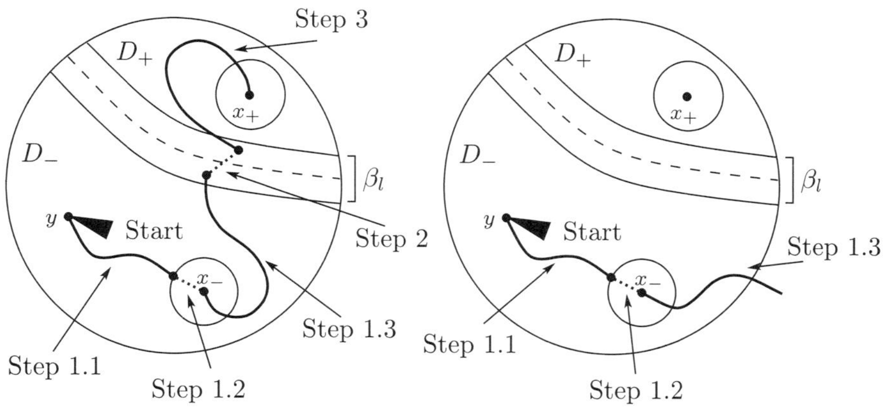

FIGURE 4.8. Steps of the proof in the two possible cases.

points $y \in D$ to points a positive distance away from D not exceeding the energy $e(s)$ by more than θ uniformly in $s \in M$ in four steps.

Step 1. Assume first that $y \in D_-$. For $l > 0$ small enough we construct $\delta_1^l > 0$, $S_1^l > 0$ and a path $\psi_1^{s,y,l}$ defined on a time interval $[0, \tau_1^{s,y,l}]$ with $\tau_1^{s,y,l} \leq S_1^l$ for all $y \in D_-$, $s \in M$ and along which we exit a δ_1^l-neighborhood of D_- at cost at most $e(s) + \frac{2}{3}\theta$.

Step 1.1. In a first step we go from y to a small neighborhood $B_l(x_-)$ of x_-, in time at most $T_1^l < \infty$, without cost.
We denote by $\varphi_1^{s,y,l}$ the trajectory starting at $\varphi_1^{s,y,l}(0) = y \in D_-$ of

$$\dot{\varphi}_1(t) = b(s, \varphi_1(t)),$$

and reaching $B_l(x_-)$ at time $\sigma_1^{y,s,l}$. Since $D_- \subset A_-$ and due to Assumption 4.17, $\sigma_1^{y,s,l}$ is finite. Moreover, since b is locally Lipschitz, stability of solutions with respect to initial conditions and smooth changes of vector fields implies that there exist open neighborhoods W_y of y and $\mathcal{W}_s$ of s and $T_1^{s,y,l} > 0$ such that, for all $z \in W_y$, $u \in \mathcal{W}_s$, $\sigma_1^{u,z,l} \leq T_1^{s,y,l}$. Recall that D_- is compact. Therefore we may find a finite cover of $D_- \times M$ by such sets, and consequently $T_1^l < \infty$ such that for all $y \in D_-$ and $s \in M$, $\sigma_1^{s,y,l} \leq T_1^l$. Denote $z^{s,y,l} = \varphi_1^{s,y,l}(\sigma_1^{s,y,l})$.

Step 1.2. In a second step, we go from a small neighborhood $B_l(x_-)$ of x_- to the equilibrium point x_-, in time at most 1, at cost at most $\frac{\theta}{3}$.

In fact, by the continuity of the cost function, for l small enough, $s \in M$, there exists a continuous path $\varphi_2^{s,y,l}$ of time length $\sigma_2^{s,y,l} \leq 1$ such that $\varphi_2^{s,y,l}(0) = z^{s,y,l}$, $\varphi_2^{s,y,l}(\sigma_2^{s,y,l}) = x_-$ and $I_{\sigma_2^{s,y,l}}(\varphi_2^{s,y,l}) \leq \theta/3$.

Step 1.3. In a third step, we exit some δ-neighborhood of D_-, starting from the equilibrium point x_-, in time at most $T_3 < \infty$, at cost at most $e(s) + \frac{\theta}{3}$ for $s \in M$.

By (4.42) and the continuity of the cost function for any $s \in M$ there exists $z_s \notin A_- \supset D_-$, $T_3^s < \infty$, some neighborhood $\mathcal{W}_s$ of s and for $u \in \mathcal{W}_s$ we have

$\varphi_3^u \in C([0,\sigma_3^u],\mathbb{R}^d)$ such that $\varphi_3^u(0)=x_-$, $\varphi_3^u(\sigma_3^u)=z_s$, $\sigma_3^u \le T_3^s$ and

$$\sup_{u\in\mathcal{W}_s} I^u_{\sigma_3^u}(\varphi_3^u) \le e(s)+\frac{\theta}{3}.$$

Use the compactness of M to find a finite cover of M by such neighborhoods, and thus some $T_3<\infty$ such that all the statements hold with $\sigma_3^s \le T_3$ for all $s\in M$. Finally remark that the exit point is at least a distance $\delta=\inf_{i\in J}|z_i|$ away from the boundary of D_-, if z_i, $i\in J$, are the exit points corresponding to the finite cover.

In order to complete **Step 1**, we now define a path $\psi_1^{s,y,l}\in C([0,\tau_1^{s,y,l}],\mathbb{R}^d)$ by concatenating $\varphi_1^{s,y,l}$, $\varphi_2^{s,y,l}$ and φ_3^s. This way, for small $l>0$ we find $S_1^l>0$ such that for all $s\in M, y\in D_-$ we have $\tau_1^{s,y,l}\le S_1^l$, $\psi_1^{s,y,l}(\tau_1^{s,y,l})=y$, $\psi_1^{s,y,l}(\tau_1^{s,y,l})\notin A_-$ and

$$I^s_{\tau_1^{s,y,l}}(\psi_1^{s,y,l}) \le\ e(s)+\frac{2}{3}\theta \quad \text{for all} \quad s\in M,\ y\in D_-.$$

At this point we can encounter two cases. In the first case $\psi_1^{s,y,l}$ exits a δ_l-neighborhood of $B_R(0)$. In this case we continue with **Step 4**. In the second case, $\psi_1^{s,y,l}$ exits D_- into β_l, and we continue with **Step 2**.

Step 2. For l small enough, we start in $y\in\beta_l$, to construct $S_2^l>0$ and a path $\psi_2^{s,y,l}$ defined on a time interval $[0,\tau_2^{s,y,l}]$ with $\tau_2^{s,y,l}\le S_2^l$ for all $y\in D_-, s\in M$ and along which we exit β_l into the interior of D_+ at cost at most $\frac{\theta}{3}$.

In fact, due to the continuity of the cost function (see Lemma 4.2), there exists $l>0$ small enough such that for any $s\in M$, $y\in\overline{\beta_l}$ there exists $z_{s,y,l}$ in the interior of D_+, such that $\psi_2^{s,y,l}(0)=y$, $\psi_2^{s,y,l}(\tau_2^{s,y,l})=z_{s,y,l}$ and $I^u_{\tau_2^{s,y,l}}(\psi_2^{s,y,l})\le\theta/3$. We may take $S_2^l=1$.

Step 3. We start in $y\in D_+$, to construct $\delta_3^l>0$, $S_3^l>0$ and a path $\psi_3^{s,y,l}$ defined on a time interval $[0,\tau_3^{s,y,l}]$ with $\tau_3^{s,y,l}\le S_3^l$ for all $y\in D_-$, $s\in M$ and along which we exit D_+ into $B_{\varrho-\delta_3^l}(x_+)$ at no cost.

Let $\delta_3^l=\varrho/2$. Since D_+ is compact and contained in the domain of attraction of x_+, stability of the solutions of the differential equation $\dot\varphi(t)=b(s,\varphi(t))$ with respect to the initial condition $y\in D_+$ and the parameter s guarantees the existence of some time $S_3^l>0$ such that the entrance time $\tau_3^{s,y,l}$ of $B_{\varrho/2}(x_+)$ by the solution starting in y is bounded by S_3^l. Therefore we may take $\psi_3^{s,y,l}$ to be defined by this solution restricted to the time interval before its entrance into $B_{\varrho/2}(x_+)$.

Step 4. For $l>0$ small enough we start in $x\in D_-$ and construct $T_0>0, \delta>0$ and a path $\zeta^{s,x}$ defined on the time interval $[0,T_0]$, exiting a δ-neighborhood of D at cost at most $e(s)+\theta$ for all $s\in M$.

For l small enough, take $T_0=S_1^l+S_2^l+S_3^l$. We just have to concatenate paths constructed in the first three steps. Recall that $\psi_1^{s,x,l}$ passes through the equilibrium x_- due to **Step 1**. In case $\psi_1^{s,x,l}$ exits a δ_1^l-neighborhood of $B_R(0)$, just let the path spend enough time in x_- without cost to obtain a path $\zeta^{s,x,l}$ defined on $[0,T_0]$, and take $\delta=\delta_1^l$. In the other case, we concatenate the three paths constructed in **Steps 1–3**, to obtain a path defined on a subinterval of $[0,T_0]$ depending on s,x,l and which exits a δ_3^l–neighborhood of D. Recall from **Step 1** that this path also passes through x_-. It remains to redefine the path by spending extra time at no cost in this equilibrium point, to complete the proof. □

We now proceed to the estimation of *uniform lower bounds* for the asymptotic law of τ_D^ε. The uniformity has to be understood in the sense of Remark 4.7.

PROPOSITION 4.21. Let K be a compact subset of D.

a) If $e(s) > \mu$, then

$$\liminf_{\varepsilon\to 0} \varepsilon \ln \inf_{x\in K} \mathbb{P}_x(\tau_D^\varepsilon < sT(\varepsilon)) \geq \mu - e(s),$$

locally uniformly on $\{(s,\mu)\colon \mu_* < \mu < \min(e(0), e(s)), 0 \leq s \leq 1\}$.

b) If $e(s) < \mu$, then

$$\lim_{\varepsilon\to 0} \varepsilon \ln \sup_{x\in K} \mathbb{P}_x(\tau_D^\varepsilon \geq sT(\varepsilon)) = -\infty,$$

locally uniformly on $\{(s,\mu)\colon e(s) < \mu < e(0), 0 \leq s \leq 1\}$.

PROOF. We choose a compact subset L of $[0,1]$ and a compact subset M of $(\mu_*, e(0))$ as well as some $\theta > 0$ such that

$$|e(s) - \mu| \geq \theta \quad \forall (s,\mu) \in L \times M.$$

Later on we shall assume that $e(s) - \mu$ is uniformly positive resp. negative in order to prove a) resp. b).

Step 1. In a first step, we apply Lemma 4.20 to approximate the energy function $e(s)$ by the rate function along a particular path, uniformly w.r.t. s. For the chosen θ it yields $T_0 > 0$ and $\delta > 0$ as well as continuous paths $\zeta^{x,s}$ indexed by $x \in D$ and $s \in [0,1]$ ending a distance at least δ away from D such that for all $x \in D$ and $s \in [0,1]$

$$I_{T_0}^s(\zeta^{x,s}) \leq e(s) + \frac{\theta}{4}.$$

Step 2. In a second step, we use the Markov property to estimate the probability of exiting D after time $sT(\varepsilon)$ by a large product of exit probabilities after time intervals of length independent of ε and μ. Since for $\varepsilon > 0$, $\mu \in M$ the interval $[0, sT(\varepsilon)]$ becomes arbitrarily large as $\varepsilon \to 0$, we introduce a splitting into intervals of length $\nu \geq T_0$ independent of ε and μ. For $k \in \mathbb{N}_0$ let $t_k = t_k(s,\varepsilon,\mu) := sT(\varepsilon) - k\nu$. Then we have for $k \in \mathbb{N}_0$ and $x \in D$

$$\begin{aligned}\mathbb{P}_x\Big(\tau_D^\varepsilon \geq t_k\Big) &= \mathbb{E}_x\Big[\mathbb{I}_{\{\tau_D^\varepsilon \geq t_k\}}\mathbb{I}_{\{\tau_D^\varepsilon \geq t_{k+1}\}}\Big] = \mathbb{E}_x\Big[\mathbb{I}_{\{\tau_D^\varepsilon \geq t_{k+1}\}}\mathbb{E}\Big[\mathbb{I}_{\{\tau_D^\varepsilon \geq t_k\}}\Big|\mathcal{F}_{t_{k+1}}\Big]\Big] \\ &\leq \mathbb{P}_x\Big(\tau_D^\varepsilon \geq t_{k+1}\Big) \cdot \sup_{y\in D} \mathbb{P}_{t_{k+1},y}\Big(\tau_D^\varepsilon \geq \nu\Big).\end{aligned}$$

Here $\mathbb{P}_{s,y}$ denotes the law of $X^{s,y,\varepsilon}$, defined by the SDE

$$dX_t^{s,y,\varepsilon} = b\Big(\frac{s+t}{T(\varepsilon)}, X_t^{s,y,\varepsilon}\Big)\,dt + \sqrt{\varepsilon}\,dW_t, \quad t \geq 0,\ X_0^{s,y,\varepsilon} = y \in \mathbb{R}^d.$$

On intervals $[0,\nu]$ it coincides with the law of the original process $X^{x,\varepsilon}$ on $[s, s+\nu]$, but of course paths may differ. Denoting $q_k(s,\varepsilon,\mu) := \sup_{y\in D} \mathbb{P}_{t_k,y}(\tau_D^\varepsilon \geq \nu)$, an iteration of the latter argument yields

$$\sup_{x\in K} \mathbb{P}_x\Big(\tau_D^\varepsilon \geq sT(\varepsilon)\Big) \leq \prod_{k=1}^{N(\varepsilon,\mu)} q_k(s,\varepsilon,\mu) \tag{4.46}$$

whenever $\nu N(\varepsilon,\mu) < sT(\varepsilon)$.

For the further estimation of the q_k we apply an LDP to the product (4.46). This relies on the following idea. If we choose $N(\varepsilon,\mu)$ of the order $\varepsilon T(\varepsilon)$, then the starting times t_k appearing in the product belong to some neighborhood of $sT(\varepsilon)$ that, compared to $T(\varepsilon)$, shrinks to a point asymptotically. Consequently, the family of diffusions underlying the product (4.46) is uniformly exponentially equivalent to the homogeneous diffusion governed by the drift $b(s,\cdot)$. This will be done in the following third step.

Step 3. For $x \in D$, $s \in [0,1]$ let

$$\Psi(x,s) := \Big\{\psi \in C([0,T_0],\mathbb{R}^d) \colon \|\psi - \zeta^{x,s}\|_{T_0} < \frac{\delta}{2}\Big\}$$

be the open $\delta/2$-neighborhood of the path chosen in the first step, and let

$$\Psi(x) := \bigcup_{s\in[0,1]} \Psi(x,s).$$

To apply our large deviations estimates in this situation, note first that conditions concerning τ_D^ε translate into constraints for the trajectories of X^ε as figuring in the preceding section: due to the definition of $\Psi(x,s)$, the choice $\nu \geq T_0$ and Lemma 4.20 we know that for $y \in D$, $k \leq N(\varepsilon,\mu)$, if $X^{t_k,y}$ belongs to $\Psi(x)$, then for sure $X^{t_k,y}$ exits D before time ν. Keeping this in mind, we may apply Proposition 4.8 to the neighborhoods

$$S^{s,0}(\varepsilon,\mu) = \Big[sT(\varepsilon) - \nu N(\varepsilon,\mu), sT(\varepsilon) + \nu\Big]$$

of $sT(\varepsilon)$. Each of the intervals $[t_k, t_k+\nu]$ is contained in $S^{s,0}(\varepsilon,\mu)$. As mentioned before, $N(\varepsilon,\mu)$ is chosen of the order $\varepsilon T(\varepsilon)$, and this can be done uniformly w.r.t. $\mu \in M$. More precisely, we assume to have constants $0 < c_1 < c_2$ such that $c_1\varepsilon T(\varepsilon) \leq N(\varepsilon,\mu) \leq c_2 \varepsilon T(\varepsilon)$. Then

$$\lim_{\varepsilon\to 0} \sup_{s\in[0,1],\mu\in M} \frac{\operatorname{diam} S^{s,0}(\varepsilon,\mu)}{T(\varepsilon)} = 0,$$

and by the uniform large deviations principle of Proposition 4.8 we obtain the lower bound

$$\begin{aligned}\liminf_{\varepsilon\to 0} \varepsilon \ln \inf_{\substack{y\in K,\mu\in M,\\ k\leq N(\varepsilon,\mu)}} \mathbb{P}_{t_k,y}(\tau_D^\varepsilon < \nu) &\geq -\sup_{y\in K} \inf_{\psi\in\Psi(y)} I_{T_0}^s(\psi)\\ &\geq -\sup_{y\in K} I_{T_0}^s(\zeta^{y,s}) \geq -e(s) - \frac{\theta}{4}.\end{aligned}$$

We stress that this bound is uniform w.r.t. s in the sense of Remark 4.7, so we can find $\varepsilon_0 > 0$ independent of s such that for $\varepsilon \leq \varepsilon_0$, $\mu \in M$ and $k \leq N(\varepsilon,\mu)$

$$\begin{aligned}1 - q_k(s,\varepsilon,\mu) &= \inf_{y\in D} \mathbb{P}_{t_k,y}(\tau_D^\varepsilon < \nu)\\ &\geq \inf_{\substack{y\in D,\mu\in M,\\ j\leq N(\varepsilon,\mu)}} \mathbb{P}_{t_j,y}(\tau_D^\varepsilon < \nu) \geq \exp\Big\{-\frac{1}{\varepsilon}\Big(e(s) + \frac{\theta}{2}\Big)\Big\}.\end{aligned}$$

From this we obtain

$$\begin{aligned}\sup_{x\in K} \mathbb{P}_x\Big(\tau_D^\varepsilon \geq sT(\varepsilon)\Big) &\leq \prod_{k=1}^{N(\varepsilon,\mu)} q_k(s,\varepsilon,\mu) \leq \Big(1 - \exp\Big\{-\frac{1}{\varepsilon}\Big(e(s)+\frac{\theta}{2}\Big)\Big\}\Big)^{N(\varepsilon,\mu)}\\ &= \exp\Big\{N(\varepsilon,\mu)\ln\Big(1 - \exp\Big\{-\frac{1}{\varepsilon}\Big(e(s)+\frac{\theta}{2}\Big)\Big\}\Big)\Big\} =: m(\varepsilon,\mu).\end{aligned}$$

Since $\ln(1-x) \le -x$ for $0 \le x < 1$ we have

$$m(\varepsilon,\mu) \le \exp\Big\{ - c_1\varepsilon \exp\Big\{\frac{\mu}{\varepsilon} - \frac{1}{\varepsilon}\Big(e(s) + \frac{\theta}{2}\Big)\Big\}\Big\}. \tag{4.47}$$

Step 4. In the fourth and last step, we exploit this bound of $m(\varepsilon,\mu)$ to obtain the claimed asymptotic bounds.

In order to prove a), assume that $\mu < e(s)$ for $(s,\mu) \in L \times M$. Then the inner exponential in (4.47) approaches 0 on $L \times M$. Using the inequality $1 - e^{-x} \ge xe^{-1}$ on $[0,1]$, we conclude that there exists $\varepsilon_1 \in (0,\varepsilon_0)$ such that for all $\varepsilon \le \varepsilon_1$ and $(s,\mu) \in L \times M$

$$\begin{aligned}
\varepsilon \ln \inf_{x\in K} \mathbb{P}_x\Big(\tau_D^\varepsilon < sT(\varepsilon)\Big) &\ge \varepsilon \ln\Big(1 - m(\varepsilon,\mu)\Big)\\
&\ge \varepsilon \ln\Big(\varepsilon c_1 \exp(-1)\exp\Big\{\frac{1}{\varepsilon}\Big(\mu - e(s) - \frac{\theta}{2}\Big)\Big\}\Big)\\
&= -\varepsilon + \varepsilon \ln c_1 + \varepsilon \ln \varepsilon + \mu - e(s) - \frac{\theta}{2}\\
&\ge \mu - e(s) - \theta.
\end{aligned}$$

For b) assume $\mu > e(s)$ on $L \times M$. Then

$$\begin{aligned}
\varepsilon \ln \sup_{x\in K} \mathbb{P}_x\Big(\tau_D^\varepsilon \ge sT(\varepsilon)\Big) &\le \varepsilon \ln m(\varepsilon,\mu)\\
&\le -c_1\varepsilon \exp\Big\{-\frac{1}{\varepsilon}\Big(\mu - e(s) - \frac{\theta}{2}\Big)\Big\} \stackrel{\varepsilon\to 0}{\to} -\infty.
\end{aligned}$$

□

As a consequence of these large deviations type results on the asymptotic distribution of τ_D^ε and the remarks preceding the statement of Lemma 4.20 and Proposition 4.21, we get the following asymptotics for the transition time of the diffusion.

PROPOSITION 4.22. Let $x \in A_-$. There exists $h_0 > 0$ such that for $h \le h_0$

$$\liminf_{\varepsilon\to 0} \varepsilon \ln \mathbb{P}_x(\tau_\varrho^\varepsilon \le (a_\mu - h)T(\varepsilon)) \ge \mu - e(a_\mu - h), \tag{4.48}$$

$$\lim_{\varepsilon\to 0} \varepsilon \ln \mathbb{P}_x(\tau_\varrho^\varepsilon \ge (a_\mu + h)T(\varepsilon)) = -\infty. \tag{4.49}$$

Moreover, these convergence statements hold uniformly w.r.t. x on compact subsets of D and w.r.t. μ on compact subsets of $(\mu_*, e(0))$.

PROOF. As the estimation (4.45) based on Theorem 2.32 at the beginning of the section shows, we may derive the required estimates for τ_D^ε instead of τ_ϱ^ε, if R is chosen large enough.

Let M be a compact subset of $(\mu_*, e(0))$. Then $0 < a_\mu < 1$ for $\mu \in M$, which yields the existence of $h_0 > 0$ such that the compact set $L_h := \{a_\mu - h \colon \mu \in M\}$ is contained in $(0,1)$ for $h \le h_0$. Moreover, we have $e(s) > \mu$ for $0 < s < a_\mu$ due to the assumptions on e, uniformly w.r.t. $(s,\mu) \in L_h \times M$ by the continuity of e. Hence by Proposition 4.21 a)

$$\liminf_{\varepsilon\to 0} \varepsilon \ln \inf_{x\in K} \mathbb{P}_x\Big(\tau_D^\varepsilon \le sT(\varepsilon)\Big) \ge \mu - e(s),$$

uniformly on $L_h \times M$ for all $h \le h_0$. By setting $s = a_\mu - h$ we obtain the first asymptotic inequality. The second one follows in a completely analoguous way from Proposition 4.21 b) since $e(a_\mu + h) < \mu$ for small enough h. □

4.3.3. Upper bound for the exit rate. Let us next derive upper bounds for the exponential exit rate which resemble the lower bounds just obtained. We need an extension of a result obtained by Freidlin and Wentzell (Lemma 5.4 in [**105**]).

LEMMA 4.23. *Let K be a compact subset of $A_-\backslash\{x_-\}$. There exist $T_0 > 0$ and $c > 0$ such that for all $T \geq T_0$, $s \in [0,1]$ and for each $\varphi \in C([0,T],\mathbb{R}^d)$ taking its values in K we have*

$$I_T^s(\varphi) \geq c(T - T_0).$$

PROOF. Let $\phi_{s,x}$ be the solution of the differential equation

$$\dot{\phi}_{s,x}(t) = b(s, \phi_{s,x}(t)), \quad \phi_{s,x}(0) = x \in K.$$

Let $\tau(s,x)$ be the first exit time of the path $\phi_{s,x}$ from the domain K. Since A_- is the domain of attraction of x_- and since K is a compact subset of $A_-\backslash\{x_-\}$, we obtain $\tau(s,x) < \infty$ for all $x \in K$. The function $\tau(s,x)$ is upper semi-continuous with respect to the variables s and x (due to the continuous dependence of $\phi_{s,x}$ on s and x). Hence the maximal value

$$T_1 := \sup_{s\in[0,1],x\in K} \tau(s,x)$$

is attained.

Let $T_0 = T_1 + 1$, and consider all functions $\varphi \in C([0,T_0],\mathbb{R}^d)$ with values in K. This set of functions is closed with respect to the maximum norm. Since there is no solution of the ordinary differential equation in this set of functions, the functional $I_{T_0}^s$ reaches a strictly positive minimum m on this set which is uniform in s. By the additivity of the functional I_T^s, we obtain for $T \geq T_0$ and $\varphi \in C([0,T],\mathbb{R}^d)$ with values in K

$$I_T^s(\varphi) \geq m\left\lfloor \frac{T}{T_0}\right\rfloor \geq m\left(\frac{T}{T_0} - 1\right) = c(T - T_0),$$

with $c = \frac{m}{T_0}$. □

Let us recall the subject of interest of this section:

$$\tau_\varrho^\varepsilon = \inf\left\{t \geq 0\colon X_t^\varepsilon \in B_\varrho(x_+)\right\},$$

the hitting time of a small neighborhood of the equilibrium point x_+. First we shall consider upper bounds for the law of this time in some window of length $\beta T(\varepsilon)$, where β is sufficiently small. The important feature of the following statement is that β is independent of s, while the uniformity of the bound again has to be understood in the sense of Remark 4.7.

PROPOSITION 4.24. For all $\theta > 0$ there exist $\beta > 0$, $\eta > 0$ such that for all $s \in [0,1]$

$$\limsup_{\varepsilon\to 0} \varepsilon \ln \sup_{x\in B_\eta(x_-)} \mathbb{P}_x\Big(sT(\varepsilon) \leq \tau_\varrho^\varepsilon \leq (s+\beta)T(\varepsilon)\Big) \leq \mu - e(s) + \theta.$$

This bound holds locally uniformly w.r.t. $\mu \in (\mu_*, e(0))$ and uniformly w.r.t. $s \in [0,1]$.

PROOF. **Step 1.** We introduce a new domain. Let M be a compact subset of $(\mu_*, e(0))$, and fix $\theta > 0$. We first introduce some parameter dependent domains the exit times of which will prove to be suitable for estimating the probability that τ_ϱ^ε is in a certain time window.

For this purpose, we define for $\delta > 0$ and $s \in [0,1]$ an open domain

$$D(\delta, s) := \Big\{ y \in \mathbb{R}^d \colon V^s(x_-, y) < \mu^* + \frac{1}{1+\delta},\ d(y, A_+) > \delta \Big\},$$

and let $D = D(\delta) = \bigcup_{s\in[0,1]} D(\delta, s)$. Then D is relatively compact in A_-, $d(y, A_+) > \delta$ for all $y \in D(\delta)$, and a transition to a ϱ-neighborhood of x_+ certainly requires an exit from $D(\delta)$.

The boundary of $D(\delta)$ consists of two hyper-surfaces one of which carries an energy strictly greater than μ^* and thus greater than $e(s)$ for all $s \in [0,1]$. The minimal energy is therefore attained on the other component of distance δ from A_+ which approaches $\chi = \partial A_-$ as $\delta \to 0$. Thus, by the joint continuity of the quasi-potential, we can choose $\delta_0 > 0$ and $\eta > 0$ such that for $\delta \le \delta_0$ and $s \in [0,1]$

$$\begin{aligned} e(s) = \inf_{z\in\chi} V^s(x_-, z) &\ge \inf_{z\in\partial D(\delta)} V^s(x_-, z) \\ &\ge \inf_{y\in B_\eta(x_-)} \inf_{z\in\partial D(\delta)} V^s(y, z) \ge e(s) - \frac{\theta}{4}. \end{aligned} \tag{4.50}$$

Step 2. We split the time interval. Let τ_D^ε be the first exit time of $X^{x,\varepsilon}$ from D. For $s \in [0,1]$ and $\beta > 0$ we introduce a covering of the interval of interest $[sT(\varepsilon), (s+\beta)T(\varepsilon)]$ into $N = N(\beta, \varepsilon, \mu)$ intervals of fixed length ν, i.e. ν is independent of ε, μ, s and β. We will have to assume that ν is sufficiently large which will be made precise later on. Thus we have $\nu N \ge \beta T(\varepsilon)$, and we can and do assume that $N \le \beta T(\varepsilon)$. For $k \in \mathbb{Z}$, $k \ge -1$, let

$$t_k = t_k(s, \varepsilon, \mu) := sT(\varepsilon) + k\nu$$

be the starting points of these intervals. We consider t_{-1} since we need some information about the past in order to ensure the diffusion to start in a neighborhood of the equilibrium x_-. Then for $x \in B_\eta(x_-)$ we get the desired estimation of probabilities of exit windows for τ_ϱ^ε by those with respect to τ_D^ε:

$$\mathbb{P}_x\Big(sT(\varepsilon) \le \tau_\varrho^\varepsilon \le (s+\beta)T(\varepsilon)\Big) \le \sum_{k=0}^{N} \mathbb{P}_x\Big(t_k \le \tau_D^\varepsilon \le t_{k+1}\Big).$$

In a second step we will fix $k \ge 0$ and estimate the probability of a first exit from D during each of the intervals $[t_k, t_{k+1}]$ separately. Here the difficulty is that we do not have any information on the location at time t_k. We therefore condition on whether or not $X^{x,\varepsilon}$ has entered the neighborhood $B_\eta(x_-)$ in the previous time interval. For that purpose, let

$$\sigma_k := \inf\Big\{t \ge t_k \vee 0 :\ X_t^\varepsilon \in B_\eta(x_-)\Big\}, \quad k \ge -1.$$

Then for $k \ge 0$

$$\begin{aligned} &\mathbb{P}_x\Big(t_k \le \tau_D^\varepsilon \le t_{k+1}\Big) \\ &\quad \le \mathbb{P}_x\Big(t_k \le \tau_D^\varepsilon \le t_{k+1} \Big| \sigma_{k-1} \le t_k\Big) + \mathbb{P}_x\Big(\tau_D^\varepsilon \wedge \sigma_{k-1} \ge t_k\Big). \end{aligned} \tag{4.51}$$

Step 3. Estimation of the second term in the right hand side of (4.51). Let $K = K(\delta, \eta) = \overline{D(\delta)} \backslash B_\eta(x_-)$. Then K is compact, and by the Markov property we have

$$\mathbb{P}_x\Big(\tau_D^\varepsilon \wedge \sigma_{k-1} \ge t_k\Big) \le \sup_{y\in K} \mathbb{P}_{t_{k-1},y}\Big(\tau_D^\varepsilon \wedge \sigma_1 \ge \nu\Big),$$

where $\mathbb{P}_{t,y}$ is as defined in the previous subsection. Now we wish to further estimate this exit probability using large deviations methods. The neighborhoods

$$S^{s,\beta}(\varepsilon,\mu) = \Big[sT(\varepsilon) - \nu, (s + \nu N(\beta,\varepsilon,\mu))T(\varepsilon)\Big]$$

of $sT(\varepsilon)$ contain each interval $[t_k, t_{k+1}]$, $-1 \leq k \leq N(\beta,\varepsilon,\mu)$, and they satisfy

$$\limsup_{\varepsilon\to 0} \sup_{\mu\in M, s\in[0,1]} \frac{\operatorname{diam}(S^{s,\beta}(\varepsilon,\mu))}{T(\varepsilon)} \leq \beta.$$

Hence by the uniform LDP of Proposition 4.6, applied to the closed set

$$\Phi_K(\delta,\eta) = \Big\{\varphi \in C([0,\nu],\mathbb{R}^d) \colon \varphi_t \in K(\delta,\eta) \quad \text{for all} \quad t \in [0,\nu]\Big\},$$

we obtain the upper bound

$$\begin{aligned}
&\limsup_{\varepsilon\to 0} \varepsilon \ln \sup_{\substack{y\in K,\mu\in M,\\ k\leq N}} \mathbb{P}_{t_{k-1},y}(\tau_D^\varepsilon \wedge \sigma_1 \geq \nu) \\
&\qquad \leq \limsup_{\varepsilon\to 0} \varepsilon \ln \sup_{\substack{y\in K,\mu\in M,\\ t\in S^{s,\beta}(\varepsilon,\mu)}} \mathbb{P}\Big(X^{t,y,\varepsilon} \in \Phi_K(\delta,\eta)\Big) \qquad (4.52)\\
&\qquad \leq -\inf_{y\in K} \inf_{\varphi\in\Phi_K(\delta,\eta)^{\gamma_0(\beta)}} I_\nu^s(\varphi),
\end{aligned}$$

where $\gamma_0(\beta) = \beta\delta(\Phi_K(\delta,\eta))$ is the *blowup-factor* induced by the diameter β. Since $\gamma_0(\beta) \to 0$ as $\beta \to 0$, we can find $\beta_0 > 0$ such that for $\beta \leq \beta_0$

$$\Phi_K(\delta,\eta)^{\gamma_0(\beta)} \subset \Phi_K\Big(\frac{\delta}{2},\frac{\eta}{2}\Big),$$

which amounts to saying that, instead of blowing up the set of paths, we consider the slightly enlarged domain $K(\frac{\delta}{2},\frac{\eta}{2})$. Thus

$$-\inf_{y\in K} \inf_{\varphi\in\Phi_K(\delta,\eta)^{\gamma_0(\beta)}} I_\nu^s(\varphi) \leq -\inf_{y\in K} \inf_{\varphi\in\Phi_K(\frac{\delta}{2},\frac{\eta}{2})} I_\nu^s(\varphi).$$

By Lemma 4.23 the latter expression, and therefore the r.h.s. of (4.52), approaches $-\infty$ as $\nu \to \infty$, uniformly w.r.t. $s \in [0,1]$. So the second term in the decomposition (4.51) of $\mathbb{P}_x(t_k \leq \tau_D^\varepsilon \leq t_{k+1})$ can be neglected since it becomes exponentially small of any desired order by choosing ν suitably large.

Step 4. We treat the first term on the r.h.s. of (4.51). It is given by the probability that, while $X^{x,\varepsilon}$ is in $B_\eta(x_-)$ at time σ_{k-1}, it exits within a time interval of length $t_{k+1} - \sigma_{k-1} \leq 2\nu$. Hence by the strong Markov property

$$\mathbb{P}_x(t_k \leq \tau_D^\varepsilon \leq t_{k+1} | \sigma_{k-1} \leq t_k) \leq \sup_{\substack{t_{k-1}\leq t\leq t_k,\\ y\in B_\eta(x_-)}} \mathbb{P}_{t,y}(\tau_D^\varepsilon \leq 2\nu).$$

Applying the uniform LDP to the closed set

$$F_D(\delta) := \Big\{\varphi \in C([0,2\nu],\mathbb{R}^d) \colon \varphi_0 \in D(\delta),\ \varphi_{t_0} \notin D(\delta) \text{ for some } t_0 \leq 2\nu\Big\},$$

yields the upper bound

$$\limsup_{\varepsilon\to 0} \varepsilon \ln \sup_{\substack{y\in B_\eta(x_-),\mu\in M,\\ t\in S^{s,\beta}(\varepsilon,\mu)}} \mathbb{P}_{t,y,}(\tau_D^\varepsilon \leq 2\nu) \leq -\inf_{y\in B_\eta(x_-)} \inf_{\varphi\in F_D(\delta)^{\gamma_0(\beta)}} I_{2\nu}^s(\varphi),$$

where $\gamma_0(\beta) = 2\beta\delta(F_D(\delta))$. By the same reasoning as before we can replace the blow-up of the path sets $F_D(\delta)$ by an enlargement of the domain $D(\delta)$. We find $\beta_1 > 0$ such that for $\beta \leq \beta_1$

$$
\begin{aligned}
-\inf_{y\in B_\eta(x_-)}\inf_{\varphi\in F_D(\delta)^{\gamma_0(\beta)}} I^s_{2\nu}(\varphi) &\leq -\inf_{y\in B_\eta(x_-)}\inf_{\varphi\in F_D(\frac{\delta}{2})} I^s_{2\nu}(\varphi)\\
&\leq -\inf_{y\in B_\eta(x_-)}\inf_{z\in\partial D(\frac{\delta}{2})} V^s(y,z).
\end{aligned}
$$

Now we apply (4.50) and recall the uniformity of the LDP w.r.t. s. We find $\varepsilon_0 > 0$ such that we have for $\varepsilon \leq \varepsilon_0$, $s \in [0,1]$, $\mu \in M$ and $\beta \leq \beta_1$

$$
\begin{aligned}
\varepsilon \ln \sup_{\substack{y\in B_\eta(x_-),\\ t\in S^{s,\beta}(\varepsilon,\mu)}} \mathbb{P}_{t,y}(\tau^\varepsilon_D \leq 2\nu) &\leq -\inf_{y\in B_\eta(x_-)}\inf_{z\in\partial D(\frac{\delta}{2})} V^s(y,z) + \frac{\theta}{4}\\
&\leq -e(s) + \frac{\theta}{2}.
\end{aligned} \tag{4.53}
$$

Step 5. We summarize our findings. We conclude that there exists $\varepsilon_1 > 0$ such that for $\varepsilon \leq \varepsilon_1$, $\mu \in M$ and $s \in [0,1]$ we have

$$
\begin{aligned}
&\varepsilon \ln \sup_{x\in B_\eta(x_-)} \mathbb{P}_x(sT(\varepsilon) \leq \tau^\varepsilon_\varrho \leq (s+\beta)T(\varepsilon))\\
&\leq \varepsilon \ln\Big(\sum_{k=0}^{N(\beta,\varepsilon,\mu)} \sup_{x\in B_\eta(x_-)} \mathbb{P}_x\Big(t_k \leq \tau^\varepsilon_D \leq t_{k+1}\Big|\sigma_{k-1} \leq t_k\Big)\Big) + \frac{\theta}{4}\\
&\leq \varepsilon \ln\Big(\beta T(\varepsilon) e^{-\frac{1}{\varepsilon}(e(s)-\frac{\theta}{2})}\Big) + \frac{\theta}{4}\\
&= \varepsilon \ln \beta + \mu - e(s) + \frac{3}{4}\theta\\
&\leq \mu - e(s) + \theta.
\end{aligned}
$$

This completes the proof. □

Remark 4.25. If we stay away from $s = 0$, in the statement of Proposition 4.24 the radius of the starting domain $B_\eta(x_-)$ can be chosen independently of the parameter θ. It may then be stated in the following somewhat different form.

Proposition 4.26. Let L and M be compact subsets of $(0,1]$ resp. $(\mu_*, e(0))$. Let $\eta > 0$ be small enough such that $B_\eta(x_-)$ belongs to the domain

$$\Big\{y \in \mathbb{R}^d \colon V^s(x_-,y) < \mu^* \text{ for all } s \in L\Big\}.$$

Then for each $\theta > 0$ there exists $\beta > 0$ such that

$$\limsup_{\varepsilon\to 0} \varepsilon \ln \sup_{x\in B_\eta(x_-)} \mathbb{P}\Big(sT(\varepsilon) \leq \tau^\varepsilon_\varrho \leq (s+\beta)T(\varepsilon)\Big) \leq \mu - e(s) + \theta,$$

uniformly w.r.t. $s \in L$ and $\mu \in M$.

Proof. To prove Proposition 4.26, one has to slightly modify the preceding proof. Instead of just η one has to choose two different parameters: η_0 for the definition of the starting domain D and some η_1 for the description of the location of the diffusion at time t_k, i.e. for the definition of the stopping times σ_k. □

In the following Proposition, we derive the upper bound for the asymptotic law of transition times, corresponding to the lower bound obtained in Proposition 4.22. Recall from (4.43) the definition $a_\mu = \inf\{t \geq 0\colon e(t) \leq \mu\}$.

PROPOSITION 4.27. Let $\mu < e(0)$. There exist $\gamma > 0$ and $h_0 > 0$ such that for all $h \leq h_0$

$$\limsup_{\varepsilon\to 0} \varepsilon \ln \sup_{x\in B_\gamma(x_-)} \mathbb{P}_x\Big(\tau_\varrho^\varepsilon \leq (a_\mu - h)T(\varepsilon)\Big) \leq \mu - e(a_\mu - h). \tag{4.54}$$

This bound is uniform w.r.t. μ on compact subsets of $(\mu_*, e(0))$.

PROOF. **Step 1**. The choice of h_0. Let M be a compact subset of $(\mu_*, e(0))$. To choose h_0, we use our assumptions on the geometry of the energy function e. Recall Assumption 4.18 according to which e is strictly monotone in the open intervals between the extrema (μ_*, μ^*). It implies that e is monotonically decreasing on the interval $[a_{e(0)}, a_\mu]$ for any $\mu \in M$. By choice of M, we further have $a_{e(0)} < \inf_{\mu\in M} a_\mu$. Now choose h_0 such that

$$\inf_{\mu\in M} a_\mu - h_0 > a_{e(0)}.$$

Then we have for $h \leq h_0$

$$\inf_{\mu\in M} a_\mu - h > 0, \tag{4.55}$$

$$e(0) > \sup_{\mu\in M, h\leq h_0} e(a_\mu - h), \tag{4.56}$$

$$e(s) \geq e(a_\mu - h) \quad \text{for all} \quad s \leq a_\mu - h. \tag{4.57}$$

To see (4.57), note that for $0 \leq s \leq a_{e(0)}$, by definition of $a_{e(0)}$, the inequality $e(s) \geq e(0) > e(a_\mu - h)$ holds, while for $a_{e(0)} \leq s \leq a_\mu - h$ by monotonicity $e(s) \geq e(a_\mu - h)$.

Step 2. We split the time interval. Fix $h \leq h_0$. For $\mu \in M$, let $\Lambda_0 = \Lambda_0(\mu) = 0$, and $\Lambda_1(\mu) \leq \inf_{\mu\in M}(a_\mu - h)T(\varepsilon)$. For $N \in \mathbb{N}$ we set

$$\Lambda_i(\mu) = \Lambda_1 + \frac{i-1}{N-1}\Big((a_\mu - h)T(\varepsilon) - \Lambda_1\Big), \quad 2 \leq i \leq N,$$

thus splitting the time interval $[0, (a_\mu - h)T(\varepsilon)]$ into the N intervals $[\Lambda_i(\mu), \Lambda_{i+1}(\mu)]$, $0 \leq i \leq N-1$. Then for $\gamma > 0$, $x \in B_\gamma(x_-)$

$$\mathbb{P}_x\Big(\tau_\varrho^\varepsilon \leq (a_\mu - h)T(\varepsilon)\Big) \leq \sum_{i=0}^{N-1} \mathbb{P}_x\Big(\tau_\varrho^\varepsilon \in [\Lambda_i(\mu), \Lambda_{i+1}(\mu)]\Big),$$

which implies

$$\begin{aligned}\limsup_{\varepsilon\to 0} \varepsilon \ln &\sup_{x\in B_\gamma(x_-)} \mathbb{P}_x\Big(\tau_\varrho^\varepsilon \leq (a_\mu - h)T(\varepsilon)\Big) \\ &\leq \max_{0\leq i\leq N-1} \limsup_{\varepsilon\to 0} \varepsilon \ln \sup_{x\in B_\gamma(x_-)} \mathbb{P}_x\Big(\tau_\varrho^\varepsilon \in [\Lambda_i(\mu), \Lambda_{i+1}(\mu)]\Big).\end{aligned}$$

Step 3. Estimation of the probabilities. Fix $\theta > 0$ such that for $h \leq h_0, \mu \in M$ we have $e(0) \geq e(a_\mu - h) + \theta$. This is guaranteed by (4.56). We shall show that

$$\limsup_{\varepsilon\to 0} \varepsilon \ln \sup_{x\in B_\gamma(x_-)} \mathbb{P}_x\Big(\tau_\varrho^\varepsilon \in [\Lambda_i(\mu), \Lambda_{i+1}(\mu)]\Big) \leq \mu - e(a_\mu - h) + \theta$$

uniformly in $0 \leq i \leq N-1$ and $\mu \in M$. Let us treat the estimation of the first term separately from the others.

First term. By Proposition 4.24, setting $s=0$, $\beta = \Lambda_1/T(\varepsilon)$, we may choose Λ, $\varepsilon_0 > 0$ and $\gamma_0 > 0$ such that for $\Lambda_1 \leq \Lambda T(\varepsilon)$, $\varepsilon \leq \varepsilon_0$, $\gamma \leq \gamma_0$, $\mu \in M$ the inequality

$$\varepsilon \ln \sup_{x \in B_\gamma(x_-)} \mathbb{P}_x\Big(\tau_\varrho^\varepsilon \in [\Lambda_0(\mu), \Lambda_1(\mu)]\Big) \leq \mu - e(0) + \theta$$

holds. Now we use the inequality $e(0) \geq e(a_\mu - h) + \theta$, valid for all $\mu \in M$. Hence there exists $\Lambda > 0$, $\varepsilon_0 > 0$ and $\gamma_0 > 0$ such that for $\Lambda_1 \leq \Lambda T(\varepsilon)$, $\varepsilon \leq \varepsilon_0$, $\gamma \leq \gamma_0$, $\mu \in M$

$$\varepsilon \ln \sup_{x \in B_\gamma(x_-)} \mathbb{P}_x\Big(\tau_\varrho^\varepsilon \in [\Lambda_0(\mu), \Lambda_1(\mu)]\Big) \leq \mu - e(a_\mu - h).$$

Next terms. Let us next estimate the contributions for the intervals $[\Lambda_i(\mu), \Lambda_{i+1}(\mu)]$ with $i \geq 1$. We use Proposition 4.24, this time with $s = \Lambda_i(\mu)/T(\varepsilon)$, $\beta = \frac{1}{N-1} \sup_{\mu \in M} a_\mu$. By the definition of a_μ, we get $e(s) > e(a_\mu)$ for all $s < a_\mu$. By (4.57), we have

$$e(s) = e(\Lambda_i(\mu)/T(\varepsilon)) \geq e(a_\mu - h).$$

By Remark 4.25,

$$\limsup_{\varepsilon \to 0} \varepsilon \ln \sup_{x \in B_\gamma(x_-)} \mathbb{P}_x\Big(\tau_\varrho^\varepsilon \in [\Lambda_i(\mu), \Lambda_{i+1}(\mu)]\Big) \leq \mu - e(a_\mu - h) + \theta$$

uniformly w.r.t. $1 \leq i \leq N$ and $\mu \in M$. Letting θ tend to 0, which implies that N tends to infinity and Λ_1 tends to zero, we obtain the desired upper bound for the exponential exit rate. □

4.4. The full dynamics: stochastic resonance in diffusions

Given the results of the previous section on the asymptotics of exit times which are uniform in the scale parameter μ, we are now in a position to consider the problem of finding a satisfactory probabilistic notion of stochastic resonance that does not suffer from the lack of robustness of physical notions such as spectral power amplification. We continue to study the SDE

$$dX_t^{x,\varepsilon} = b\Big(\frac{t}{T(\varepsilon)}, X_t^{x,\varepsilon}\Big)\,dt + \sqrt{\varepsilon}\,dW_t, \quad t \geq 0,\ X_0^{x,\varepsilon} = x \in \mathbb{R}^d,$$

introduced before, thereby recalling that the drift term b satisfies the local Lipschitz conditions (4.11) and (4.10) in space and time, as well as the growth condition (2.44). Moreover, b is assumed to be one-periodic in time such that $T(\varepsilon)$ is the period of the deterministic input of the randomly perturbed dynamical system described by X^ε.

In typical applications $b = -\nabla_x U$ is given by the (spatial) gradient of some time periodic double well potential U.

The potential possesses at all times two local minima well separated by a barrier (see Figure 4.11 on p. 174 for the 1-dimensional case). The depth of the wells and the roles of being the deep and shallow one change periodically. The diffusion X^ε then roughly describes the motion of a Brownian particle of intensity ε in a double well landscape. Its attempts to get close to the energetically most favorable deep position in the landscape makes it move along random trajectories which exhibit randomly periodic hopping between the wells. The average time the trajectories need to leave a potential well of depth $\frac{v}{2}$ being given by the Kramers–Eyring law $T(\varepsilon) = \exp(\frac{v}{\varepsilon})$ motivates our choice of time scales $T(\varepsilon) = \exp(\frac{\mu}{\varepsilon})$ and also our convention to measure time scales in their corresponding free energy units μ.

The problem of stochastic resonance consists in characterizing the *optimal tuning* of the noise, i.e. the best relation between the noise amplitude ε and the input period $T(\varepsilon)$ — or, in our units the energy parameter μ — of the deterministic system which makes the diffusion trajectories look as periodic as possible. Of course, the optimality criterion must be based upon a *quality measure* for periodicity in random trajectories.

In this section we shall design a measure of quality based on the transition probabilities investigated in Section 4.3. Its counterpart for the Markov chains describing the reduced dynamics was discussed in Section 4.2. With respect to this measure we identify a resonance energy $\mu_0(\varepsilon)$ for which the diffusion trajectories follow the periodic forcing of the system at intensity ε in an optimal way. We shall study the problem in the general situation, which includes the double well potential gradient case as an important example, and draws its intuition from it. The deterministic system

$$\dot{\varphi}_s(t) = b(s, \varphi_s(t)), \quad t \geq 0,$$

has to satisfy Assumption 4.17, i.e. it possesses two well separated domains of attraction the common boundary of which is time invariant.

In the first subsection we shall describe the *resonance interval*, namely the set of scale parameters μ for which trivial behavior (constant or continuously jumping trajectories) is excluded. The second subsection shows that a quality measure of periodic tuning is given by the exponential rate at which the first transition from one domain of attraction to the other one happens within a fixed time window around $a_\mu T(\varepsilon)$. This quality measure is robust, as demonstrated in the final subsection: in the small noise limit the diffusion and its reduced model, a Markov chain jumping between the domains of attraction reduced to the equilibrium points (see Section 4.2), display the same resonance pattern.

4.4.1. Resonance interval. According to Freidlin [**39**], quasi-periodic hopping behavior of the trajectories of our diffusion in the small noise limit of course requires that the energies needed to leave the domains of attraction of the two equilibria switch their order periodically: if $e_\pm$ denotes the energy needed to leave $A_\pm$, then e_+ needs to be bigger than e_- during part of one period, and vice versa for the rest. We assume that $e_\pm$ both satisfy Assumption 4.18 and associate to each of these functions the transition time

$$a_\mu^\pm(s) = \inf\Big\{t \geq s \colon e_\pm(t) \leq \mu\Big\}.$$

The scale parameters μ for which relevant behavior of the system is expected clearly belong to the intervals

$$I_i = \Big(\inf_{t\geq 0} e_i(t), \sup_{t\geq 0} e_i(t)\Big), \quad i \in \{-,+\}.$$

Our aim being the observation of periodicity, we have to make sure that the process can travel back and forth between the domains of attraction on the time scales considered, but not instantaneously. So, on the one hand, in these scales it should not *get stuck* in one of the domains. On the other hand, they should not allow

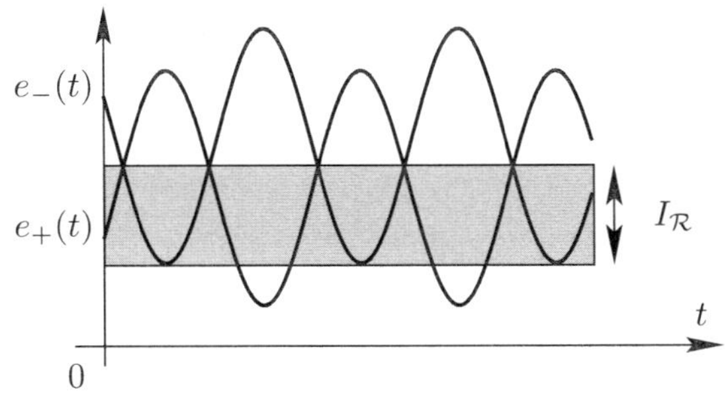

FIGURE 4.9. Resonance interval $I_{\mathcal{R}}$.

for *chaotic behavior*, i.e. immediate re-bouncing after leaving a domain has to be avoided.

To make these conditions mathematically precise, recall that transitions from A_i to A_{-i} become possible as soon as the energy e_i needed to exit from A_i falls below μ, which represents the available energy. Not to get stuck in one of $A_\pm$, we therefore have to guarantee

$$\mu > \max_{i=\pm} \inf_{t\geq 0} e_i(t).$$

To avoid immediate re-bouncing, we have to assure that the diffusion cannot leave A_{-i} at the moment it reaches it, coming from A_i. Suppose we consider the dynamics after time $s \geq 0$, and the diffusion is near i at that time. Its first transition to A_{-i} occurs at time $a_\mu^i(s)T(\varepsilon)$ where $a_\mu^i(s)$ is the first time in the original scale at which e_i falls below μ after s. Provided $e_{-i}(a_\mu^i(s))$ is bigger than μ, it stays there for at least a little while. This is equivalent to saying that for all $s \geq 0$ there exists $\delta > 0$ such that on $[a_\mu^i(s), a_\mu^i(s) + \delta]$ we have $\mu < e_{-i}$. Since by definition for t shortly after $a_\mu^i(s)$, we always have $e_i(t) \leq \mu$, our condition may be paraphrased by: for all $s \geq 0$ there exists $\delta > 0$ such that on $[a_\mu^i(s), a_\mu^i(s) + \delta]$ we have $\mu < \max_{i=\pm} e_i$. This in turn is more elegantly expressed by

$$\mu < \inf_{t\geq 0} \max_{i=\pm} e_i(t).$$

Our search for a set of scales μ for which the diffusion exhibits non-trivial transition behavior may be summarized in the following definition. The interval

$$I_{\mathcal{R}} = \left(\max_{i=\pm} \inf_{t\geq 0} e_i(t), \inf_{t\geq 0} \max_{i=\pm} e_i(t) \right)$$

is called *resonance interval* (see Figure 4.9).

In this interval, for small ε, we have to look for an optimal energy scale $\mu_0(\varepsilon)$. See Section 4.2 for the definition of the corresponding interval in the case of two state Markov chains. In Freidlin's [**39**] terms, stochastic resonance in the sense of quasi-deterministic periodic motion is given if the parameter μ exceeds the lower boundary of our resonance interval.

Let us briefly comment on the potential gradient case.

Assume $b = -\nabla_x U$, where U is a time periodic double well potential with time invariant local minima $x_\pm$, the wells A_- and A_+, and and a separatrix χ. The energy $e_\pm$ is essentially the energy a Brownian particle needs to cross χ.

Freidlin and Wentzell [**40**] give the link between this energy and the depth of the well (see Section 3.1).

LEMMA 4.28. *If* $D_\pm(t) = \inf_{y\in\chi} U(t,y) - U(t,x_\pm)$ *denote the depths of the wells, then* $e_\pm(t) = 2D_\pm(t)$ *for all* $t \geq 0$.

This link is the origin of the name *quasipotential.* The minimal energy e is reached by some path which intersects the level sets of the potential with orthogonal tangents. This path satisfies an equation of the form

$$\dot\varphi_s = \nabla_x U(t,\varphi_s), \quad s \in (-\infty, T), \quad \varphi_T \in \chi.$$

The resonance interval is given by

$$I_{\mathcal{R}} = \Big(\max_{i=\pm}\inf_{t\geq 0} 2D_i(t), \inf_{t\geq 0}\max_{i=\pm 1} 2D_i(t)\Big).$$

4.4.2. Transition rates as quality measure. Let us now explain in detail our measure of quality designed to give a concept of optimal tuning which, as opposed to physical measures (see Section 3.4.3), is robust for model reduction to Markov chains just retaining the jump dynamics between the meta-stable equilibria of the diffusion. We shall use a notion that is based just on this rough transition mechanism. In fact, generalizing the approach for two state Markov chain models (see Section 4.2), we measure the quality of tuning by computing for varying energy parameters μ the probability that, starting in x_i, the diffusion is transferred to x_{-i} within the time window $[(a^i_\mu(0)-h)T(\varepsilon), (a^i_\mu(0)+h)T(\varepsilon)]$ of width $2hT(\varepsilon)$. To find the *stochastic resonance point* for large $T(\varepsilon)$ (small ε) we have to maximize this measure of quality w.r.t. $\mu \in I_{\mathcal{R}}$ and finally let the window width parameter h tend to 0. The probability for transition within this window will be approximated using the estimates of Section 4.3. Uniformity of convergence to the exponential rates will enable us to maximize in μ for fixed small ε.

Let us now make these ideas precise. To make sure that the transition window makes sense at least for small h, we have to suppose that $a^i_\mu > 0$ for $i = \pm$ and $\mu \in I_{\mathcal{R}}$. This is guaranteed if

$$e_i(0) > \inf_{t\geq 0}\max_{j=\pm} e_j(t), \quad i = \pm.$$

If this is not granted from the beginning, it suffices to start the diffusion a little later. For ϱ small enough such that $B_\varrho(x_\pm) \subset A_\pm$, for $\varepsilon > 0$, and $\mu \in I_{\mathcal{R}}$, we call

$$\mathcal{M}^h(\varepsilon,\mu,\varrho) = \min_{i=\pm}\sup_{x\in B_\varrho(x_i)} \mathbb{P}_x\Big(\tau^{\varepsilon,-i}_\varrho \in [(a^i_\mu - h)T(\varepsilon), (a^i_\mu + h)T(\varepsilon)]\Big) \tag{4.58}$$

transition probability for a time window of width h. Here

$$\tau^\varepsilon_{i,\varrho} = \inf\Big\{t \geq 0 \colon X^\varepsilon_t \in B_\varrho(x_i)\Big\},$$

and a^i_μ abbreviates $a^i_\mu(0)$, the transition time at zero. We are ready to state our main result on the asymptotic law of transition time windows. This is an obvious consequence of Theorem 4.19.

THEOREM 4.29. *Let* M *be a compact subset of* $I_{\mathcal{R}}$, $h_0 > 0$ *and* ϱ *be given according to Theorem* 4.19. *Then for all* $h \leq h_0$

$$\lim_{\varepsilon\to 0}\varepsilon\ln\Big(1 - \mathcal{M}^h(\varepsilon,\mu,\varrho)\Big) = \max_{i=\pm}\Big\{\mu - e_i(a^i_\mu - h)\Big\} \tag{4.59}$$

uniformly w.r.t. $\mu \in M$.

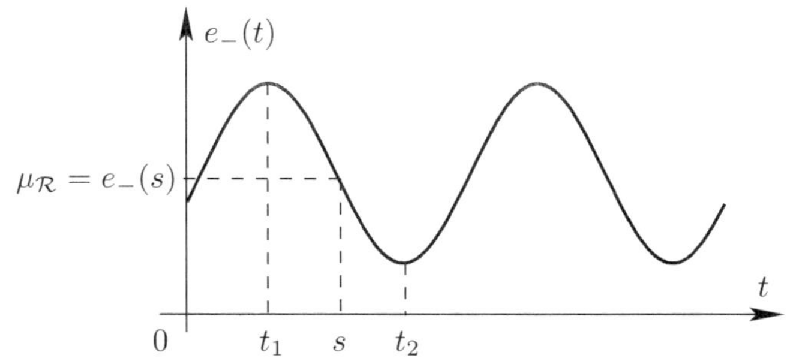

FIGURE 4.10. The point of maximal decrease.

It is clear that for h small, the global minimizer $\mu_{\mathcal{R}}(h)$ of

$$I_{\mathcal{R}} \ni \mu \mapsto \max_{i=\pm 1} \left\{ \mu - e_i(a_\mu^i - h) \right\}$$

if it exists, would be a good candidate for our resonance point. To get rid of the dependence on h, we shall consider the limit of $\mu_{\mathcal{R}}(h)$ as $h \to 0$.

DEFINITION 4.30. Suppose that

$$I_{\mathcal{R}} \ni \mu \mapsto \max_{i=\pm} \left\{ \mu - e_i(a_\mu^i - h) \right\}$$

possesses a global minimum $\mu_{\mathcal{R}}(h)$. Suppose further that

$$\mu_{\mathcal{R}} = \lim_{h \to 0} \mu_{\mathcal{R}}(h)$$

exists in $I_{\mathcal{R}}$. We call $\mu_{\mathcal{R}}$ the *stochastic resonance point* of the diffusion $X^{x,\varepsilon}$.

Let us now illustrate this resonance notion in a situation in which the energy functions are related by a phase lag $\phi \in (0,1)$, i.e. $e_-(t) = e_+(t+\phi)$ for all $t \geq 0$. We shall show that in this case the stochastic resonance point exists if one of the energy functions, and thus both, has a unique point of maximal decrease on the interval where it is strictly decreasing, see Figure 4.10.

THEOREM 4.31. *Suppose that e_- is twice continuously differentiable and has its global maximum at t_1, and its global minimum at t_2, where $t_1 < t_2$. Suppose further that there is a unique point $t_1 < s < t_2$ such that $e_-(\cdot)$ is strictly concave on (t_1, s) and strictly convex on (s, t_2). Then $\mu_{\mathcal{R}} = e_-(s)$ is the stochastic resonance point.*

PROOF. As a consequence of the phase lag of the energy functions,

$$\max_{i=\pm} \left\{ \mu - e_i(a_\mu^i - h) \right\} = \mu - e_-(a_\mu^- - h).$$

Write $a_\mu = a_\mu^-$, and recall that on the interval of decrease of e_- we have $a_\mu = e_-^{-1}(\mu)$. Therefore, the differentiability assumption yields

$$1 = e_-'(a_\mu - h) \cdot a_\mu' = e_-'(a_\mu - h) \frac{1}{e_-'(a_\mu)}.$$

Our hypothesis concerning convexity and concavity of e_- essentially means that $e_-''(s) = 0$, and $e_-'' < 0$ on $(t_1, s) < 0$, $e_-'' > 0$ on (s, t_2), which may be stated alternatively by saying that $\mu \mapsto e'_-(a_\mu)$ has a local maximum at $a_\mu = s$. Hence for h small there exists a unique point $a_\mu(h)$ such that

$$e'_-(a_\mu(h) - h) = e'_-(a_\mu(h))$$

and

$$\lim_{h \to 0} a_\mu(h) = s.$$

To show that $a_\mu(h)$ corresponds to a minimum of the function

$$\mu \mapsto \mu - e_-(a_\mu - h),$$

we take the second derivative of this function at $a_\mu(h)$, which is given by

$$\frac{e'_-(a_\mu(h) - h)e''_-(a_\mu(h)) - e''_-(a_\mu(h) - h)e'_-(a_\mu(h))}{e'_-(a_\mu(h))}.$$

But $e'_-(a_\mu(h)), e'_-(a_\mu(h) - h) < 0$, whereas $e''_-(a_\mu(h) - h) > 0$, $e''_-(a_\mu(h)) < 0$. This clearly implies that $a_\mu(h)$ corresponds to a minimum of the function. By definition, as $h \to 0$, $a_\mu(h) \to s$. Therefore, finally, $e_-(s)$ is the stochastic resonance point. □

4.4.3. Example in dimension 1. Let us consider the double-well potential

$$U(t,x) = \frac{x^6}{6} - \cos\left\{2\pi\Big(t - \frac{1}{4} + \psi\,\mathrm{sign}(x)\Big)\right\}\Big(\frac{x^5}{5} - \frac{x^3}{3}\Big) - \frac{x^2}{2}, \quad x \in \mathbb{R},\ t \geq 0.$$

with $\psi \in [0, \frac{1}{4})$ (see Figure 4.11). We shall compute the resonance interval and the resonance point for the following diffusion:

$$dX_t^{x,\varepsilon} = -\frac{\partial U}{\partial x}\Big(\frac{t}{T(\varepsilon)}, X_t^{x,\varepsilon}\Big)\,dt + \sqrt{\varepsilon}\,dW_t, \quad t \geq 0,\ X_0^{x,\varepsilon} = x \in \mathbb{R}.$$

with $T(\varepsilon) = \exp(\frac{\mu}{\varepsilon})$.

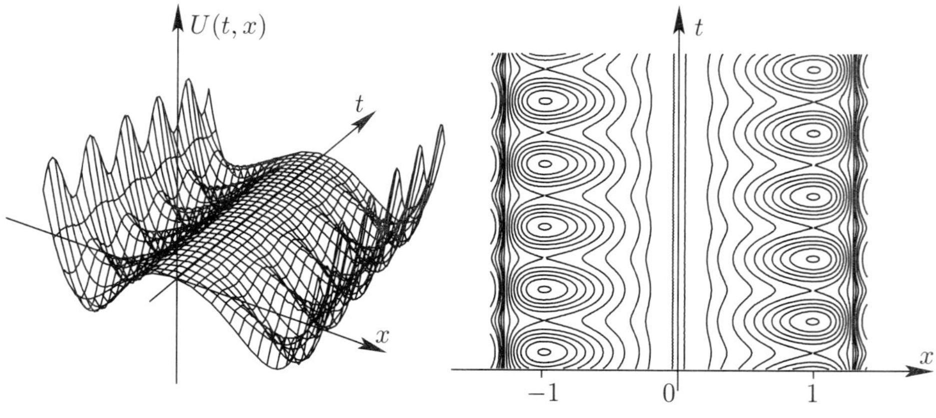

FIGURE 4.11. The time-periodic double-well potential U for $\psi = 0$ and its level sets.

The potential function U satisfies all the assumptions required for potentials above, in particular

$$\frac{\partial U}{\partial x}(t,x) = 0 \text{ iff } x \in \{-1, 0, 1\}.$$

-1 and 1 are the meta-stable states of the potential, and 0 the saddle point. We can then compute the barrier height of the two wells. For $i \in \{-1, 1\}$,

$$D_i(t) = \frac{2}{3} - i\frac{4}{15}\cos\Big(2\pi t + i2\pi\psi - \frac{\pi}{2}\Big).$$

Let us note that $D_+(t) = D_-(t + 2\psi + \frac{1}{2})$. Since $\psi \in [0, \frac{1}{4})$, the phase between the function $e_- = 2D_-$ and $e_+ = 2D_+$ is given by $\phi = 2\psi + 1/2$. The resonance interval is then given by

$$I_{\mathcal{R}} = \Big(\frac{4}{5}, \frac{4}{3} - \frac{8}{15}\cos\Big(\frac{\pi}{2} - 2\pi\psi\Big)\Big).$$

In the symmetric case, that is if $\psi = 0$ we obtain $I_{\mathcal{R}} = (4/5, 4/3)$. Let us now compute the optimal tuning scale. We obtain

$$a_\mu^- = \frac{1}{2\pi}\arccos\Big(\frac{15}{2}\Big(\frac{\mu}{4} - \frac{1}{3}\Big)\Big) + \psi + \frac{1}{4}.$$

Hence, for $h > 0$ small enough, we define

$$\begin{aligned} F(\mu) &= \mu - e_-(a_\mu^- - h) \\ &= \mu - \frac{4}{3} - \frac{8}{15}\cos\Big(\arccos\Big\{\frac{15}{2}\Big(\frac{\mu}{4} - \frac{1}{3}\Big)\Big\} - 2\pi h\Big) \\ &= \Big(\mu - \frac{4}{3}\Big)(1 - \cos 2\pi h) - \frac{8}{15}\sin(2\pi h)\sqrt{1 - \Big(\frac{15}{2}\Big(\frac{\mu}{4} - \frac{1}{3}\Big)\Big)^2}. \end{aligned}$$

Let us recall that F does not depend on the phase which implies

$$\mu - e_-(a_\mu^- - h) = \mu - e_+(a_\mu^+ - h).$$

Hence, to obtain optimal tuning, it suffices to compute the minimum of F for $\mu \in I_{\mathcal{R}}$. Differentiating F we obtain

$$F'(\mu) = 1 - \cos(2\pi h) + \frac{15}{2}\sin(2\pi h)\frac{\frac{\mu}{4} - \frac{1}{3}}{\sqrt{1 - \left(\frac{15\mu}{8} - \frac{15}{6}\right)^2}}.$$

Hence F attains its minimum for

$$\mu_{\mathcal{R}}(h) = \frac{4}{3} - \frac{4\sqrt{2}}{15}\sqrt{1 - \cos(2\pi h)}$$

and

$$\mu_{\mathcal{R}} = \lim_{h \to 0}\mu_{\mathcal{R}}(h) = \frac{4}{3}.$$

Thus we obtain that $\mu_{\mathcal{R}}$ is the stochastic resonance point if $\mu_{\mathcal{R}} \in I_{\mathcal{R}}$, that is, if the phase is near to $\frac{1}{2}$, i.e. if ψ is close to 0. In the other case, the optimal tuning rate on every interval $[a, b] \subset I_{\mathcal{R}}$ is given by the upper bound b.

4.4.4. Robustness of the notion of optimal rates for transition windows. In this last subsection we shall conclude the study of our new quality measure for periodic tuning based on transitions between the meta-stable states, by showing that it is robust with respect to model reduction. First we recall the definition of the reduced model and the related analysis of optimal tuning (see Section 4.2). Then we compare the results obtained in the different models.

To describe the reduced model, let e_- and e_+ be the energy functions corresponding to transitions from A_- to A_+ (resp. from A_+ to A_-) as before. Let us consider a continuous time Markov chain $(Y_t^\varepsilon)_{t \geq 0}$ taking values in the state space

$\{-1,+1\}$ with initial data $Y_0^\varepsilon = -1$. Suppose the infinitesimal generator is given by

$$Q(t) = \begin{pmatrix} -\varphi(t) & \varphi(t) \\ \psi(t) & -\psi(t) \end{pmatrix},$$

where $\varphi(t)$ and $\psi(t)$, $t \geq 0$, are a 1-periodic functions describing the rates which just produce the transition dynamics of the diffusion between the equilibria $x_\pm$, i.e.

$$\varphi(t) = \exp\Big(- \frac{e_-(t/T(\varepsilon))}{\varepsilon}\Big), \quad t \geq 0,$$

and

$$\psi(t) = \exp\Big(- \frac{e_+(t/T(\varepsilon))}{\varepsilon}\Big), \quad t \geq 0.$$

This reduced model was studied in Section 4.2. As a quality measure for the Markov chain Y^ε, we investigate the corresponding analogue of (4.58) for the Markov chain Y^ε.

Let τ_i^ε denote the first transition time of Y^ε. For $h > 0$ we consider the following measure of quality of periodic tuning

$$\mathcal{N}^h(\varepsilon, \mu) = \min_{i=\pm} \mathbb{P}_i\Big(\tau_{-i}^\varepsilon \in [(a_\mu^i - h)T(\varepsilon), (a_\mu^i + h)T(\varepsilon)]\Big), \quad \varepsilon > 0, \ \mu \in I_{\mathcal{R}},$$

which is called *transition probability for a time window of width h* for the Markov chain. We summarize the key result of Section 4.2:

Let M be a compact subset of $I_{\mathcal{R}}$ and $h_0 < a_\mu^-$. Then for $0 < h \leq h_0$

$$\lim_{\varepsilon \to 0} \varepsilon \ln(1 - \mathcal{N}^h(\varepsilon, \mu)) = \max_{i=\pm 1} \Big\{\mu - e_-(a_\mu^i - h)\Big\}$$

uniformly w.r.t. $\mu \in M$.

It is clear that the Markov chain Y^ε and the diffusion process X^ε have exactly the same resonance behavior: $\mathcal{M}^h(\varepsilon, \mu)$ and $\mathcal{N}^h(\varepsilon, \mu)$ have the same logarithmic asymptotic scale. Of course, we may define the *stochastic resonance point* for Y^ε just as we did for X^ε. So the following final robustness result holds true.

THEOREM 4.32. *The resonance points of X^ε with time periodic drift b and of Y^ε with exponential transition rate functions $e_\pm$ coincide.*

APPENDIX A

Supplementary tools

LEMMA A.1 (Gronwall). *Suppose* $f, \phi\colon [0,T] \to [0,\infty)$ *satisfy* $\phi, \phi f \in L^1([0,T],\mathbb{R})$ *and*

$$f(t) \le a + \int_0^t \phi(s) f(s)\, ds, \quad 0 \le t \le T, \tag{A.1}$$

for some constant $a > 0$. *Then for* $0 \le t \le T$

$$f(t) \le a \exp\Big\{ \int_0^t \phi(s)\, ds \Big\}.$$

PROOF. For $0 \le t \le T$, let

$$F(t) = \exp\Big\{ - \int_0^t \phi(s)\, ds \Big\} \int_0^t \phi(s) f(s)\, ds.$$

Using the assumption (A.1) on f, we see that $F'(t) \le a\phi(t) \exp\Big\{ - \int_0^t \phi(s)\, ds \Big\}$. Since $F(0) = 0$, this implies

$$F(t) \le a\Big(1 - \exp\Big\{ - \int_0^t \phi(s)\, ds \Big\}\Big).$$

Multiplying the latter inequality by $\exp\{\int_0^t \phi(s)\, ds\}$ and using (A.1) once again yields the claimed estimate. $\square$

APPENDIX B

Laplace's method

In this Appendix Laplace's method of asymptotic evaluation of integrals depending on parameter is explained. In our exposition we follow Erdelyi [**36**] and Olver [**84**].

Consider the integral

$$I(\varepsilon) = \int_a^b e^{-\frac{2U(x)}{\varepsilon}} w(x)\, dx, \tag{B.1}$$

in which $a, b \in [-\infty, +\infty]$, U and w are smooth functions on $\mathbb{R}$, $\varepsilon > 0$. The following powerful method for approximating $I(\varepsilon)$, $\varepsilon \to 0$, goes back to Laplace [**66**]. According to Laplace, the major contribution to the value of the integral arises from the immediate vicinity of those points of the interval $[a, b]$ at which U assumes its smallest value. Let the minimum of U occur, say, at $x = x_0$. If ε is small, the graph of the integrand has a very sharp peak at x_0. It suggests that the overwhelming contribution to the integral comes from the neighbourhood of x_0. Accordingly, we replace U and w by the leading terms in their series expansions in $x - x_0$, and then extend the integration limits to $\pm\infty$. The evaluation of the resulting integral yields the required approximation.

We consider two major cases. Suppose first that a is finite, $x_0 = a$, $U'(a) > 0$ and $w(a) \neq 0$. Then Laplace's estimation reads

$$\begin{aligned} I(\varepsilon) &= \int_a^b e^{-\frac{2U(x)}{\varepsilon}} w(x)\, dx \doteqdot \int_a^b e^{-\frac{2}{\varepsilon}(U(a)+(x-a)U'(a))} w(a)\, dx \\ &\doteqdot w(a) e^{-\frac{2U(a)}{\varepsilon}} \int_a^\infty e^{-\frac{2}{\varepsilon}(x-a)U'(a)}\, dx = \frac{\varepsilon w(a) e^{-\frac{2U(a)}{\varepsilon}}}{2U'(a)}. \end{aligned}$$

The second major case arises if U has a simple minimum at an interior point x_0 of (a, b) and $w(x_0) \neq 0$. Then

$$\begin{aligned} I(\varepsilon) &= \int_a^b e^{-\frac{2U(x)}{\varepsilon}} w(x)\, dx \doteqdot \int_a^b e^{-\frac{2}{\varepsilon}(U(x_0)+\frac{1}{2}(x-x_0)^2 U''(x_0))} w(x_0)\, dx \\ &\doteqdot w(x_0) e^{-\frac{2U(x_0)}{\varepsilon}} \int_{-\infty}^\infty e^{-\frac{U''(x_0)}{\varepsilon}(x-x_0)^2}\, dx = w(x_0) e^{-\frac{2U(x_0)}{\varepsilon}} \sqrt{\frac{\pi\varepsilon}{U''(x_0)}}. \end{aligned}$$

If U has a finite number of minima, we may break up the integral (B.1) into a finite number of integrals so that in each interval U reaches its minimum at one of the end-points and at no other point. Accordingly, we shall assume that U reaches its minimum at $x = a$ and that $U(x) > U(a)$, $a < x \leq b$. Now we precisely formulate the theorem about Laplace's approximation, see [**84**, Chapters 7,9].

THEOREM B.1. *Let $a \in \mathbb{R}$, $b \in \mathbb{R} \cup \{+\infty\}$, $a < b$. Let $U\colon \mathbb{R} \to \mathbb{R}$ be differentiable, and $w : \mathbb{R} \to \mathbb{R}$ or $\mathbb{C}$ be measurable.*

Suppose in addition that

i) the minimum of U *is attained only at* a*;*

ii) U' *and* w *are continuous in a neighbourhood of* a*;*

iii) as $x \downarrow a$,

$$U(x) = U(a) + P(x-a)^{\mu} + \mathcal{O}((x-a)^{\mu+1}),$$
$$w(x) = Q(x-a)^{\lambda-1} + \mathcal{O}((x-a)^{\lambda}),$$

and the first of these relations is differentiable; here P, μ *and* λ *are positive constants, and* Q *is a real or complex constant;*

iv)

$$I(\varepsilon) = \int_a^b e^{-\frac{2U(x)}{\varepsilon}} w(x)\,dx$$

exists in the Lebesgue sense throughout its range for all sufficiently small ε.

Then

$$I(\varepsilon) = \frac{Q}{\mu} \cdot \Gamma\Big(\frac{\lambda}{\mu}\Big) \cdot \Big(\frac{\varepsilon}{2P}\Big)^{\frac{\lambda}{\mu}} \cdot e^{-\frac{2U(a)}{\varepsilon}} \Big(1 + \mathcal{O}(\varepsilon^{\frac{1}{\mu}})\Big).$$

If the asymptotic expansions in ascending powers of $x-a$ exist for U and w, the expansion of the integral $I(\varepsilon)$ can be also obtained. Although there is no general formula for this expansion, we determine its first three terms.

THEOREM B.2. *Let conditions (i), (ii) and (iv) of Theorem* B.1 *be satisfied and the expansions*

$$U(x) = U(a) + \sum_{s=0}^{n-1} p_s (x-a)^{\mu+s} + \mathcal{O}\left((x-a)^{\mu+n}\right),$$
$$w(x) = \sum_{s=0}^{n-1} q_s (x-a)^{\lambda-1+s} + \mathcal{O}\left((x-a)^{\lambda+n}\right)$$

hold. Suppose that $p_0 \neq 0$, $q_0 \neq 0$. *Then*

$$I(\varepsilon) = e^{-\frac{2U(a)}{\varepsilon}} \left[\sum_{s=0}^{n-1} \Gamma\Big(\frac{\lambda+s}{\mu}\Big) a_s \Big(\frac{\varepsilon}{2}\Big)^{\frac{\lambda+s}{\mu}} + \mathcal{O}(\varepsilon^{\frac{\lambda+n}{\mu}}) \right], \tag{B.2}$$

where

$$a_0 = \frac{q_0}{\mu p_0^{\lambda/\mu}},$$
$$a_1 = \left\{ \frac{q_1}{\mu} - \frac{(\lambda+1)p_1 q_0}{\mu^2 p_0} \right\} \frac{1}{p_0^{(\lambda+1)/\mu}}, \tag{B.3}$$
$$a_2 = \left[\frac{q_2}{\mu} - \frac{(\lambda+2)p_1 q_1}{\mu^2 p_0} + \left\{ (\lambda+\mu+2)p_1^2 - 2\mu p_0 p_2 \right\} \frac{(\lambda+2)q_0}{2\mu^3 p_0^2} \right] \frac{1}{p_0^{(\lambda+2)/\mu}}.$$

Let us apply Theorems B.1 and B.2 to the double-well potential U from Chapters 3.3 and 3.4 to find the asymptotics of the integral $\int_{\mathbb{R}} e^{-\frac{2U(x)}{\varepsilon}}\,dx$ for $\varepsilon \to 0$.

The function U is supposed to be infinitely differentiable and to possess a unique global minimum at -1 such that $U(-1) = -\frac{V}{2}$. We break the interval $(-\infty, +\infty)$

into two intervals $(-\infty,-1]$ and $[-1,+\infty)$, and note that

(B.4)
$$\begin{aligned}\int_{\mathbb{R}} e^{-\frac{2U(x)}{\varepsilon}}\,dx &= \int_{-\infty}^{-1} e^{-\frac{2U(x)}{\varepsilon}}\,dx + \int_{-1}^{\infty} e^{-\frac{2U(x)}{\varepsilon}}\,dx\\ &= \int_{-1}^{+\infty} e^{-\frac{2U(x)}{\varepsilon}}\,dx + \int_{1}^{+\infty} e^{-\frac{2\overline{U}(x)}{\varepsilon}}\,dx,\end{aligned}$$

where $\overline{U}(x) = U(-x)$, $x \in \mathbb{R}$. Both integrals in the last line of (B.4) satisfy the conditions of Theorem B.2. To determine the coefficients p_k, $k = 0, 1, 2$, we expand U near -1 and $\overline{U}$ near 1 to get

(B.5)
$$U(x) = -\frac{V}{2} + \frac{\omega_-}{2}(x+1)^2 + \frac{U^{(3)}(-1)}{6}(x+1)^3 + \frac{U^{(4)}(-1)}{24}(x+1)^4 + \mathcal{O}((x+1)^5),$$

(B.6)
$$\overline{U}(x) = -\frac{V}{2} + \frac{\omega_-}{2}(x-1)^2 - \frac{U^{(3)}(-1)}{6}(x-1)^3 + \frac{U^{(4)}(-1)}{24}(x-1)^4 + \mathcal{O}((x-1)^5).$$

Thus, $\mu = 2$, $\lambda = 1$, $q_0 = 1$ and $q_k = 0$, $k \geq 1$. A direct application of (B.3) and (B.2) yields

(B.7)
$$\begin{aligned}&\int_{-1}^{+\infty} e^{-\frac{2U(x)}{\varepsilon}}\,dx = e^{\frac{V}{\varepsilon}}\frac{1}{2}\sqrt{\frac{\pi\varepsilon}{\omega_-}}\\ &\times\left[1 - \frac{U^{(3)}(-1)}{3\omega_-^{3/2}\sqrt{\pi}}\sqrt{\varepsilon} + \frac{1}{16\omega_-^3}\left(\frac{5U^{(3)}(-1)^2}{3} - \omega_- U^{(4)}(-1)\right)\varepsilon + \mathcal{O}(\varepsilon^{3/2})\right]\\ &\int_{1}^{+\infty} e^{-\frac{2\overline{U}(x)}{\varepsilon}}\,dx = e^{\frac{V}{\varepsilon}}\frac{1}{2}\sqrt{\frac{\pi\varepsilon}{\omega_-}}\\ &\times\left[1 + \frac{U^{(3)}(-1)}{3\omega_-^{3/2}\sqrt{\pi}}\sqrt{\varepsilon} + \frac{1}{16\omega_-^3}\left(\frac{5U^{(3)}(-1)^2}{3} - \omega_- U^{(4)}(-1)\right)\varepsilon + \mathcal{O}(\varepsilon^{3/2})\right],\end{aligned}$$

and, consequently,

(B.8)
$$\int_{\mathbb{R}} e^{-\frac{2U(x)}{\varepsilon}}\,dx = e^{\frac{V}{\varepsilon}}\sqrt{\frac{\pi\varepsilon}{\omega_-}}\left[1 + \frac{1}{16\omega_-^3}\left(\frac{5U^{(3)}(-1)^2}{3} - \omega_- U^{(4)}(-1)\right)\varepsilon + \mathcal{O}(\varepsilon^{3/2})\right].$$

The error term $\mathcal{O}(\varepsilon^{3/2})$ is in fact of order ε^2, since due to the infinite differentiability of U all terms a_k with odd indices k vanish. This variant of the asymptotics is used in Lemma 3.49. The less exact asymptotics

(B.9)
$$\int_{\mathbb{R}} e^{-\frac{2U(x)}{\varepsilon}}\,dx = e^{\frac{V}{\varepsilon}}\sqrt{\frac{\pi\varepsilon}{\omega_-}}(1 + \mathcal{O}(\varepsilon)).$$

is also used.

Analogously, one evaluates the integral

(B.10)
$$\begin{aligned}&\int_{\mathbb{R}} x\, e^{-\frac{2U(x)}{\varepsilon}}\,dx = \int_{-1}^{\infty} x\, e^{-\frac{2U(x)}{\varepsilon}}\,dx - \int_{1}^{\infty} x\, e^{-\frac{2\overline{U}(x)}{\varepsilon}}\,dx =\\ &- e^{\frac{V}{\varepsilon}}\sqrt{\frac{\pi\varepsilon}{\omega_-}}\left[1 + \left\{\frac{U^{(3)}(-1)}{4\omega_-^2} + \frac{1}{16\omega_-^3}\left(\frac{5U^{(3)}(-1)^2}{3} - \omega_- U^{(4)}(-1)\right)\right\}\varepsilon + \mathcal{O}(\varepsilon^2)\right].\end{aligned}$$

Bibliography

1. V. S. Anishchenko, A. B. Neiman, F. Moss, and L. Schimansky-Geier, *Stochastic resonance: noise-enhanced order*, Physics–Uspekhi **42** (1999), no. 1, 7–36.
2. R. Azencott, *Grandes déviations et applications.*, Ecole d'ete de probabilites de Saint-Flour VIII-1978, Lect. Notes Math. 774, 1-176, Springer, 1980 (French).
3. P. Baldi and M. Chaleyat-Maurel, *An extension of Ventsel-Freidlin estimates.*, Stochastic analysis and related topics, Proc. Workshop, Silivri/Turk., Lect. Notes Math. 1316, 305-327, Springer, 1988 (English).
4. P. Baldi and B. Roynette, *Some exact equivalents for the Brownian motion in Hölder norm*, Probab. Theory Related Fields **93** (1992), no. 4, 457–484. MR1183887 (94a:60117)
5. R. Benzi, G. Parisi, A. Sutera, and A. Vulpiani, *A theory of stochastic resonance in climatic change.*, SIAM J. Appl. Math. **43** (1983), 563–578.
6. R. Benzi, A. Sutera, and A. Vulpiani, *The Mechanism of Stochastic Resonance*, J. Phy. A **14** (1981), L453–L457.
7. F. A. Berezin and M. A. Shubin, *The Schrödinger equation*, Mathematics and Its Applications. Soviet Series., vol. 66, Kluwer Academic Publishers, Dordrecht etc., 1991.
8. N. Berglund and B. Gentz, *A sample-paths approach to noise-induced synchronization: Stochastic resonance in a double-well potential*, Ann. Appl. Probab. **12** (2002), 1419–1470.
9. ———, *Metastability in simple climate models: Pathwise analysis of slowly driven Langevin equations*, Stoch. Dyn. **2** (2002), 327–356.
10. ———, *Geometric singular perturbation theory for stochastic differential equations*, J. Differential Equations **191** (2003), no. 1, 1–54.
11. ———, *On the noise-induced passage through an unstable periodic orbit I: Two-level model*, J. Statist. Phys. **114** (2004), no. 5/6, 1577–1618.
12. ———, *Universality of first-passage- and residence-time distributions in non-adiabatic stochastic resonance*, Europhys. Letters **70** (2005), no. 1, 1–7.
13. ———, *Noise-induced phenomena in slow-fast dynamical systems. A sample-paths approach.*, Probability and its Applications. London: Springer. xiv, 276 p. , 2006 (English).
14. A. Bovier, M. Eckhoff, V. Gayrard, and M. Klein, *Metastability and low lying spectra in reversible Markov chains*, Comm. Math. Phys. **228** (2002), no. 2, 219–255. MR1911735 (2004g:60102)
15. ———, *Metastability in reversible diffusion processes I: Sharp asymptotics for capacities and exit times*, Journal of the European Mathematical Society **6** (2004), no. 4, 399–424.
16. A. Bovier, V. Gayrard, and M. Klein, *Metastability in reversible diffusion processes II: Precise asymptotics for small eigenvalues*, Journal of the European Mathematical Society **7** (2005), no. 1, 69–99.
17. A. Bulsara, E. W. Jacobs, T. Zhou, F. Moss, and L. Kiss, *Stochastic resonance in a single neuron model: Theory and analog simulation*, Journal of Theoretical Biology **152** (1991), 531–555.
18. V. A. Buslov and K. A. Makarov, *Hierarchy of time scales in the case of weak diffusion.*, Theor. Math. Phys. **76** (1988), no. 2, 818–826.
19. ———, *Life times and lower eigenvalues of an operator of small diffusion.*, Mat. Zametki **51** (1992), no. 1, 20–31.
20. Z. Ciesielski, *On Haar functions and on the Schauder basis of the space* $C_{\langle 0,1\rangle}$, Bull. Acad. Polon. Sci. Sér. Sci. Math. Astronom. Phys. **7** (1959), 227–232. MR0131751 (24 #A1599)
21. J. J. Collins, T. T. Imhoff, and P. Grigg, *Noise-mediated enhancements and decrements in human tactile sensation*, Phys. Rev. E **56** (1997), 923–926 (English).

22. M. V. Day, *On the exponential exit law in the small parameter exit problem.*, Stochastics **8** (1983), 297–323.
23. ______, *Recent progress on the small parameter exit problem.*, Stochastics **20** (1987), 121–150.
24. ______, *Mathematical approaches to the problem of noise-induced exit.*, Stochastic analysis, control, optimization and applications. A volume in honor of W. H. Fleming, on the occasion of his 70th birthday (McEneaney, W. M. et al., eds.), Birkhäuser, Boston, 1999, pp. 269–287.
25. A. Dembo and O. Zeitouni, *Large deviations techniques and applications*, second ed., Springer-Verlag, New York, 1998. MR99d:60030
26. R. E. Lee Deville and E. Vanden-Eijnden, *Self-induced stochastic resonance for Brownian ratchets under load*, Communications in Mathematical Sciences **5** (2007), no. 2, 431–446.
27. C. R. Doering, *Stochastic ratchets*, Physica A **254** (1998), 1–6.
28. C. R. Doering and T. C. Elston, *Random walks, resonance, and ratchets*, Nonequilibrium Statistical Mechanics in One Dimension (V. Privman, ed.), Cambridge University Press, 1997.
29. J. K. Douglass, L. Wilkens, E. Pantazelou, and F. Moss, *Noise enhancement of information transfer in crayfish mechanoreceptors by stochastic resonance*, Nature **365** (1993), 337–340.
30. M. I. Dykman, A. L. Velikovich, G. P. Golubev, D. G. Luchinskii, and S. V. Tsuprikov, *Stochastic resonance in an all-optical passive bistable system*, Journal of Experimental and Theoretical Physics Letters **53** (1991), 193–197.
31. Weinan E, W. Ren, and E. Vanden-Eijnden, *Minimum action method for the study of rare events*, Communications on Pure and Applied Mathematics **57** (2004), no. 5, 637–656.
32. J.-P. Eckmann and L. E. Thomas, *Remarks on stochastic resonance.*, J. Phys. A **15** (1982), 261–266.
33. D. E. Edmunds and W. D. Evans, *Spectral theory and differential operators.*, Oxford Mathematical Monographs, Clarendon Press, Oxford, 1989.
34. Yu. V. Egorov and M. A. Shubin, *Foundations of the classical theory of partial differential equations.*, Springer, Berlin, 1998.
35. T. C. Elston and C. S. Peskin, *Protein flexibility and the correlation ratchet*, SIAM Journal on Applied Mathematics **61** (2000), no. 3, 776–791.
36. A. Erdélyi, *Asymptotic expansions.*, Dover Publications, Inc., New York, 1956 (English).
37. H. Eyring, *The activated complex in chemical reactions.*, The Journal of Chemical PHysics **3** (1935), 107–115.
38. S. Fauve and F. Heslot, *Stochastic resonance in a bistable system*, Physics Letters A **97** (1983), 5–7.
39. M. Freidlin, *Quasi-deterministic approximation, metastability and stochastic resonance*, Physica D **137** (2000), 333–352.
40. M. Freidlin and A. Wentzell, *Random perturbations of dynamical systems*, second ed., Springer-Verlag, New York, 1998.
41. J. A. Freund, J. Kienert, L. Schimansky-Geier, B. Beisner, A. Neiman, D. F. Russell, T. Yakusheva, and F. Moss, *Behavioral stochastic resonance: How a noisy army betrays its outpost*, Physical Review E **63** (2001), 031910+.
42. J. A. Freund, L. Schimansky-Geier, B. Beisner, A. Neiman, D. F. Russell, T. Yakusheva, and F. Moss, *Behavioral stochastic resonance: How the noise from a Daphnia swarm enhances individual prey capture by juvenile paddlefish*, Journal of Theoretical Biology **214** (2002), no. 1, 71–83.
43. L. Gammaitoni, P. Hänggi, P. Jung, and F. Marchesoni, *Stochastic resonance*, Reviews of Modern Physics **70** (1998), 223–287.
44. L. Gammaitoni, P. Hänggi, P. Jung, and F. Marchesoni, *Stochastic resonance: a remarkable idea that changed our perception of noise.*, Eur. Phys. J. B **69** (2009), 1–3 (English).
45. A. Ganopolski and S. Rahmstorf, *Abrupt glacial climate changes due to stochastic resonance.*, Physical Review Letters **88** (2001), no. 3, 038501.
46. C. W. Gardiner, *Handbook of stochastic methods for physics, chemistry and the natural sciences.*, Springer Series in Synergetics, vol. 13 (3rd ed.). Springer, Berlin, 2004 (English).
47. P. E. Greenwood, L. M. Ward, D. F. Russel, A. Neiman, and F. Moss, *Stochastic Resonance Enhances the Electrosensory Information Available to Paddlefish for Prey Capture*, Phys. Rev. Letters **84** (2000), 4773–4776 (English).

48. R. Z. Has′minskiĭ, *Stochastic stability of differential equations*, Monographs and Textbooks on Mechanics of Solids and Fluids: Mechanics and Analysis, vol. 7, Sijthoff & Noordhoff, 1980.
49. S. Herrmann and P. Imkeller, *Barrier crossings characterize stochastic resonance*, Stochastics and Dynamics **2** (2002), no. 3, 413–436. MR2004i:60082
50. ______, *The exit problem for diffusions with time-periodic drift and stochastic resonance.*, The Annals of Applied Probability **15** (2005), no. 1A, 39–68 (English).
51. S. Herrmann, P. Imkeller, and D. Peithmann, *Transition times and stochastic resonance for multidimensional diffusions with time periodic drift: a large deviations approach*, The Annals of Applied Probability **16** (2006), no. 4, 1851–1892.
52. S. Herrmann and B. Roynette, *Boundedness and convergence of some self-attracting diffusions*, Mathematische Annalen **325** (2003), no. 1, 81–96.
53. M. Heymann and E. Vanden-Eijnden, *The geometric minimum action method: A least action principle on the space of curves*, Communications on Pure and Applied Mathematics **61** (2008), no. 8, 1052–1117.
54. I. Hidaka, D. Nozaki, and Y. Yamamoto, *Functional Stochastic Resonance in the Human Brain: Noise Induced Sensitization of Baroreflex System*, Phys. Rev. Lett. **85** (2000), 3740–3743.
55. P. D. Hislop and I. M. Sigal, *Introduction to spectral theory. With applications to Schrödinger operators*, Applied Mathematical Sciences., vol. 113, Springer-Verlag, New York, NY, 1996.
56. R. A. Holley, S. Kusuoka, and D. W. Stroock, *Asymptotics of the spectral gap with applications to the theory of simulated annealing.*, J. Funct. Anal. **83** (1989), no. 2, 333–347.
57. C.-R. Hwang and S.-J. Sheu, *Large-time behavior of perturbed diffusion Markov processes with applications to the second eigenvalue problem for Fokker-Planck operators and simulated annealing.*, Acta Appl. Math. **19** (1990), no. 3, 253–295.
58. P. Imkeller and I. Pavlyukevich, *Stochastic resonance in two-state Markov chains*, Archiv der Mathematik **77** (2001), no. 1, 107–115.
59. ______, *Model reduction and stochastic resonance*, Stochastics and Dynamics **02** (2002), no. 04, 463–506.
60. C. Ivey, A. V. Apkarian, and D. R. Chialvo, *Noise-induced tuning curve changes in mechanoreceptors*, J. Neurophysiol. **79** (1998), 1879–1890.
61. S. Jacquot, *Comportement asymptotique de la seconde valeur propre des processus de Kolmogorov.*, J. Multivariate Anal. **40** (1992), no. 2, 335–347.
62. P. Jung, *Periodically driven stochastic systems.*, Physics Reports **234** (1993), 175–295 (English).
63. T. Kanamaru, T. Horita, and Y. Okabe, *Theoretical analysis of array-enhanced stochastic resonance in the diffusively couple FitzHugh-Nagumo equation.*, Physical Review E **64** (2001), 031908 (English).
64. V. N. Kolokoltsov, *Semiclassical analysis for diffusions and stochastic processes*, Lecture Notes in Mathematics, vol. 1724, Springer, Berlin, 2000.
65. H. A. Kramers, *Brownian motion in a field of force and the diffusion model of chemical reactions.*, Physica **7** (1940), 284–304.
66. Laplace, le Marquis de, *Théorie analytique des probabilités*, third ed., Courcier, Paris, 1820, Reprinted in *Complete works*, Vol. 7. Gautier-Villars, Paris, 1886.
67. D. S. Leonard and L. E. Reichl, *Stochastic resonance in a chemical reaction*, Physical Review E **49** (1994), 1734–1737.
68. B. M. Levitan and I. S. Sargsjan, *Introduction to spectral theory: Selfadjoint ordinary differential operators*, Translations of Mathematical Monographs., vol. 39, American Mathematical Society., Providence, R.I., 1975.
69. F. Marchesoni, F. Apostolico, L. Gammaitoni, and S. Santucci, *Color effects in a near-threshold Schmitt trigger.*, Physical Review E **58** (1998), no. 6, 7079–7084.
70. F. Marchesoni, F. Apostolico, and S. Santucci, *Stochastic resonance in an asymmetric Schmitt trigger.*, Physical Review E **59** (1999), no. 4, 3958–3963.
71. B. J. Matkowsky and Z. Schuss, *Eigenvalues of the Fokker-Planck operator and the approach to equilibrium for diffusions in potential fields.*, SIAM J. Appl. Math. **40** (1981), 242–254.
72. R. M. Mazo, *Brownian motion: fluctuations, dynamics, and applications.*, International Series of Monography on Physics, vol. 112. Oxford University Press, New York, 2002 (English).

73. M. D. McDonnell, N. G. Stocks, C. E. M. Pearce, and D. Abbott, *Stochastic resonance*, Cambridge University Press, 2008.
74. B. McNamara and K. Wiesenfeld, *Theory of stochastic resonance*, Physical Review A (General Physics) **39** (1989), 4854–4869.
75. B. McNamara, K. Wiesenfeld, and R. Roy, *Observation of stochastic resonance in a ring laser*, Physical Review Letters **60** (1988), 2626–2629.
76. V. I. Melnikov, *Schmitt trigger: A solvable model of stochastic resonance*, Physical Review E **48** (1993), no. 4, 2481–2489.
77. G. N. Milstein and M. V. Tretyakov, *Numerical analysis of noise-induced regular oscillations*, Physica D **140** (2000), no. 3-4, 244–256 (English).
78. T. Mori and S. Kai, *Noise-induced entrainment and stochastic resonance in human brain waves*, Phys. Rev. Lett. **88** (2002), 218101 (English).
79. F. Moss, L. M. Ward, and Walter G. Sannita, *Stochastic resonance and sensory information processing: a tutorial and review of application.*, Clinical Neurophysiology **115** (2004), 267–281 (English).
80. C. B. Muratov and E. Vanden-Eijnden, *Self-induced stochastic resonance in excitable systems*, Physica D **210** (2005), 227–240.
81. A. Neiman, B. Shulgin, V. Anishchenko, W. Ebeling, L. Schimansky-Geier, and J. Freund, *Dynamical entropies applied to stochastic resonance*, Physical Review Letters **76** (1996), no. 23, 4299–4302.
82. E. Nelson, *Dynamical theories of Brownian motion.*, Princeton University Press, Princeton, NJ, 1967 (English).
83. C. Nicolis, *Stochastic aspects of climatic transitions — responses to periodic forcing.*, Tellus **34** (1982), 1–9.
84. F. W. J. Olver, *Asymptotics and special functions*, Computer Science and Applied Mathematics., Academic Press, a subsidiary of Harcourt Brace Jovanovich, Publishers, 1974.
85. A. Patel and B. Kosko, *Stochastic resonance in continuous and spiking neuron models with Lévy noise*, IEEE Transactions on Neural Networks **19** (2008), no. 12, 1993–2008.
86. I. E. Pavlyukevich, *Stochastic Resonance*, Ph.D. thesis, Humboldt University Berlin, Berlin, 2002, Logos–Verlag.
87. C. Penland and P. D. Sardeshmukh, *The optimal growth of tropical sea surface temperature anomalies.*, J. Climate **8** (1995), 1999–2024 (English).
88. P. Priouret, *Remarques sur les petites perturbations de systèmes dynamiques.*, Seminaire de probabilites XVI, Univ. Strasbourg 1980/81, Lect. Notes Math. 920, 184-200, Springer, 1982 (French).
89. A. Priplata, J. Niemi, M. Salen, J. Harry, L. A. Lipsitz, and J. J. Collins, *Noise-Enhanced Human Balance Control*, Phys. Rev. Lett. **89** (2002), 238101.
90. O. Raimond, *Self-attracting diffusions: case of the constant interaction*, Probab. Theory Rel. Fields **107** (1997), no. 2, 177–196.
91. P. Reimann, *Brownian motors: noisy transport far from equilibrium*, Physics Reports **361** (2002), no. 2–4, 57–265.
92. M. Renardy and R. C. Rogers, *An introduction to partial differential equations*, Texts in Applied Mathematics., vol. 13, Springer-Verlag, New York, 1993.
93. D. Revuz and M. Yor, *Continuous martingales and Brownian motion*, third ed., Grundlehren der Mathematischen Wissenschaften [Fundamental Principles of Mathematical Sciences], vol. 293, Springer-Verlag, Berlin, 1999. MR2000h:60050
94. M. Riani and E. Simonotto, *Stochastic resonance in the perceptual interpretation of ambiguous figures: A neural network model*, Physical Review Letters **72** (1994), 3120–3123.
95. D. F. Russel, L. A. Wilkens, and F. Moss, *Use of behavioural stochastic resonance by paddle fish for feeding*, Nature **402** (1999), 291 (English).
96. Z. Schuss, *Theory and applications of stochastic differential equations.*, Wiley Series in Probability and Mathematical Statistics., John Wiley & Sons, New York etc., 1980.
97. F. Schweitzer, *Brownian agents and active particles.*, Springer Series in Synergetics, Collective dynamics in the natural and social sciences. Springer, Berlin, 2003 (English).
98. D. B. Sears, *Note on the uniqueness of the Greens functions associated with certain differential equations.*, Can. J. Math. **2** (1950), 314–325.
99. E. Simonotto, M. Riani, C. Seife, M. Roberts, J. Twitty, and F. Moss, *Visual perception of stochastic resonance*, Physical Review Letters **78** (1997), 1186.

100. M. L. Spano, M. Wun-Fogle, and W. L. Ditto, *Experimental observation of stochastic resonance in a magnetoelastic ribbon*, Physical Review A **46** (1992), 5253–5256.
101. N. G. Stocks, *Suprathreshold stochastic resonance in multilevel threshold systems*, Physical Review Letters **84** (2000), no. 11, 2310–2313.
102. D. W. Stroock and S. R. S. Varadhan, *Multidimensional diffusion processes*, Grundlehren der Mathematischen Wissenschaften [Fundamental Principles of Mathematical Sciences], vol. 233, Springer-Verlag, Berlin, 1979. MR532498 (81f:60108)
103. M. Usher and M. Feingold, *Stochastic resonance in the speed of memory retrieval*, Biological Cybernetcs **83** (2000), L011–L016 (English).
104. G. Vemuri and R. Roy, *Stochastic resonance in a bistable ring laser.*, Phys. Review A **39** (1989), 4668–4674 (English).
105. A. D. Ventcel and M. I. Freidlin, *Small random perturbations of dynamical systems*, Uspehi Mat. Nauk **25** (1970), no. 1 (151), 3–55. MR42#2123
106. B. Wang, A. Barcilon, and Z. Fang, *Stochastic dynamics of El Nino-Southern Oscillation.*, J. Atmos. Sci. **56** (1999), 5–23 (English).
107. B. Wang and Z. Fang, *Chaotic oscillations of tropical climate: A dynamic system theory for ENSO.*, J. Atmos. Sci. **53** (1996), 2786–2802 (English).
108. T. Wellens, V. Shatokhin, and A. Buchleitner, *Stochastic resonance.*, Rep. Prog. Phys. **67** (2004), 45–105 (English).
109. K. Wiesenfeld, D. Pierson, E. Pantazelou, C. Dames, and F. Moss, *Stochastic resonance on a circle*, Phys. Rev. Lett. **72** (1994), 2125–2129.
110. M. Williams, *Asymptotic exit time distributions.*, SIAM J. Appl. Math. **42** (1982), 149–154.

Index

Selected Published Titles in This Series

194 **Samuel Herrmann, Peter Imkeller, Ilya Pavlyukevich, and Dierk Peithmann,** Stochastic Resonance, 2014

192 **Messoud Efendiev,** Attractors for Degenerate Parabolic Type Equations, 2013

191 **Grégory Berhuy and Frédérique Oggier,** An Introduction to Central Simple Algebras and Their Applications to Wireless Communication, 2013

190 **Aleksandr Pukhlikov,** Birationally Rigid Varieties, 2013

189 **Alberto Elduque and Mikhail Kochetov,** Gradings on Simple Lie Algebras, 2013

188 **David Lannes,** The Water Waves Problem, 2013

187 **Nassif Ghoussoub and Amir Moradifam,** Functional Inequalities: New Perspectives and New Applications, 2013

186 **Gregory Berkolaiko and Peter Kuchment,** Introduction to Quantum Graphs, 2013

185 **Patrick Iglesias-Zemmour,** Diffeology, 2013

184 **Frederick W. Gehring and Kari Hag,** The Ubiquitous Quasidisk, 2012

183 **Gershon Kresin and Vladimir Maz'ya,** Maximum Principles and Sharp Constants for Solutions of Elliptic and Parabolic Systems, 2012

182 **Neil A. Watson,** Introduction to Heat Potential Theory, 2012

181 **Graham J. Leuschke and Roger Wiegand,** Cohen-Macaulay Representations, 2012

180 **Martin W. Liebeck and Gary M. Seitz,** Unipotent and Nilpotent Classes in Simple Algebraic Groups and Lie Algebras, 2012

179 **Stephen D. Smith,** Subgroup complexes, 2011

178 **Helmut Brass and Knut Petras,** Quadrature Theory, 2011

177 **Alexei Myasnikov, Vladimir Shpilrain, and Alexander Ushakov,** Non-commutative Cryptography and Complexity of Group-theoretic Problems, 2011

176 **Peter E. Kloeden and Martin Rasmussen,** Nonautonomous Dynamical Systems, 2011

175 **Warwick de Launey and Dane Flannery,** Algebraic Design Theory, 2011

174 **Lawrence S. Levy and J. Chris Robson,** Hereditary Noetherian Prime Rings and Idealizers, 2011

173 **Sariel Har-Peled,** Geometric Approximation Algorithms, 2011

172 **Michael Aschbacher, Richard Lyons, Stephen D. Smith, and Ronald Solomon,** The Classification of Finite Simple Groups, 2011

171 **Leonid Pastur and Mariya Shcherbina,** Eigenvalue Distribution of Large Random Matrices, 2011

170 **Kevin Costello,** Renormalization and Effective Field Theory, 2011

169 **Robert R. Bruner and J. P. C. Greenlees,** Connective Real K-Theory of Finite Groups, 2010

168 **Michiel Hazewinkel, Nadiya Gubareni, and V. V. Kirichenko,** Algebras, Rings and Modules, 2010

167 **Michael Gekhtman, Michael Shapiro, and Alek Vainshtein,** Cluster Algebras and Poisson Geometry, 2010

166 **Kyung Bai Lee and Frank Raymond,** Seifert Fiberings, 2010

165 **Fuensanta Andreu-Vaillo, José M. Mazón, Julio D. Rossi, and J. Julián Toledo-Melero,** Nonlocal Diffusion Problems, 2010

164 **Vladimir I. Bogachev,** Differentiable Measures and the Malliavin Calculus, 2010

163 **Bennett Chow, Sun-Chin Chu, David Glickenstein, Christine Guenther, James Isenberg, Tom Ivey, Dan Knopf, Peng Lu, Feng Luo, and Lei Ni,** The Ricci Flow: Techniques and Applications: Part III: Geometric-Analytic Aspects, 2010